Advances in Intelligent Systems and Computing

Volume 278

Series editor

Janusz Kacprzyk, Polish Academy of Sciences, Warsaw, Poland
e-mail: kacprzyk@ibspan.waw.pl

For further volumes:
http://www.springer.com/series/11156

About this Series

The series "Advances in Intelligent Systems and Computing" contains publications on theory, applications, and design methods of Intelligent Systems and Intelligent Computing. Virtually all disciplines such as engineering, natural sciences, computer and information science, ICT, economics, business, e-commerce, environment, healthcare, life science are covered. The list of topics spans all the areas of modern intelligent systems and computing.

The publications within "Advances in Intelligent Systems and Computing" are primarily textbooks and proceedings of important conferences, symposia and congresses. They cover significant recent developments in the field, both of a foundational and applicable character. An important characteristic feature of the series is the short publication time and world-wide distribution. This permits a rapid and broad dissemination of research results.

Advisory Board

Chairman

Nikhil R. Pal, Indian Statistical Institute, Kolkata, India
e-mail: nikhil@isical.ac.in

Members

Rafael Bello, Universidad Central "Marta Abreu" de Las Villas, Santa Clara, Cuba
e-mail: rbellop@uclv.edu.cu

Emilio S. Corchado, University of Salamanca, Salamanca, Spain
e-mail: escorchado@usal.es

Hani Hagras, University of Essex, Colchester, UK
e-mail: hani@essex.ac.uk

László T. Kóczy, Széchenyi István University, Győr, Hungary
e-mail: koczy@sze.hu

Vladik Kreinovich, University of Texas at El Paso, El Paso, USA
e-mail: vladik@utep.edu

Chin-Teng Lin, National Chiao Tung University, Hsinchu, Taiwan
e-mail: ctlin@mail.nctu.edu.tw

Jie Lu, University of Technology, Sydney, Australia
e-mail: Jie.Lu@uts.edu.au

Patricia Melin, Tijuana Institute of Technology, Tijuana, Mexico
e-mail: epmelin@hafsamx.org

Nadia Nedjah, State University of Rio de Janeiro, Rio de Janeiro, Brazil
e-mail: nadia@eng.uerj.br

Ngoc Thanh Nguyen, Wroclaw University of Technology, Wroclaw, Poland
e-mail: Ngoc-Thanh.Nguyen@pwr.edu.pl

Jun Wang, The Chinese University of Hong Kong, Shatin, Hong Kong
e-mail: jwang@mae.cuhk.edu.hk

Zhenkun Wen · Tianrui Li
Editors

Knowledge Engineering and Management

Proceedings of the Eighth International
Conference on Intelligent Systems
and Knowledge Engineering, Shenzhen,
China, Nov 2013 (ISKE 2013)

 Springer

Editors
Zhenkun Wen
College of Computer and Software
  Engineering
Shenzhen University
Shenzhen
China

Tianrui Li
School of Information Science
  and Technology
Southwest Jiaotong University
Chengdu
China

ISSN 2194-5357   ISSN 2194-5365 (electronic)
ISBN 978-3-642-54929-8   ISBN 978-3-642-54930-4 (eBook)
DOI 10.1007/978-3-642-54930-4
Springer Heidelberg New York Dordrecht London

Library of Congress Control Number: 2013942994

Printed on acid-free paper

Springer is part of Springer Science+Business Media (www.springer.com)

Preface

This book is a part of the Proceedings of the Eighth International Conference on Intelligent Systems and Knowledge Engineering (ISKE 2013) held in Shenzhen, China, during November 20–23, 2013. ISKE is a prestigious annual conference on ISKE with the past events held in Shanghai (2006, 2011), Chengdu (2007), Xiamen (2008), Hasselt, Belgium (2009), Hangzhou (2010), and Beijing (2012). Over the past few years, ISKE has matured into a well-established series of international conferences on ISKE and related fields over the world.

ISKE 2013 received 609 submissions in total from about 1,434 authors from 18 countries (United States of American, Singapore, Russian Federation, Saudi Arabia, Spain, Sudan, Sweden, Tunisia, United Kingdom, Portugal, Norway, Korea, Japan, Germany, Finland, France, China, Argentina, Australia, and Belgium). Based on rigorous reviews by the Program Committee members and reviewers, among 263 papers contributed to ISKE 2013, high-quality papers were selected for publication in the proceedings with the acceptance rate of 43 %. The papers were organized into 22 cohesive sections covering all major topics of intelligent and cognitive science and applications. In addition to the contributed papers, the technical program includes four plenary speeches by Ronald R. Yager (Iona College, USA), Yu Zheng (Microsoft Research Asia), and Hamido Fujita (Iwate Prefectural University, Japan).

As organizers of this conference, we are grateful to Shenzhen University, Science in China Press, Chinese Academy of Sciences for their sponsorship, grateful to IEEE Computational Intelligence Society, Chinese Association for Artificial Intelligence, State Key Laboratory on Complex Electronic System Simulation, Science and Technology on Integrated Information System Laboratory, Southwest Jiaotong University, University of Technology, Sydney for their technical co-sponsorship. We would also like to thank the members of the Advisory Committee for their guidance, the members of the International Program Committee and additional reviewers for reviewing the papers, and members of the Publications Committee for checking the accepted papers in a short period of time. Particularly, we are grateful to the publisher, Springer, for publishing the proceedings in the prestigious series of *Advances in Intelligent Systems and Computing*. Meanwhile, we wish to express our heartfelt appreciation to the plenary speakers, special session organizers, session chairs, and student helpers. In addition, there are still many colleagues, associates, and friends who helped us in immeasurable ways. We are also grateful to them all. Last but not the least, we are thankful to all the authors and participants for their great contributions that made ISKE 2013 successful.

November 2013   Zhenkun Wen
                                                                                      Tianrui Li

Contents

Parallel Approaches to Neighborhood Rough Sets: Classification and Feature Selection

Junbo Zhang, Chizheng Wang, Yi Pan and Tianrui Li

Abstract In these days, the ever-increasing volume of data requires that data mining algorithms should not only have high accuracy but also have high performance, which is really a challenge for the existing data analysis methods. Traditional algorithms, such as classification and feature selection under neighborhood rough sets, have been proved to be very effective in real applications. Parallel approach to these traditional algorithms could be a way to take the challenge. This is what we present in this paper, the design of parallel approaches to neighborhood rough sets and the implementation of classification and feature selection. Two optimizing strategies are proposed to improve the performance of the approaches: (1) The distributed cache is used to reduce I/O time. (2) Most of computations are put into the Map phase which helps reduce the overhead of communication. The experimental results show that the proposed algorithms scale pretty well and the speedup is getting higher with the increasing size of data.

J. Zhang · T. Li (✉)
School of Information Science and Technology, Southwest Jiaotong University,
Chengdu 610031, China
e-mail: trli@swjtu.edu.cn

J. Zhang
e-mail: JunboZhang86@163.com; jbzhang@cs.gsu.edu

J. Zhang · Y. Pan
Department of Computer Science, Georgia State University, Atlanta, GA 30303, USA
e-mail: pan@cs.gsu.edu

C. Wang
Department of Computer Science, University of Illinois at Urbana-Champaign,
Urbana, IL 61801, USA
e-mail: cwang86@illinois.edu

Z. Wen and T. Li (eds.), *Knowledge Engineering and Management*,
Advances in Intelligent Systems and Computing 278,
DOI: 10.1007/978-3-642-54930-4_1, © Springer-Verlag Berlin Heidelberg 2014

1

1 Introduction

People always say we are living in the information age [3]. What they say is actually we are living in *big data* age. Growing amount and kinds of data play important roles in uncountable fields, including business, astronomy, and bioinformatics. Data mining focuses on discovering knowledge from different kinds of data. Classification is one of the hot topics in data mining. There are basically two kinds of classification algorithms, namely eager learner and lazy learner [3]. Eager learner algorithm usually contains two steps, learning step and classification step. In the learning step, the classifier will create a model based on a set of data, which is usually called training data set. This model will be used to classify the test data. ID3, C4.5, and SVM are all eager learner algorithms.

The lazy algorithms also need a set of training data. However, unlike eager learner algorithm, lazy algorithms do not use the training set to create a model, instead they use the test data to compare with training data set to make decisions. The neighborhood classifiers (NEC), which is introduced in [5], and k-nearest-neighbor classifiers are both lazy learner algorithms. As the development of the research in data mining, many mathematical and statistical tools are involved to help to increase the performance of data mining. Rough set theory, introduced by Pawlak in 1982 [9], is one of the powerful mathematical techniques in data mining. It works well in decision situations which involve inconsistent information [6, 7, 13].

Nowadays, people not only concern about the accuracy but also the performance of a data mining program. There are usually just two ways to improve the performance, either increasing the speed of computers or decreasing the complexity of algorithms. However, both ways are tough. In the information age, the volume of data grows exponentially. The growing speed of CPU performance is much slower than the growing speed of data. It is even harder to decrease the complexity of algorithms that already exist. To solve this problem, many parallel techniques are developed. MapReduce is a new program model proposed by Google [1]. Its basic idea is to split a large task into small pieces and assigns them to different machines. It is easy to use and allows a big task to be split into small pieces and be done in different machines. The model abstracts the processes into two parallel processes, Map and Reduce. The two algorithms proposed in this paper are implemented with Hadoop [11], which is a framework driving from Google MapReduce. To process massive data with rough sets, Zhang et al. proposed a parallel method for computing rough set approximations [12, 14]. However, that method can only process categorical data rather than numerical data. In real-life applications, most of data are heterogeneous, including categorical and numerical. Therefore, in this paper, we discuss how to process numerical data in parallel for feature selection and classification.

The rest of the paper is organized as follows. The basic concepts and definitions are provided in Sect. 2. Section 3 presents two new parallel rough set-based algorithms. The experimental results are shown in Sect. 4. Section 5 describes the conclusion and future research directions.

2 Preliminaries

In this section, we review some basic concepts of neighborhood rough sets [4, 5] and the MapReduce model [1].

2.1 Neighborhood Rough Sets

An information system is defined as a 4-tuple (U, A, V, f), where U is a non-empty finite set that contains all objects. A is a non-empty finite set of features. $V = \bigcup_{a \in A} V_a$, V_a is a domain of the feature a. $f : U \times A \rightarrow V$ is an information function. The information system is a decision table if $A = C \cup D$, where $C \cap D = \emptyset$. D is usually called the decision, while C is usually called features.

Definition 1 Let $B \subseteq C$ be a subset of attributes $x \in U$. The neighborhood $\delta_B(x)$ of x in the feature space B is defined as

$$\delta_B(x) = \{y \in U | \Delta_B(x, y) \leq \delta\} \tag{1}$$

where $\Delta_B(x, y)$ represents the distance between x and y.
The distance between x and y in B is defined as

$$\Delta_B(x, y) = \left(\sum_{i=1}^{m} |f(x, a_i) - f(y, a_i)|^P \right)^{1/P} \tag{2}$$

where $a_i \in B$.

Definition 2 Given a neighborhood decision table $NDT = (U, C \cup D, V, f)$, suppose $U/D = \{D_1, D_2, \cdots, D_N\}$, $\delta_B(x)$ is the neighborhood information granules including x and generated by attributes of $B \subseteq C$. Then the lower and upper approximations of the decision D with respect to B are defined as

$$\underline{N}_B(D) = \bigcup_{i=1}^{N} \underline{N}_B(D_i) \tag{3}$$

$$\overline{N}_B(D) = \bigcup_{i=1}^{N} \overline{N}_B(D_i) \tag{4}$$

where

$$\underline{N}_B(D_i) = \{x \in U | \delta_B(x) \subseteq D_i\}$$

$$\overline{N}_B(D_i) = \{x \in U | \delta_B(x) \cap D_i \neq \emptyset\}$$

The boundary region of D with respect to B is defined as

$$BN(D) = \overline{N}_B(D) - \underline{N}_B(D) \tag{5}$$

Definition 3 The lower approximations of the decision, also called the positive region of the decision, denoted by $POS_B(D)$, is the subset of the object set whose neighborhood granules consistently belong to one of the decision classes. It is easy to show that all elements in $\delta_B(x_i)$ have the same decision, $\forall x_i \in POS_B(D)$.

Definition 4 The dependency degree of B with respect to D is defined as

$$\gamma_B(D) = \frac{|POS_B(D)|}{|U|} \tag{6}$$

Definition 5 Given the decision table $(U, C \cup D), B \subseteq C$, the significance of the attribute $a \in B$ is defined as

$$\text{Sig}(a, B, D) = \gamma_B(D) - \gamma_{B-\{a\}}(D) \tag{7}$$

Another definition of the significance of an attribute is denoted as follows.

Definition 6 Given the decision table $(U, C \cup D), B \subseteq C$, the significance of the attribute $a \in C, a \notin B$ is defined as

$$\text{Sig}(a, B, D) = \gamma_{B \cup \{a\}}(D) - \gamma_B(D) \tag{8}$$

2.2 MapReduce Model

MapReduce is designed to handle and generate large scale, no dependency, data sets in distributed environment [1]. It provides a convenient way to parallelize data analysis process. Its advantages include conveniences, robustness, and scalability. The basic idea of MapReduce is to split the large input data set into many small pieces and assign small tasks to different devices. To work in the cluster, MapReduce model usually works with the distributed file system which provides storage [2].

3 Parallel Approaches Based on MapReduce

3.1 Parallel Neighborhood Classification

The basic idea of the neighborhood classification algorithm is to make decisions for the data based on its neighbor. For each test data, the program will calculate the distance between itself and each training data to determine the neighborhood and use the majority voting to make decisions. Algorithm 1 shows the sequential algorithm [5]. The value of δ is important to the accuracy. The program cannot make decisions if δ is too small because all the training data will not be included. The program could make wrong decisions if δ is too big because all the training data will be included. Hu et al. [5] calculated δ as

$$\delta = \min(\Delta(x_i, s)) + w \cdot \text{range}(\Delta(x_i, s)), w \leq 1 \qquad (9)$$

Algorithm 1: NEighborhood Classifiers (NEC)

---

input : Training set: $(U, C \cup D)$, Test sample: s; Threshold δ; Specify the norm used.
output: Class of s
1 Compute the distance between s and $x_i \in U$ with the used norm;
2 Find the samples in the neighborhood $\delta(s)$ of s;
3 Find the class d_j with the majority training samples in $\delta(s)$;
4 Assign d_j to the test s.

Algorithm 2: PNEC-Map

---

input : $< key, value >$
            // key is the byte offset of the line and value is the information of test data
output: $< key', value' >$
            // key' is the ID of test data and value' is the decision of test data
1 **begin**
2 | U = GetFileFromCache(); *// get the training data set from*
    | `distributed cache`
3 | **for** $x_i \in U$ **do**
4 | | distance[i] = CalcDistance(value, x_i); *// calculate all the distances*
5 | **end**
6 | value'=Vote(distance); *// majority vote*
7 | key'=GetID(value);
8 | output.collect(key',value');
9 **end**

The parallel NEC proposed by this paper is designed to handle large test data sets which means the program will make decisions not just for one test data but for huge number of test data. The whole test data set will be put into the input file, one item per line. The training data set will be put in another file and added into the distributed cache. The test data will be separated into many Map tasks and assigned to different machines. The Map function will get the training data from the distributed cache and find out the neighborhood of the test data. After that majority voting will be employed to make decisions for the test data. The Parallel Neighborhood Classification (PNEC) is originally a Map-only algorithm.

3.2 Parallel Neighborhood Feature Selection

Even though many features have been recorded, some of them may not relate to the decision. Some of them may even guide the program to make wrong decisions. Too many features will make the program too time consuming. The purpose of feature selection algorithms is to improve the accuracy and performance of classifiers. The total number of possible combination is $2^{|C|}$, where $|C|$ is the number of features. It will be too time consuming to check all the possible combination, so a greedy feature selection algorithm was proposed [5]. The basic idea of the algorithm is to find the optimized combination in each iteration. There is a reduction set which is initially empty in the program. In each iteration, each candidate is firstly combined with the features in the reduction set, and then the significance of this combination is calculated. The feature which leads to maximal significance will be added into the reduction set. The program will terminate when the maximal significance is less than ε (a small positive real number used to control the convergence) or all candidates are added into the reduction set. Different from the neighborhood classifiers, the threshold of the neighborhood is a constant.

Algorithms 3, 4, 5 show the pseudo codes and Fig. 1 shows the process. In Algorithm 5, ε is a small positive real number used to control the convergence.

Algorithm 3: PNFS-Map

---

    input : $< key, value >$, reduction set: *red*, candidate set: C, threshold δ
        // *key is the byte offset of the line and value is the information of training data*
    output: $< key', value' >$
        // *key' is the name or ID of the candidates. Its value will be 1 if this training data is in the positive region when the combination contains this feature and 0 means it is not in the positive region*

1 **begin**
2      $U =$ GetFileFromCache(); // get the training data set from the distributed cache
3   **for** $a_i \in C$ **do**
4   boolean result = isPositive($value, U, \delta, red \cup a_i$);
5   **if** *result* == *True* **then**
6              $value' = 1$;
7   **else**
8              $value' = 0$;
9   **end**
10          $key' = a_i$;
11   output.collect($key', value'$);
12   **end**
13 **end**

Algorithm 4: PNFS-Reduce

---

    input : $< key, V >$ // *key is the name or ID of candidates and V is a list of 1 or 0*
    output: $< key', value' >$ // *key' is the name or ID of candidates and value' is number of data in the positive region*

1 **begin**
2      $key' = key, value' = 0$;
3   **for** $val \in V$ **do**
4          $value' + = val$;
5   **end**
6   output.collect($key', value'$);
7 **end**

Fig. 1 Process of parallel neighborhood feature selection

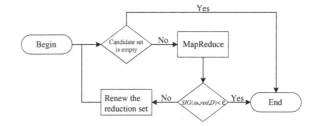

Algorithm 5: Parallel Neighborhood based Feature Selection-Main

---

    input : $(U, C \cup D)$ and δ
    output: Reduction set: *red*
1 **begin**
2     $\emptyset \rightarrow red$;
3   **while** $C \neq \emptyset$ **do**
4   Call MapReduce(U, red, C, δ); // Calculate $SIG(a_i, red, D)$, $\forall a_i \in C$
5   Select the attribute a_k which satisfies $SIG(a_k, red, D) = \max(SIG(a_i, red, D))$;
6   **if** $SIG(a_k, red, D) < \varepsilon$ **then** // ε is a little positive real number
7   break;
8   **else**
9             $C - a_k \rightarrow C$;
10            $red \cup a_k \rightarrow red$;
11   **end**
12   **end**
13   Return *red*;
14 **end**

---

4 Experimental Analysis

4.1 Experimental Environment

Our experiments were conducted on large clusters of computing machines. In detail, the task nodes consist of two kinds of machines. One kind of machines has 16 GB main memory and uses AMD Opteron Processor 2376 with two Quad-Core CPUs. The other kind of machines has 8 GB main memory and uses Intel Xeon CPU E5410, comprising two Quad-Core CPUs. Hence, there are 64 working cores in all. The operating system in these machines is Linux CentOS 5.2 Kernel 2.6.18. All experiments run on the Apache Hadoop platform [11].

4.2 A Comparison of Sequential and Parallel Algorithms

Here, we give a comparison of sequential and parallel algorithms. To simplify the process, Euclidean distance is used in all the experiments. To compare with the sequential algorithms in [5], seven data sets from machine learning data repository,

Table 1 Data sets

	Data sets	Records	Features	Decision
1	Cardiotocography	2126	23	3
2	Credit	690	14	2
3	Ionosphere	351	34	2
4	Iris	150	4	3
5	MAGIC-gamma-telescope	19020	11	2
6	Sonar-mines-versus-rocks	208	60	2
7	WDBC	569	31	2

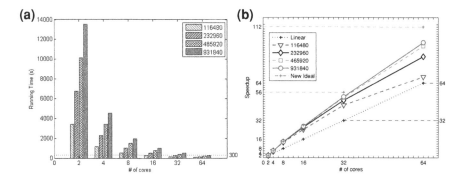

Fig. 2 a Running time and **b** Speedup plots of PNEC versus the number of cores used for 116480, 232960, 465920, 931840 records/60 features

University of California at Irvine [8], are selected in the experiments. Table 1 shows the basic information of data sets, all of which are numerical data.

First, we use sequential algorithm NFS and parallel algorithm parallel neighborhood feature selection (PNFS) to select relevant feature subsets with two different sizes of the neighborhood. The experiments demonstrate that the sequential and parallel algorithms have the same results.

4.3 Performance Analysis

In the experiment, to test the performance of PNEC and PNFS, the data set Sonar-Mines-versus-Rocks in Table 1 is used as input many times to stimulate the large input condition.

We mainly measured the speedup of both parallel algorithms [10]. Figure 2 shows the running time and speedup of different size data with different numbers of cores. Using the same number of cores, every time the input size is double, and the running time is double, and the running time is double. It is easy to find that the speedup is superlinear.

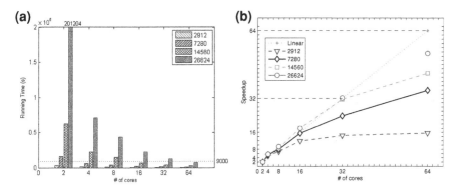

Fig. 3 Running time and speedup plots of PNFS versus the number of cores used for 2912, 7280, 14560, 26624 records/60 features

Figure 3 shows the running time and speedup of the PNFS algorithm with different input data sizes and different number of cores. Here, we find that PNFS is neither superlinear speedup nor as excellent as PNEC. In spite of this, the results show that PNFS still has a good speedup with different input data sizes. And the speedup is getting higher with the increasing size of data. When the input size is double, the running time is quadruple if the same number of cores is used.

5 Conclusions

There is no doubt that data plays an extremely important role nowadays. Discovering precise knowledge from data is one of the hot topics. However, the amount of data grows exponentially that the traditional techniques could not keep up with the rhythm. Parallelizing the sequential algorithm is one of the best ways to solve this problem. We mainly parallelized two algorithms in neighborhood rough sets: classification and feature selection. Similarly, the proposed parallel approaches are also applicable to other classification and feature selection algorithms based on neighborhood rough sets. The experimental results showed both of them not only had the same result with the sequential algorithm, but also had good scalability. Through implementing the algorithms in Hadoop and lots of experiments, we acquire the following optimizing strategies which can help process data efficiently.

- We find that the partial data will be used every time. Hence, we store this part of the data to the distributed cache, which greatly reduces I/O time.
- According to the characteristic of Hadoop and MapReduce, we put most of computation into the Map phase and only integrate results in Reduce phase, which greatly reduces the overhead of communication.

More issues about load balancing will be discussed in future. Besides, different kinds of data could be processed through the developing algorithms since the proposed algorithms in this paper can only process the numerical data.

Acknowledgments This work is supported by the National Science Foundation of China (Nos. 61175047, 61100117, 61202043), the US National Science Foundation (No. OCI-1156733), the Science and Technology Planning Project of Sichuan Province (No. 2012RZ0009), China, and the Fostering Foundation for the Excellent Ph.D. Dissertation of Southwest Jiaotong University (No. 2012ZJB), China.

References

1. Dean J, Ghemawat S (2008) MapReduce: simplified data processing on large clusters. Commun ACM 51(1):107–113
2. Ghemawat S, Gobioff H, Leung ST (2003) The google file system. In: ACM symposium on operating systems principles, pp 29–43
3. Han J, Kamber M, Pei J (2011) Data mining: concepts and techniques, 3rd edn. Morgan Kaufmann, San Francisco
4. Hu Q, Yu D, Liu J, Wu C (2008) Neighborhood rough set based heterogeneous feature subset selection. Inf Sci 178(18):3577–3594
5. Hu Q, Yu D, Xie Z (2008) Neighborhood classifiers. Expert Syst Appl 34(2):866–876
6. Hu Q, Pedrycz W, Yu D, Lang J (2010) Selecting discrete and continuous features based on neighborhood decision error minimization. IEEE Trans Syst Man Cybern B Cybern 40(1):137–150
7. Li T, Ruan D, Geert W, Song J, Xu Y (2007) A rough sets based characteristic relation approach for dynamic attribute generalization in data mining. Knowl Based Syst 20:485–494
8. Newman D, Hettich S, Blake C, Merz C (1998) UCI repository of machine learning databases. University of California, Department of information and computer science, Irvine, CA. (http://archive.ics.uci.edu/ml/)
9. Pawlak Z (1991) Rough sets: theoretical aspects of reasoning about data, system theory, knowledge engineering and problem solving, vol 9. Kluwer Academic Publishers, Dordrecht
10. Ruan Y, Guo Z, Zhou Y, Qiu J, Fox G (2012) Hymr: a hybrid mapreduce workflow system. In: Proceedings of ECMLS 12
11. White T (2010) Hadoop: the definitive guide, 2nd edn. O'Reilly Media/Yahoo Press, Sebastopol, USA
12. Zhang J, Li T, Ruan D, Gao Z, Zhao C (2012) A parallel method for computing rough set approximations. Inf Sci 194:209–223
13. Zhang J, Li T, Ruan D, Liu D (2012) Rough sets based matrix approaches with dynamic attribute variation in set-valued information systems. Int J Approximate Reasoning 53(4):620–635
14. Zhang J, Wong JS, Li T, Pan Y (2014) A comparison of parallel large-scale knowledge acquisition using rough set theory on different mapreduce runtime systems. Int J Approximate Reasoning 55(3):896–907

Research on the Intensity of Subjective and Objective Vocabulary in Interactive Text Based on E-Learning

Wansen Wang and Peishen Li

Abstract Based on the text subjective judgment algorithm based on the rough set, we proposed an improved logarithmic linear model and fuzzy set combining the subjective intensity of learning method Chinese words and lexical subjectivity recognition, which is applied in the E-learning interactive text, and achieved better recognition results.

Keywords Log-linear model · Fuzzy set · E-Learning interactive text · Subjectivity intensity

1 Introduction

With the development of network information technology, E-Learning has become an effective form of school education, enterprise training, organization training. However, the traditional E-Learning system lack of emotion generally, in order to increase the emotional functions of E-Learning system, people began to study the emotion of learners studying with E-Learning. Approaches commonly used are: facial expression recognition, text sentiment analysis, speech emotion analysis. In fact, to mine the learners' ideas from academic texts, and then analyzed the

Supported by the National Natural Science Foundation of China under No. 60970052, Beijing National Natural Science Foundation (The Study of Personalized E-learning Community Education based on Emotional Psychology 4112014).

W. Wang (✉) · P. Li
Department of Information Engineering Institute, University of Capital Normal, Beijing, China
e-mail: wansenw@126.com

P. Li
e-mail: leeps2013@gmail.com

psychological condition is in the premise of academic interactive text subjectivity classification [1].

Subjectivity of Chinese words is a basic problem in text sentiment analysis. Its accuracy will directly affect the follow-ups; it is the basis of the sentiment analysis of the phrase level, sentence level, and paper-level. Although many studies have done [2, 3], the existing analysis methods in dealing with large-scale texts still face the following difficulties: For example, different words in the expression of Opinion may have different subjective intensity, and thus have different effects on subjective analysis of sentences or articles. Moreover, the same words in different language environments may have different subjective intensity, a major problem we are faced is distinguish the subjectivity of words according to the current language environments, but there is less research on the words subjective intensity. Also, a subjective sentence may include two or more subjectivity of the words, but the roles they play to express their opinions are different.

This article firstly introduces the rough set theory for reduction of the text, and secondly extracts corpus from E-leaning platform, and the use of rough set theory to reduce; then extracts emotional candidate words and views indicator words, and calculates their subjective weights; Finally, we combine fuzzy set theory, inspecting the impact of the intensity of subjective word on Chinese sentence subjectivity classification.

2 The Text Subjectivity Judgment Based on Rough Set

The so-called subjective text refers to the objective facts described in the text. Its main contents are based on the allegations or arguments, and with one's personal feelings and intentions. A sentence, no matter what form it expressed, as long as the sentence including a subjective component, then it is defined as the subjective sentence. Based on the subjective sentences these features, undoubtedly it is difficult to achieve the purpose of distinct subjective and objective words with methods of analysis of sentence constitutes.

Under the premise of maintaining the same sentence subjectivity, you can arbitrarily change the organization of the sentence, add modifiers. However, regardless of the form of subjective sentences changes the expression of subjective thinking will not change. For example, "I particularly like the movie" change its expression, under the premise of maintaining the same sentence subjectivity, can be said to be: No matter how they evaluate, I like the movie "or" the story of the drama develops more reasonable, but I still like the movie, these three sentences eventually to express personal views are "I like the movie", result of word software process is R R V I R R. So, in the sentence, it can be judged as long as it contains R R V I R R this model, they think this sentence is subjective sentence. This is what we determine the basis of subjective sentence. If a sentence contains the subjective sentences model, then the sentence is subjective sentence; otherwise, this sentence is the objective sentence.

Based on the above ideas, the first thing to do is collecting 1,000 subjective sentences from the E-Learning platform, then extract the structure model of the subjective sentences using word software. To this end, firstly, set these models to a certain threshold, if the threshold is reached, then save this mode; otherwise not retained. In this way, under the premise of ensuring the precision rounding part of the sentence patterns. In the remaining modes may contain redundant elements of sentence or mode of redundancy, for example: model one: R R V I N and model two: R R D V I N, of this case, it is necessary to consider the rough set theory to reduction experimental initial parameters, that is, the use of the knowledge of rough set areas reduction of redundant sentence elements, for example, mode: R R D V I N reduction off "D", the result is: R R V I N. And then re-use rough set attribute reduction, reduction of redundant attributes; here is the reduction redundant mode, the mode I and mode II reduction for the same mode: R R V I R R. Thus, from the original two modes, and seven sentence elements compare match to the final one mode, and three sentences Comparison of components, and this can improve the efficiency of the implementation of the program to a large extent [4].

Rough Set Theory plays a fundamental role in text for the Reduction and reducing the time cost of the system. The experiments show that either the precision or time cost of the research of the Chinese text based on rough set has a lot of improvements.

3 The Subjective Words Extraction Based on E-Learning

The next three steps are subjective words extracted from the training corpus. First of all, regard the verb as the candidate words of potential views indicator, adjectives and adverbs as underlying emotional candidate words. Then we use log-linear model to calculate the relevance of the words and subjective categories, as the weight of subjective words. Finally, we exclude those who cannot directive as candidate words of advice and emotional according to the weight of words.

In general, a view sentence [5] often contains comments indication words and emotional words to express their views. In Chinese, generally, opinions indication words are some verb, for example, regard, say advocates, these verb and opinions holders jointly published some comments. Emotional information often expressed with some polarity words or phrase, such as the adjective 'beautiful', 'ugly', the adverb 'but', 'may'. For convenience of presentation, we regard the words related with emotions expressed as emotional words. For log-linear model can be well predict variables and variables, and the degree of correlation between the variables and categories, so we use the log-linear model to predict the weight of words subjectivity, for it can better describe the degree of correlation between words and words, words and subjective categories in our training corpus.

Firstly, let's calculate the probability and frequency of candidate words in the training corpus. Table 1: The probability and frequency of subjectivity words in

Table 1 Contingency tables of frequencies and probabilities for weighting subjective words

W	C					
	Sub	Obj	$\sum_j$	Sub	Obj	$\sum_j$
W_1	n_{11}	n_{12}	$n_1.$	p_{11}	p_{12}	$p_1.$
$\vdots$	$\vdots$	$\vdots$	$\vdots$	$\vdots$	$\vdots$	$\vdots$
W_k	n_{k1}	n_{k2}	$n_k.$	p_{k1}	p_{k2}	$p_k.$
$\sum_i$	$n_{.1}$	$n_{.2}$	n	$p_{.1}$	$p_{.2}$	p
	Frequency table			Probability table		

the training corpus. Here W represents the words in the training corpus, C represents the subjective and objective categories, namely {subjective sentences, the objective sentences} n_{ij} means that frequencies of a subjective words $(w_i(1 \leq i \leq k))$ in a subjective and objective category $(c_j(j = 1, 2))$, its corresponding probability is $p_{ij} = n_{ij}/n$, n is sum of all the n_{ij}.

As shown in Eq. (1), the probability table is expressed as logarithmic form.

$$\eta_{ij} = \ln p_{ij} = \ln\left(p_i.p._j \frac{p_{ij}}{p_i.p._j}\right) = \ln p_i. + \ln p._j + \ln\frac{p_{ij}}{p_i.p._j}. \tag{1}$$

Make $\eta_i. = \sum_{j=1}^{2} \eta_{ij}, \eta._j = \sum_{i=1}^{k} \eta_{ij}, \eta.. = \sum_{i=1}^{k}\sum_{j}^{2} \eta_{ij}.$

So, the average logarithmic probability can be calculated by the following formula (2, 3, and 4).

$$\bar{\eta}_i. = \frac{1}{2}\sum_{j=1}^{2} \eta_{ij} \tag{2}$$

$$\bar{\eta}._j = \frac{1}{k}\sum_{i=1}^{k} \eta_{ij} \tag{3}$$

$$\bar{\eta}.. = \frac{1}{2 \times k}\sum_{i=1}^{k}\sum_{j}^{2} \eta_{ij}. \tag{4}$$

Make $\gamma_{ij} = \eta_{ij} - \bar{\eta}_i. - \bar{\eta}._j + \bar{\eta}.., \hat{p}_{1j} = n_{1j}/n, \hat{p}_i. = n_i./n, \hat{p}._j = n._j/n.$

And γ_{ij} is the interaction between words w_i and subjectivity Category C_j. $\gamma_{ij} > 0$ presents there is positive interaction, and $\gamma_{ij} < 0$ presents they have a reverse effect on the interaction, and when $\gamma_{ij} = 0$, there is no interaction between them. We also define $\hat{\eta}_{ij}, \hat{\bar{\eta}}_i., \hat{\bar{\eta}}._j$ 和 $\hat{\bar{\eta}}..$ as follows:

$$\hat{\eta}_{ij} = \ln \hat{p}_{ij} = \ln n_{ij} - \ln n \tag{5}$$

$$\hat{\bar{\eta}}_{i\cdot} = \ln\frac{1}{2}\sum_{j=1}^{2}\eta_{ij} = \frac{1}{2}\sum_{j=1}^{2}\left(\ln\frac{n_{ij}}{n}\right) = \frac{1}{2}\sum_{j=1}^{2}\left(\ln n_{ij}\right) - \ln n \qquad (6)$$

$$\hat{\bar{\eta}}_{\cdot j} = \frac{1}{k}\sum_{i=1}^{k}\eta_{ij} = \frac{1}{k}\sum_{i=1}^{k}\left(\ln\frac{n_{ij}}{n}\right) = \frac{1}{k}\sum_{j=1}^{2}\left(\ln n_{ij}\right) - \ln n \qquad (7)$$

$$\hat{\bar{\eta}}_{\cdot\cdot} = \frac{1}{2k}\sum_{i=1}^{k}\sum_{j=1}^{2}\eta_{ij} = \frac{1}{2k}\sum_{i=1}^{k}\sum_{j=1}^{2}\left(\ln\frac{n_{ij}}{n}\right) = \frac{1}{2k}\sum_{i=1}^{k}\sum_{j=1}^{2}\left(\ln n_{ij}\right) - \ln n. \qquad (8)$$

And further we can calculate the estimated value of γ_{ij} by the Eq. (9)

$$\hat{\gamma}_{ij} = \hat{\eta}_{ij} - \hat{\bar{\eta}}_{i\cdot} - \hat{\bar{\eta}}_{\cdot j} + \hat{\bar{\eta}}_{\cdot\cdot}$$

$$= \ln n_{ij} - \frac{1}{2}\sum_{j=1}^{2}\left(\ln n_{ij}\right) - \frac{1}{k}\sum_{i=1}^{k}\left(\ln n_{ij}\right) + \frac{1}{2\times k}\sum_{i=1}^{k}\sum_{j=1}^{2}\left(\ln n_{ij}\right). \qquad (9)$$

We use $\hat{\gamma}_{ij}$ to measure the contribution of candidate words (w_i) to subjective category (C_j), $\hat{\gamma}_{ij}$ shows the words' subjectivity weight. Table 2 shows the value ($\hat{\gamma}_{ij}$) of candidate subjective words in the training corpus.

4 The Identification of Subjective Words in Fuzzy Sets

Depending on the weight of the subjectivity of the words, we will divide them into fuzzy sets, namely: 'high subjective intensity', 'moderate subjective intensity', 'low subjective intensity', then we construct the membership function of each collection, according to the membership function to determine the subjective intensity of the unknown words.

4.1 Membership Function of the Intensity of the Subjective Words

We selected trigonometric functions as membership function to describe the distribution. Firstly, make cluster centers of three-level collection $M = \{m_1, m_2, m_3\}$, and then we defined the membership function as follows:

$$T_l(x) = \begin{cases} 1 & x \leq m_1 \\ \frac{m_2 - x}{m_2 - m_1} & m_1 < x < m_2 \\ 0 & x \geq m_2 \end{cases} \qquad (10)$$

Category of subjective words	Examples	$\hat{\gamma}_{ij}$
Opinion indicator words	'Feel'	3.6310
	'Indicate'	3.6042
	'Assert'	2.3900
	'Forecast'	2.3233
	'Report'	−1.7848
	'Cassette'	−2.1916
Sentiment words	'Inevitable'	2.3510
	'Satisfied'	2.2740
	'Afraid'	2.0715
	'Pollution'	1.9279
	'Accept'	−0.5214
	'Issue'	−0.5389

Table 2 The weight of some opinion indicators and sentiment words under log-linear modeling

$$T_{med}(x) = \begin{cases} 0 & x \leq m_1 \\ \frac{x-m_1}{m_2-m_1} & m_1 < x < m_2 \\ \frac{m_3-x}{m_3-m_2} & m_2 < x < m_3 \\ 0 & x \geq m_3 \end{cases} \tag{11}$$

$$T_h(x) = \begin{cases} 1 & x \geq m_3 \\ \frac{x-m_2}{m_3-m_2} & m_2 < x < m_3 \\ 0 & x \leq m_2 \end{cases} \tag{12}$$

In this paper, we use the method of self-organizing feature maps to determine the center collection M, the method of SOM has corrected the distance of sample point to the center point by the method of error propagation and via an iterative convergence ultimately determine the cluster center. According to the SOM algorithm we can calculate the weight set of cluster centers of opinion indicator words and sentiment words. Among the membership function of the views indicator words $m\_1 = -1.226$, $m\_2 = 1.035$, $m\_3 = 3.890$, in the membership function of the sentiment words $m\_1 = -0.854$, $m\_2 = 1.205$, $m\_3 = 3.114$ [6].

4.2 Subjective and Objective Classification Based on Complex Rules

To test the impact of the words subjective intensity on Chinese sentence subjectivity classification, we use a set of classifier based on rules, which mainly determined sentence's subjectivity by looking for the different subjective intensity of the subjectivity of the words in the sentences. Unlike (Riloff and Wiebe 2003) [7] using the single rule classifier, it mainly contains the following rules.

- If the sentence contains lots of high intensity or medium intensity view indicator words whose intensity is greater than a given threshold value δ, then sentence is subjectivity sentence.
- If the sentence contains lots of high intensity or medium intensity sentiment words whose intensity is greater than a given threshold value δ, then sentence is subjectivity sentence.
- If the sentence contains lots of high-intensity or medium intensity view indicator words whose intensity is greater than a given threshold value δ, then sentence is subjectivity sentence; at the same time, the intensity of high intensity or medium intensity sentiment words contained in the sentence is greater than a given threshold value μ, then the sentence is subjectivity sentence.

δ and μ are two experienced threshold value which can be determined by experiment. As it can be seen from the above rules, rule 1 and rule 2 are view indicator words' and sentiment words' single effect on the subjective identification of sentence, rules 3 taking into account both.

5 Experiment and Analysis

5.1 Experiment Setting

Data we used in this paper are from the E-Learning platform academic text.We constructor own training sets and test sets by extracting the data from these texts and the evaluation criteria we use is Lenient-AWK: recall rate (R), precision rate (P) and their harmonic mean (F) [8].

5.2 Experiment Result

5.2.1 Effect of View Indicator Words and Sentiment Words on Sentence Subjectivity Identification

The first sets of experiments were to test the effect of different types of subjective words on the recognition of subjectivity of Chinese sentences, including views indicator and sentiment words. The experiments in this article uses the following three types of rules to evaluate the impact of words in the sentence subjectivity classification, the experimental results are given in Table 3.

As can be seen from Table 4, a system was constructed according to the rules 1 and rule 2 to obtain a higher precision, but the recall rate is lower. Rule 3 has achieved the best overall performance, it embodies an active role in considering the views indicator words and sentiment words in Chinese Sentences subjectivity

Table 3 Basic statistics of the experimental data

Text type	Training data	Test data
Theme	35	14
Document	901	188
Sentence	13416	4655

Table 4 Evaluation results for different classifier using different rules with $\delta = 1$ and $\mu = 1$

Subjectivity and objectivity classifier	P	R	F
Rule 1	0.7085	0.5551	0.6225
Rule 2	0.8144	0.5363	0.6447
Rule 3	0.7175	0.8006	0.7567

recognition. Rule 2 creates the classifier that has obtained the highest accuracy rate to 81.44 %, but the recall rate of only 53.63 %, the main reason may be the opinion sentences containing sentiment words in text, but the subjectivity of these sentiment words' intensity is weak, so the classifier constructed in rule 2 does not recognize too many subjective sentences.

5.2.2 Distinguish the Effect of Subjective Words' Intensity on Subjective and Objective Recognition of the Sentences

The second set of experiments was designed to test the subjectivity intensity of the words in the subjective and objective classification of the sentence [9]. We compared the experimental results with the Baseline system in the NTCIR-7MOAT and with the best WIA-Opinmine system. Our sentiment dictionary contains 8,596 subjectivity words, mainly from the CUHK and NTU sentiment dictionary. In this experiment, the Baseline system did not distinguish sentence subjectivity intensity, namely, the subjectivity of the level of intensity of the sentence is completely ignored, the WIA-Opinmine with a fine-grained to coarse-grained strategy to explore the composite characteristics sentence, and in the recognition of the subjectivity of the sentence, and ultimately achieve very good results.

Table 5 shows the results of the second set of experiments. We found that this system's precision and F values are higher than the WIA-Opinmine system, but lower in recall rate. Perhaps we use less subjective characteristics than WIA-Opinmine system, and therefore cannot recognize more subjective sentences in corpus. Our system is beyond the Baseline system about 10 %, which indicates that distinguish the subjectivity of the words in the intensity of subjective identification of the sentence has a very important role.

In order to study the key role of words' subjective intensity in subjectivity identification of Chinese sentence, the paper improved a subjective intensity learning method based on the log-linear model and fuzzy sets. Including:

Table 5 Comparison of our system with the best system at NTCIR-7 under the lenient standard

System	P	R	F
Baseline system	0.5288	0.8511	0.6523
WIA-Opinmine system	0.6520	0.8698	0.7453
Our system	0.7175	0.8006	0.7567

(1) candidate subjective words extraction and weighting terms; (2) constructed and parameters determined of membership function of subjective words with different intensity; (3) methods with the subjective and objective classification of sentences based complex rules. Experimental results show that the view indicator Verbs and sentiment words play a very important role in the classification of subjective sentences, and each of them are in a different way to boot opinions expressed in the sentence.

Even though the distinction the subjective intensity of words in the sentence can make subjective and objective classification results significantly improved, there is still insufficient during rough set reduction, for those with subjective and objective sentence model, whether to retain or reduction will affect the system in the implementation of judgment subjective or objective of recall and precision. Therefore, our follow-up work will improve it.

Acknowledgment First of all, I would like to thank my mentor the Professor Wang. Without his help I could not complete it successfully. Secondly, I sincerely thank the support and encouragement that my classmates give me. Thirdly, I would like to thank the support of the National Natural Science Foundation of China, and Beijing Natural Science Foundation. And finally I would show special thanks to my family for their support understanding and encouragement.

References

1. Qing Y, Zi-qiong Z, Zhen-xiong L (2007) Study of methods of subjectivity automatically discriminating of Chinese on sentiment analysis in Internet comments. China J Inf Syst 1(1):79–91
2. Ellen R, Siddhart P, Janyee W (2006) Feature subsumptionf or opinion analysis. In: Proceedings of EMNLP'06, Sydney, Australia, pp 440–448
3. Xin-fan M, Hou-feng W (2009) Research about effect of the context of factors on Subjective and objective recognition. China Acad J, 594–599
4. Long-shu L, Xiao-hong Z, Zhi-wei Z (2011) The subjective and objectivity Research of the Chinese text based on rough set theory. Comput Technol Dev, 114–115
5. Bo Z (2011) Chinese views sentence extraction based on SVM. The Academy of Computer of Beijing University of Posts and Telecommunications, Beijing
6. Xi W (2011) Research of methods on multi-granularity fusion of Chinese sentences' subjectivity and sentiment classification. The Academy of Computer Science of Heilongjiang University, Harbin
7. Ellen R, Janyce W, Phillips W (2003) Learning subjective nouns using extraction pattern bootstrapping. In: Proceedings of CoNLL'03:25-3

8. Yohei S, David KE, Lun-Wei K, Hsin-His C, Noriko K, Chin-Yew L (2007) Overview of opinion analysis pilot task at NTCIR-6. In: Proceedings of NTCIR-6 workshop meeting, Tokyo, Japan, pp 265–278
9. Yohei S, David KE, Lun-Wei K, Le S, Hsin-His C, Noriko K (2008) Overview of multilingual opinion analysis task at NTCIR-7. In: Proceedings of NTCIR-7 workshop meeting, Tokyo, Japan, pp 185–203

Geometry Knowledge Base Learning from Theorem Proofs

Hongguang Fu, Xiuqin Zhong, Qunan Li, Huadong Xia and Jie Li

Abstract Geometry theorem proofs like propositions in Euclid's geometry elements contain fruitful geometry knowledge, and the statements of geometry proofs are almost structural mathematics language. Hence, it is possible to let computer understand geometry theorem proofs. Based on the process ontology, a novel geometry knowledge base (GKB) in this paper is built by letting computer learn from theorem proofs. The resulting process ontology is automatically constructed by extracting abstract and instance models (IMS) from proofs. The abstract model displays the causal relations of conditions with conclusions, and the instance model (IM) holds the formal relationship of abstract model so that the deduction can be reused. Thus, two kinds of models completely describe the proving process of geometry theorem. Furthermore, GKB based on the process ontology can be gradually extended by learning from more and more proofs. Finally, GKB learning from about 200 examples is implemented, and an application in automated theorem proving is given.

Keywords Knowledge base · Process ontology · Abstract model · Instance model · Automated theorem proving

H. Fu (✉) · X. Zhong (✉) · Q. Li · J. Li
School of Computer Science and Engineering, University of Electronic Science and Technology of China, Chengdu 610054, China
e-mail: fu_hongguang@hotmail.com

X. Zhong
e-mail: zhongxiuqin2009@gmail.com

H. Xia
Department of Computer Science, Virginia Tech, Blacksburg, USA

Z. Wen and T. Li (eds.), *Knowledge Engineering and Management*,
Advances in Intelligent Systems and Computing 278,
DOI: 10.1007/978-3-642-54930-4_3, © Springer-Verlag Berlin Heidelberg 2014

1 Introduction

Knowledge base (KB) is a special kind of database for knowledge management, which aims at machine-readable and knowledge sharing. With the development of KB, the construction can be performed by reusing existing knowledge components, which mainly includes pieces of domain knowledge [1]. Moreover, KB based on ontology can supply semantic information efficiently [2].

Ontology has been a frontier in information systems and artificial intelligence, as a conceptual model tool which can describe the information in semantic and knowledge level [3]. As Studer defined in 1998, "An ontology is a formal, explicit specification of a shared conceptualization" [4], it can accurately regulate the domain concepts and clearly define their relations [5].

Existing common ontologies to express our knowledge have only focused on deriving static knowledge, namely entity models, and can depict classes, instances, and their hierarchical relations [6]. Because the construction of ontology is very difficult and costly, the automatic ontology construction or semiautomatic ontology construction becomes more important. Recently, most ontologies have been built by automatic approaches, for example, YAGO is just a core of semantic knowledge unifying WordNet and Wikipedia [7], which has the advantage of automated construction and growth [8].

According to the type of thinking, the knowledge is divided into static knowledge and procedural knowledge, then domain ontology is divided into static ontology and process ontology correspondingly. In recent years, because the static one has shortcomings to describe domain knowledge with processes, process thinking has gradually been accepted by most scholars [9]. Heraclitus, a Greek philosopher, is one of the representatives who firstly used process thinking to explore ontology, and his ontology named "process flow" laid a solid philosophical foundation for the development of process ontology [10]. Whitehead put forward "Process Philosophy" clearly and discussed it systematically. His interpretation of "process ontology" promoted the ontological exploration turning from the static to dynamic ontology.

Furthermore, many researchers focus on process ontology, such as Heravi et al. [11] proposed a process ontology for ebXML Business Process Specification Schema, enabling knowledge deduction and reasoning over the knowledge; Ko et al. [12] introduced the Business-OWL, and modeled as a hierarchical task network for the dynamic formation of business processes; Liang et al. [13] conceptualized the customization of service-based business processes leveraging the existing knowledge of Web services and business processes. Besides, Harrison et al. [14] introduced the use of ontology in the implementation phase of a manufacturing system and applied process ontology to hardware-in-the-loop and hybrid process simulation.

2 Related Work

Gelernter [15] produced the earliest geometry knowledge base (GKB) of merit in proving geometry theorems, whose method is primarily based on the traditional Euclidean proof. Tarski [16] provided an algebraic decision procedure for "elementary geometry." Tarski's work was further improved upon by others, yet they still failed in practice to mechanically prove any nontrivial theorems. The breakthrough year came in 1977, when Wu [17] introduced an algebraic method capable of providing mechanical proofs to many nontrivial traditional geometry theorems. In 1994, Chou et al. [18] put forward a geometry invariants approach, so that the resulting proving processes are readable.

After many years, though a machine can prove the geometry theorems by algebraic methods, it seems that the machine proofs cannot substitute the traditional Euclidean proofs twinkling man's wonderful wisdom yet. So it is time to go back the start point of Gelernter for learning from traditional Euclidean proofs again.

The authors in this paper proposed a schema of applying the ontology of geometry to prove geometry theorems [19]. But the former geometry ontologies are built manually by domain experts.

The automatic construction of ontology concerns semantic analysis on textual data. It is fortunate that geometry theorem proofs are written in almost structural mathematics language, so the semantic templates can be produced by learning problem processes. Hence, the characteristic makes it possible to let computer understand geometry theorem proofs.

Process ontology is a description of the components and their relationships that make up the process. It is usually designed to describe domain-related process model in a declarative way. A geometry theorem proof is the typical process thinking, so it can be depicted by the process ontology clearly.

In this paper, we will study the automatic construction method of geometry process ontology by learning from problem proving processes, i.e., theorem proofs. In Sect. 3, we will illustrate the process ontology of geometry. In Sect. 4, a GKB will be built on process ontology. In Sect. 5, construction of abstract models (AMS) and instance models (IMS) will be shown, and automated theorem proving will be given.

3 Process Ontology Learning

A theorem proving process means that there exists a feasible path (i.e., workflow) from given conditions to conclusions according to rules such as definitions and theorems. Moreover, the proving process can be divided into a series of logical segments with causality, so that each logical segment is composed of conditions, conclusion, and reason. Therein, the conditions and conclusions are relative, that is, the conditions of a logical segment may be the conclusions of former ones.

Before the process ontology is constructed, initial geometry conceptual classes (such as point, segment and triangle, etc.) and properties (such as parallel, perpendicular, equal, etc.) need to be defined ahead. Here, classes and properties which have been constructed in literature [19] can be reused.

But the defined classes and properties cannot adequately describe the process ontology. Therefore, classes and properties need to be extended, for example, add new relation classes such as Segment_Parallel class and Segment_Equal class, and new properties such as abstract known theorem properties and abstract unknown theorem properties. Then a basic geometry conceptual ontology is yielded, so that it is possible to understand sentences in proofs by simple semantic matching.

In the following, after extending the classes and properties, extracting abstract model and instance model (IM) from problem solving process will be illustrated for automatic learning of process ontology.

3.1 Abstract Model

The process ontology learning is the kernel to construct geometry ontology. Our main approach is based on abstract model and IMS extracting from proofs.

Definition 1 Abstract model is a directed graph structure which describes one rule of the problem proving process by learning, and the structure is defined by AM: = {sub, pre, obj}, therein, sub is abstract description of the set of conditions about the rule; pre is the concrete theorem about the rule, furthermore, it is the reason from conditions to conclusion; and obj is abstract description of the conclusion about the rule.

Naturally, the conditions, conclusion, and reason in a logical segment are corresponding to condition individuals, conclusion individual, and reasoning property, respectively. Besides, a condition individual is taken as a subject (i.e., sub), a reasoning property as a predicate (i.e., pre), and a conclusion individual as an object (i.e., obj). Then, triples ⟨sub, pre, obj⟩ composed by them will be created and added into the ontology database.

In order to extract the abstract model from logical segment, it is necessary to match patterns in conditions and conclusions by regular expressions. Hence, the regular expressions and their classes need to be summarized and stored in advance.

Here, let A be the set of conditions in the AMS extracted from problem proving process. Then, an abstract model which takes the conclusion's individual as its object will be queried by SPARQL. For example, "select ?sub ?pre where {?sub ?pre ns:segment_parallel_1}" can be used to query subject and predicate of the abstract model whose conclusion is the segment_parallel_1 class. Therein, subject is condition of the model, and the set of conditions in the corresponding models queried from the ontology database is denoted by B. Table 1 describes the construction of abstract model algorithm in detail.

Table 1 Construction of abstract model algorithm

Algorithm. Construction of Abstract model

Input: The logical segment L_j , The Abstract model database *AMs*

Output: *NULL*

1. Extract the conditions from L_j , and convert its conditions to the set of abstract relation's individual A:$\{ a_1 , a_2 ,..., a_m \}$ (a_i is abstract relation's individual). Extract the reasoning property **pr** from the predicate of logical segment L_j .. Extract the conclusion from L_j and convert it to the abstract relation's individual **ob**. Combination of A, **pr** and **ob** into an abstract model
    AM:$\{ N_1 , N_2 ,..., N_m \}$(N_i :=$\{ a_i$, **pr**, **ob** \}triples,i=1,2,...,m).

2. Query the abstract model X:$\{ X_1 , X_2 ,..., X_k \}$(X_i :=\{*sub, pre, obj*\}triples, i=1,2,...,k) by the condition of the object **ob** in *AMs*.

3. *if* X is not *NULL*

4. *then*   Extract the subjects of the abstract model in X and use   the extract results to construct the set of abstract relation's individual B$\{ b_1 , b_2 ,..., b_k \}$(b_i is abstract relation's individual).

5.   *if* A⊆B

6.   *then return*

7.   *end if*

8.   *else if* A=B

9.   *then if* **pr** is known reasoning property *and* the   reasoning property in X is unknown

10.   *then* update the reasoning property in X

11.   *return*

12.   *end if*

13.   *end if*

14.   *else if* B⊆A

15.   *then* complete the subjects of the abstract model X

16.   *return*

17.   *end if*

18.   *else*

19.   construct abstract model *AM* in *AMs*

20. *end if*

If the abstract model which takes the conclusion individual as its object does not exist, then it will create an abstract model initially. Otherwise, it will compare the relationship between set A and set B.

If A contains B, it means that the extracted model is more perfect than the model in ontology database, therefore, the model's condition in ontology database needs to be improved.

If A is equal to B, then computer would recognize whether the relations need to be updated, that is to say, if the extracted reasoning property is known (i.e., the reasoning property can be learned from the text) and the reasoning property of the ontology is unknown, then the relations need to be updated.

Else if B contains A, it means that the extracted model is not better than the model of ontology database, which can be ignored. Of course, if there is no relationship between A and B, then it will continue to check other AMS which take the conclusion individual as its object. Therefore, AMS can be created or improved in this way.

3.2 Instance Model

Abstract model can clearly display the structure and sequence of relation individuals. However, abstract model has coarse granularity, and its relation individuals cannot express the relation of their elements. In order to make the model contain more detailed information, it needs to refine the extracted granularity greatly.

Definition 2 Instance model is a directed graph structure which describes one rule of the problem proving process by learning, and the structure is defined by IM: = {sub, pre, obj}, therein, sub is concrete instance description of the set of conditions about the rule; pre is the concrete theorem about the rule, furthermore, it is the reason from conditions to conclusion; and obj is concrete instance description of the conclusion about the rule.

Here, IM can be used to explain the corresponding abstract model. In this paper, the extracted granularity of IM is refined to points of elementary geometry, and each point is marked with alphabet.

In the ontology, some mathematical symbols cannot be used to construct individuals, so some symbols will be replaced by text. For example, basic relation unit "AB = CD" can be denoted by the relation individual "AB Segment_equal CD"; the element "$\triangle ABC$" can be denoted by "Triangle ABC."

The construction of IM algorithm is similar to but different from the algorithm in Table 1. First, it will adopt regular expressions to match patterns for the extracted basic conditions and basic conclusions, then try to find their relation class and convert it into relation individual of instance (for example, AB Segment_Equal CD); If an abstract model corresponding to this IM existed in the ontology database, then it can decide how to construct or improve this IM according to operations on the abstract model.

If it needs to improve the conditions of the abstract model, we should delete all IMS (IM is associated with concrete points, and one abstract model corresponds to many IMS, so it is difficult to improve the IMS with the existed conditions), and then construct the extracted IM. Furthermore, if an abstract model only needs to update relation, it should update the relation of IM and then construct the extracted IM. If improving conditions and updating relations are both needed, it should delete all IMS and construct the extracted IM.

Construction or improvement of IM mainly includes that of relation individuals, element individuals, properties, the relations between element individuals, and relation individuals.

3.3 Process Ontology

By then, the AMS and the IMS are extracted by learning from problem proving processes of geometry problems (i.e., geometry exercises with solution process).

The process ontology learning is a process of machine learning, which is composed of two steps and the detail algorithm is summarized in Table 2.

First, a computer would parse the problem proving process and then deal with its logical segmentations. In the problem proving process, a logical segmentation is regarded as the segmentation of a series of conditions and conclusions which contains "∵" and "∴" as the keywords. If the keywords are omitted, the sections cannot be used as the learning materials.

Second, the abstract and IMS are extracted from logical segments by the semantic analysis. Here the abstract model displays the causal relations of conditions with conclusions, i.e., it is the conceptual description of a definition or theorem, and the IM holds the formal relations of the abstract model, which is used to illustrate the relationship of individuals.

Finally, process ontology will be constructed by sequencing AMS and instantiating IMS. The sequence of the problem solving process is made up of many items. Each item is defined by {SNO, C, R, CSNO}, in which SNO is the current item's sequence number, C is the formal description of current item, R is the corresponding rule of current item, and CSNO is the set of sequence numbers of conditions. The sequence has strict hierarchical relationships. That is to say, the first level consists of the whole conditions of the problem, the second level consists of all conclusions based on the conditions of first level, and the N level of the problem-solving process must contain at least one process in $N - 1$ level. Of course, the sequence numbers in the same level may be different but in disorder.

3.4 GKB on Process Ontology

Hence, an GKB can be built based on the process ontology. As is shown in Fig. 1, the GKB is composed of geometry problem (exercise) database, process ontology of geometry, and reasoning engine. Therein, process ontology consists of three components: core geometry ontology (i.e., reused geometry ontology (RGO)) which is constructed in literature [19], AMS, and IMS which are extracted from problem proving process. The process ontology is expressed by RDF triples and is stored in AllegroGraph (AG) database. Therefore, applications such as automated theorem proving, intelligent tutoring, and so on can be executed based on the GKB.

Table 2 Algorithm of Process Ontology Learning

Algorithm. Process ontology learning
Input: the text of praxis content and the proving process **P**.
Output: process ontology **PO**
1.   Extract the logical segments **L**:{ L_1 , L_2 ,..., L_r } (L_i :={condition→conclusion} logical relation, i=1,2,...,r) by analyzing **P**.
2.   *for* L_i *in* **L**
3.   *do* construct the abstract model **AM**
4.   construct the instance model **IM**
5.   *end for*
6.   *return* **PO**

Fig. 1 Structure of GKB

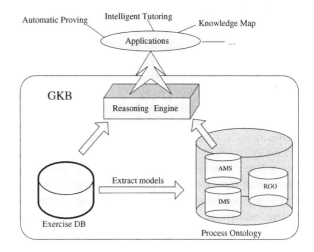

4 Implementation

In this paper, the experiment system environment is Windows XP; the tool for ontology storage and compilation is AG Store 3.3.4; the tool for demonstrating ontological data is AG Gruff, and the domain is elementary geometry.

4.1 Construction of Abstract Models and Instance Models

In the following, we will describe the construction of abstract model by illustration. Logic segmentations in textual data can be managed according to process ontology learning algorithm in Table 2, and the extracted AMS are shown in Table 3. The relation individuals such as "Segment_Equal_1," "Angle_Equal_1," and "Segment_Equal_2" point to "Triangles_Congruent_1" through the property "SAS" (Side Angle Side Congruence Theorem), which constitutes an abstract model. Each abstract model corresponds to several triples which will be stored into process ontology as RDF.

But relation individuals in abstract model cannot display concrete relations. For example, the segments in "Segment_Equal_1" are actually edges of triangles in "Triangles_Congruent_1", which cannot be represented by the abstract model. Therefore, it needs to extract IM from the segmentations.

Partial IMS corresponding to AMS in the first example in Table 3 can be shown in Fig. 2. There are three relation individuals "AB Equal AC," "BAD Equal CAD," and "AD Equal AD" as for known conditions, and one relation individual "ABD and ACD Congruent" as for conclusion in the IMS. In addition, the IM contains element individuals, just like "Segment AB," "Segment AC," "Angle BAD," and so on.

Furthermore, it has relation between element individuals. For example, "Triangle ABD" points to "Angle BAD" through the property "Has Angle," and points to "Segment AB" and "Segment AD" through the property "Has Edge," which can be shown in Fig. 2. It indicates that IM can fully represent relations between element individuals.

4.2 Elementary Geometry Process Ontology

At this moment, the process ontology can be constructed through AMS and IMS. All the models are represented with RDF triples and stored as xml files in AG.

The experiment platform deals with 200 problems with proving process. The AMS are constructed according to algorithm in Table 1, and the IMS are constructed according to algorithm in 3.2. In Table 4, there list the number of problems, the number of produced AMS and IMS, the number of covered theorems, and the number of updated AMS and IMS during learning. By statistics, there are 239 AMS and 347 IMS produced from 200 problems, which cover 178 theorems. And the corresponding column graph is generated in Fig. 3.

In Fig. 3, it can be shown that the newly constructed AMS and IMS will be reduced gradually with the increasing learning problems. When a number of problems are learned, the AMS and the IMS which constitute the process ontology will cover the entire elementary geometry field.

By sequencing the AMS and IMS, the process ontology can be generated for each problem. For example, the following problem "Statement: BE = CD, AB = AD, ∠ABE = ∠ACE, ∠ADC = ∠ACE. Proof: △ABE≅△ACD," whose corresponding abstract description of condition is "(1) Segment_Equal_1 (2) Segment_Equal_2 (3) Angle_Equal_1 (4) Angle_Equal_2."

The process ontology can be constructed by sequencing the abstracted model from known conditions [(3), (4)] in Level 0 to newly produced result (Angle_Equal_3) in Level 1 through the theorem of Transitive Property. The corresponding

Table 3 The examples of abstract models

NO.	Input logical unit	Abstract models
1	∵AB = AC, ∠BAD = ∠CAD, AD = AD, ∴△ABD≅△ACD	⟨Segment_Equal_1, SAS, Triangles_Congruent_1⟩ ⟨Segment_Equal_2, SAS, Triangles_Congruent_1⟩ ⟨Angle_Equal_1, SAS, Triangles_Congruent_1⟩
2	∵∠MAI = ∠GAC, ∠XIB = ∠MAI, ∴∠XIB = ∠GAC	⟨Angle_Equal_1, Transitive Property, Angle_Equal_3⟩ ⟨Angle_Equal_2, Transitive Property, Angle_Equal_3⟩
3	∵∠DIU and ∠IUQ are supplementary, ∠IUQ and ∠EHZ are supplementary, ∴ ∠DIU = ∠EHZ	⟨Angle_Supplementary_1, Congurent Supplements Theorem, *Angle*_Equal_1⟩ ⟨Angle_Supplementary_2, Congurent Supplements Theorem, *Angle*_Equal_1⟩

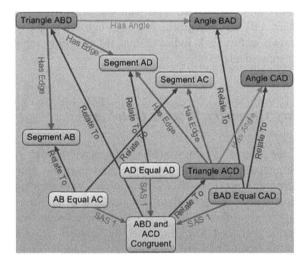

Fig. 2 Display of partial instance models

Table 4 The corresponding model number of problems

Problems	AMs	IMs	Theorems	Updated AMs	Updated IMs
10	63	80	51	2	1
50	160	193	125	6	14
100	196	241	152	7	12
150	221	295	166	5	10
200	239	347	178	6	8

Fig. 3 The histogram of models

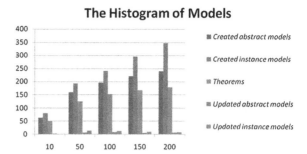

formal description is ⟨Angle_Equal_1, Angle_Equal_2; Transitive Property; Angle_Equal_3⟩. And the corresponding item of the problem solving process is described by: ⟨(5), ∠ABE = ∠ADC, Transitive Property(3)(4)⟩.

Forthermore, the abstracted model from known conditions [(1), (2)] in Level 0 and newly produced result [(5)] in Level 1 to the conclusion (Triangle_Congruent_1) in Level 2 through the theorem of SAS. The corresponding formal description is ⟨Segment_Equal_1, Segment_Equal_2, Angle_Equal_1; SAS; Triangles_Congruent_1⟩. And the new item of problem-solving process is described by: ⟨(6), ΔABE ≅ ΔACD, SAS(1)(2)(5)⟩. Therefore, the conclusion of the problem is achieved. The whole process flow is shown in Fig. 4.

For each problem, there exists a sequence of the AMS and IMS. Therefore, the process ontology is constructed by AMS and IMS abstracted from the problems according to the method proposed in this paper.

4.3 Automated Theorem Proving Based on GKB

The approach of automated theorem proving based on process ontology is querying whether it can reach the conclusion according to the given conditions. If other conditions required by the conclusion are also given, then the conclusion can become a new condition, until the final conclusion is achieved. So, GKB is built based on the process ontology. Automated theorem proving can be executed based on the GKB.

Figure 5 is an example of proving "triangle congruence theorem" based on the process ontology by inputting the textual data.

Therefore, the process ontology of elementary geometry can fully express the reusability and sharing of ontology, and provide a new method for automated theorem proving. In addition, the ability of automated proving can be enhanced with the growing number of machine learning problems.

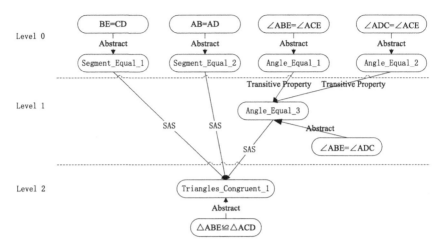

Fig. 4 The example of process ontology

Fig. 5 An example of automated theorem proving

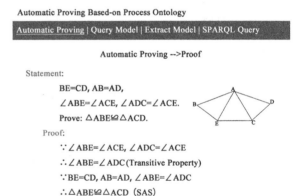

5 Conclusions

The resulting process ontology has good structure, rigorous logic, and formalization description. Therefore, it can implement automated theorem proving based on the elementary geometry process ontology. In the future, the process ontology should be extended by updating relations or improving conditions of AMS and IMS, so that the reasoning efficiency can be enhanced, and the machine can solve more complicated problems. Moreover, the method is helpful for constructing other disciplinary knowledge bases.

Acknowledgments The work in this paper was supported by National Natural Science Foundation of China (No. 61073099 and No. 61202257) and the Sichuan Provincial Science and Technology Department (No. 2012FZ0120).

References

1. Chaudhri VK, Farquhar A, Fikes R et al (1998) OKBC: A programmatic foundation for knowledge base interoperability. In: Proceedings of AAAI-98, pp 600–607
2. Chenjian H (2011) Research on knowledge model for ontology-based knowledge base. In: 2011 international conference on business computing and global informatization (BCGIN), pp 397–399
3. Antoniou G, van Harmelen F (2008) A semantic web primer. China Machine Press, Beijing, pp 47–62
4. Lu J-J, Zhang Y-F (2007) Semantic web principles and technology. Science Press, Beijing, pp 48–52
5. Gao Z, Yue P, Li M (2009) Principle and application of the semantic web. China Machine Press, Beijing, pp 1–15
6. Berners-Lee T, Hendler J, Lassila O (2001) The semantic web. Sci Am 284(5):34–43
7. Suchanek FM, Kasneci G, Weikum G (2008) YAGO—A large ontology from Wikipedia and WordNet. Elsevier J Web Seman 6(3):203–217
8. Suchanek FM (2008) Automated construction and growth of a large ontology. PhD thesis, Saarland University, Germany
9. Li J (2010) The evolution from material ontology to process ontology. J Yichun Coll 32(1):16–18
10. History of Foreign Philosophy, Peking University Department of Philosophy Teaching (1982) Ancient Greek and Roman philosophy. Commercial Press, Beijing
11. Heravi BR, Bell D, Lycett M, Green SD (2010) Towards an ontology for automating collaborative business processes. In: 2010 14th IEEE international enterprise distributed object computing conference workshops, pp 311–319
12. Ko RKL, Lee EW, Lee SG (2012) Business-OWL (BOWL)—a hierarchical task network ontology for dynamic business process decomposition an formulation. IEEE Trans Serv Comput 5(2):246–259
13. Liang Q, Wu X, Park EK, Khoshgoftaar TM, Chi C-H (2011) Ontology-based business process customization for composite web services. IEEE Trans Syst Man Cybern A Syst Humans. 41(4):717–729
14. Harrison WS, Tilbury DM, Yuan C (2012) From hardware-in-the-loop to hybrid process simulation: an ontology for the implementation phase of a manufacturing system. IEEE Trans Autom Sci Eng 9(1):96–109
15. Gelernter H (1959) Realization of a geometry-theorem proving machine. In: Proceedings of International Conference on Information Processing, Paris, pp 273–282
16. Tarski A (1951) A decision method for elementary algebra and geometry. University of California Press, Berkeley
17. Wu W-T (1986) Basic principles of mechanical theorem proving in elementary geometries. J Autom Reasoning 2(3):221–252
18. Chou SC, Gao XS, Zhang JZ (1994) Machine proofs in geometry. World Scientific, Singapore
19. Zhong X-Q, Fu H-G, She L, Huang B (2010) Geometry knowledge acquisition and representation on ontology. Chin J Comput 33(1):167–174

Return Forecast of Subscription for New Shares in Growth Enterprise Market Using Simulation Method

Xiutian Zheng and Yongbin Xu

Abstract This paper studies return of subscribing for new shares in Chinese Growth Enterprise Market (GEM) based on random simulation, and draws some useful conclusions. One is that investors can make more profit by selling their new shares at closing price than at opening price in the first trading day, the other is that percentage increase of a new stock in the first trading day has a bigger impact than the rate of successful subscription for new shares on return of subscribing for new shares in the Chinese GEM. The paper also finds both offering price to earnings ratio and amount of capital raised by initial public offerings (IPO) have significant negative effect on return of subscription for new shares by building a linear regression model. The paper is useful for investors to forecast return and control risk when they subscribe for new shares in Chinese GEM.

Keywords Return forecast · Subscription for new shares · Chinese Growth Enterprise Market · Simulation · Influencing factors

1 Introduction

The launch of the Growth Enterprise Market (GEM) is of significant importance to the development of China's capital market. As one part of China's multi-level capital market system, GEM is an important measure to improve the structure of China's capital market and expand the market's depth and width. Firstly, GEM

X. Zheng
Business Administration School of Zhejiang Gongshang University, Hangzhou 310018, China

X. Zheng (✉)
Qianjiang College, Hangzhou Normal University, Hangzhou 310036, China
e-mail: zxt1369@163.com

Y. Xu
Finance and Accounting College, Zhejiang Gongshang University, Hangzhou, China

Z. Wen and T. Li (eds.), *Knowledge Engineering and Management*,
Advances in Intelligent Systems and Computing 278,
DOI: 10.1007/978-3-642-54930-4_4, © Springer-Verlag Berlin Heidelberg 2014

enables growth enterprises to capitalize on the growth opportunities of the region by raising expansion capital. The growth enterprises that have good business ideas and growth potential, however, may not always be able to take advantage of these opportunities. A great number of them do not fulfill the profitability or track record requirements of main board of the Exchanges in China and are therefore unable to obtain a listing. The GEM is designed to bridge this gap. Because GEM does not require growth enterprises to have achieved a record of profitability as a condition of listing. Secondly, GEM is opened to growth companies. Technology companies in particular should find it attractive to align themselves with the strong growth theme of the market. In providing a fund raising venue and a strong identity to technology companies, GEM complements and supports the Chinese Government's initiative to promote the development of technology industries. Thirdly, GEM promotes the development of venture capital investments, because it provides both an exit ground and a venue for further fund raising for investments made by venture capitalists. This facilitates more and earlier investments to be made by the venture capitalists in support of the growth of the industry. Additionally, GEM offers investors an alternative of investing in "high growth, high risk" businesses.

After the establishment of GEM, a great many enterprises become listed companies by initial public offerings (IPOs) of shares. An initial public offering is defined as an existing private enterprise being listed on the GEM in China. From October 30th, 2009 to July seventh, 2011, there are total 243 companies issued their new stocks in GEM in Shenzhen Stock Exchanges. These Listed companies are mainly innovative companies, from manufacturing, telecommunications, electronics, electricity and power to chemicals sectors.

It is difficult to predict the stock price movement in the secondary market, especially for small investors and those who have less time and skill, so many of them subscribe for new shares in the GEM. Because there is a gross spread which is the difference between the underwriting price received by the issuing company and the actual price offered to the public generally.

Among the total 243 IPOs in Chinese GEM, almost every new stock has a percentage increase on first trading day. For example, Beijing Toread Outdoor Products Company obtained a percentage increase of 153 % on the first trading day, which is listed in October 30th, 2009; Chengdu Geeya Technology Company got a 210 % percentage increase on its first trading day. The high percentage increase of most companies in first trading day has attracted investors' attention. Both individual and institutional investors has been actively engaging in subscription for new shares in GEM. Although return of subscription for new shares are attractive, losses also may take place.

A substantial body of literature focuses on the initial return to IPOs. Ibbotson [1] finds that IPOs have significantly positive initial return and names it the mystery of IPOs. Beatty and Ritter predict that the difference between the market price and the offering price, i.e., the initial return, will be systematically related to this information asymmetry between firms and the market as in [2]. Baron and Holmstrom find there is information asymmetry between underwriters and issuers, with underwriters having superior information to the issuers as in [3]. By using several benchmarks, [4] examines the stock return performance of the IPO stocks which are listed on the GEM in Hong Kong, and finds two factors causing the underperformance of GEM stocks are the "technology boom" and "IPO effects". Weiya and Xiutian [5] research the return and risk of purchasing new issues in Chinese main-board stock markets.

In this paper, we research return of subscribing for new shares in the GEM and their influencing factors based on random simulation and an econometric model. The conclusions are useful for small and short-run investors who are engaging in subscription for new shares in GEM in China. The rest of the paper is organized as follows. Section 2 introduces the variables and data. Section 3 conducts simulation and analyzes the result. Section 4 is about factors influencing the return of subscribing for new shares in GEM using an econometric model. Conclusions are presented in Sect. 5.

2 Variables and Data

The sample used in this study consists of 243 A-share IPOs issued and listed in Chinese GEM from October 30th, 2009 to July fourth, 2011. The closing prices, the opening prices, and the offering prices are collected from the http://www.stockstar.com. The Chinese IPOs markets have some special characteristics when compared to IPOs markets in other countries. Chinese regulations require investors who want to subscribe for shares in an IPO to put funds up front which are frozen for four days. Potential investors are then selected randomly in a computerized lottery. Unlucky investors who are unsuccessful in winning the right to subscribe for new shares have their funds returned within three or four days. New shares' prices often jump a lot in the first trading day and investors who successfully subscribe for new shares in an IPO can make tidy profits by quickly selling their shares in the first trading day.

Most of the existing studies concerning Chinese IPO market just consider the IPO initial return is defined as the first day closing price minus the offering price, and then divided by the offering price, however the rates of successfully subscription for new shares attracts little attention. An innovation of this paper is to research the return of subscribing for new shares in the GEM by taking both IPO initial return and the rates of successful subscription into account.

We assume that investors who are lucky enough to win the right to buy new shares in the GEM would sell their shares on the first trading day at the opening price or the closing price. Therefore two kinds of percentage increase should be considered. ZFo is defined as the first day opening price minus the offering price, and then divided by the offering price, while ZFc is defined as the first day closing price minus the offering price, and then divided by the offering price. ZQL stands for the rate of successful subscription for new shares in GEM. So the paper should consider two rates of return of subscription: $Ro = ZFo * ZQL$ and

$Rc = ZFc * ZQL$. It is obvious that rates of return of subscription are random and uncertainty, because percentage increase and rates of successful subscription are random variables.

3 Simulation and Results

According to [6], Crystal Ball is one of the easiest software to perform random simulations. Crystal Ball automatically calculates thousands of different "what if" cases, saving the inputs and results of each calculation. It can reveal the range of possible outcomes, their probability of occurring, which input has the most effect on our model and where we should focus our efforts.

Firstly we should find the best probability distribution between a standard probability distribution and a data set's distribution by goodness-of-fit test using Crystal Ball software. A probability distribution describes the likelihood of specific values occurring out of a range or set of values. After knowing the probability distributions of the percentage increase and the rate of successful subscription respectively, we will use them to forecast the rates of return of purchasing new shares in GEM according to $Ro = ZFo * ZQL$ and $Rc = ZFc * ZQL$. At last, an analysis of conclusions will be conducted.

3.1 Goodness-of-Fit Test on Variables

We consider Goodness-of-fit test as a set of mathematical tests performed to find the best fit between a standard probability distribution and a data set's distribution. There are three data sets need to fit. Goodness-of-fit test shows that ZQL is subject to Lognormal distribution with mean of 1.08 and standard deviation equals to 0.79; ZFo is Lognormal distribution with mean of 0.32 and standard deviation equals to 0.28; ZFc is Gamma distribution with location of -2.52, scale of 0.06, and shape of 46.84.

3.2 Return Forecast of Subscribing for New Shares

We now begin to simulate the rate of return of purchasing new shares. After 1,000 trials, We can get the results of the simulation. The value of Ro and Rc is shown in Table 1.

The above table shows that the mean of Ro is 0.32 %, median is 0.23 % and standard deviation is 0.53, while mean of Rc is 0.35 %, median is 0.23 % and standard deviation is 0.55. So we attain a conclusion that it is better for investors to sell new shares at the closing price rather than opening price in the first trading day.

Table 1 The result of simulation about Ro and Rc	Statistics	Ro (%)	Rc (%)
	Mean	0.32	0.35
	Median	0.23	0.23
	Standard deviation	0.53	0.55
	Variance	0.28	0.31

We also can get the frequency charts of Ro and Rc, which are shown in the following Figs. 1 and 2 respectively.

Figure 1 shows the probability of $Ro > 0$ is 83.4 %, in other words, the probability of making profit by subscribing for new shares and selling them on the first trading day at opening price is 83.4 %.

As shown in Fig. 2 the probability of $Rc > 0$ is 81.6 %, that is, the probability of making profit by subscribing for new shares and selling them on the first trading day at closing price is 81.6 %. Meanwhile, the probabilities of the other rates of return of subscribing for new shares can be attained, which are listed in Table 2.

The above table shows that the probability of Ro or $Rc > 1.5$ % is very low, because probability of $Ro > 1.5$ % equals to 2.3 % and probability of $Rc > 1$ % equals to 3.5 %.

3.3 The Sensitivity Analysis of the Return

The Sensitivity Chart feature provides you with the ability to quickly and easily judge the influence each factor has on rate of return of purchasing a new issue. During a simulation, Crystal Ball ranks the factors according to their importance to rate of return. The factor with the highest sensitivity ranking can be considered the most important one in the model, while the factor with the lowest sensitivity ranking is the least important one in the model. We get the sensitivity charts of Ro and Rc measured by contribution to variance. We find that rates of successful subscription are more important than percentage increase at opening price in the first trading day in determining the rates of return of subscribing for new shares in GEM. Rates of successful subscription account for 20.1 %, and percentage increase accounts for 79.9 % in determining Ro. We also find percentage increase at closing price in the first trading day is more important than the rate of successful subscription in determining the rate of return of subscribing new shares in GEM. Rates of successful subscription account for 17.2 %, and percentage increase accounts for 82.8 % in determining Rc.

4 Analysis of Influencing Factors

In the above sections, this paper finds percentage increase in the first trading day has a bigger impact on return of subscribing new shares. Many factors can affect percentage increase of a stock in the first trading day, such as offering price

Fig. 1 Frequency charts of
Ro

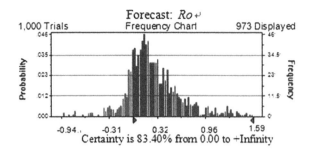

Fig. 2 Frequency charts of
Rc

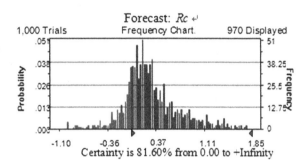

Table 2 Rate of return and its probability

Rate of return (%)	Probability of *Ro* (%)	Probability of *Rc* (%)	Rate of return (%)	Probability of *Ro* (%)	Probability of *Rc* (%)
0.05	77.5	75.9	0.5	23.2	26.7
0.1	71.3	69.3	0.7	13.3	18.1
0.2	54.8	55.9	1	6.9	9.6
0.3	41.6	42.7	1.2	5.2	6.5
0.4	31.8	33.0	1.5	2.3	3.5

to earnings ratio, amount of capital raised in initial public offerings. This section
will study how these factors influence return of subscribing for new shares in GEM
using a linear regression model. So the results will be useful for investors to make
profit and control risk. We firstly take augmented Dickey-Fuller Test (Table 3).

Table 3 shows that each ADF statistic is smaller than corresponding test critical
value in 1 % significant level, so four variables are stationary series, which are
suitable to construct linear regression models. *Rc* is dependent variable, and the
others are explanatory variables.

Table 3 Result of ADF test

Variables	(C, T, L)	ADF statistic	Critical value	P
Ro (%)	(C, 0, 0)	−6.3824	−3.4574	0.0000
Rc (%)	(C, 0, 0)	−6.4568	−3.4574	0.0000
PE (¥)	(C, 0, 0)	−3.7748	−3.4574	0.0036
CZ (billion ¥)	(C, 0, 0)	−7.7568	−3.4574	0.0000

Notes C, T, L stands for constant, time trend and lag length. We select Lag length by Akaike info criterion and Schwarz criterion

The linear regression model can be obtained by using Eviews5.0.

$$Rc = 0.0994 + 1.1562 * Ro - 0.0011 * PE - 0.084 * CZ$$
$$t = 2.4101\ 25.6919 - 1.8233\ -2.1585 \tag{1}$$
$$R^2 = 0.7359\ F = 222.0607\ DW = 1.7956.$$

According to statistic theory, the above equation is feasible.

Meanwhile, a residual series can be generated. ADF test shows the residual series is a stationary series. So the model is reasonable.

Equation (1) shows both offering price to earnings ratio (*PE*) and amount of capital raised (*CZ*) by IPO have significant negative effect on return of subscribing new shares which would be sold at closing price in first trading day (*Rc*). The return of subscribing new shares which would be sold at opening price in first trading day (*Ro*) has a positive impact on *Rc*. When *Ro* increases 0.1 %, *Rc* will goes up 0.115 %. This conclusion supports the result of above simulation that it is better for investors to sell the new shares at the closing price rather than opening price in the first trading day in order to control risk.

5 Conclusion

This paper research return of subscribing for new shares in the GEM and their influencing factors based on random simulation and an econometric model. The paper reaches some useful conclusions. Firstly, investors can make more profit by selling new shares at closing price than at opening price in the first trading day. Secondly, compare with the rate of successful subscription for new shares, percentage increase in the first trading day has a bigger impact on the return of subscribing new shares in the Chinese GEM. Besides, the linear regression model shows that both offering price to earnings ratio (*PE*) and amount of capital raised (*CZ*) have significant negative effect on return of subscribing new shares. Our conclusions are useful for small and short-run investors who are engaging in subscription for new shares in GEM in China to forecast return and control risk.

Acknowledgments The research is supported by Foundation of Humanities and Social Sciences of Chinese Ministry of Education (11YJA630171); Zhejiang Nature Science Foundation (Y13G020028); Zhejiang Nature Science Foundation (Q13G020070); Social Science Foundation of Hangzhou (C13YJ01).

References

1. Ibbotson RG (1975) Price performance of common stock new issues. J Financ Econ 02(03):235–272
2. Beatty RP, Ritter JR (1985) Investment banking, reputation, and the underpricing of initial public offerings. J Financ Econ 15(1–2):213–232
3. Baron DP, Holmstrom B (1980) The investment banking contract for new issues under asymmetric information: delegation and the incentive problem. J Financ 35(05):1115–1138
4. Pak TC, Fariborz M, David N, Eliza W (2007) The underperformance of the growth enterprise market in Hong Kong. Res Int Bus Financ 21(03):428–446
5. Weiya Z, Xiutian Z (2008) Return forecasting and risk analysis of purchasing new issues in Chinese Primary Market based on simulation. In: Proceedings of fifth international conference on fuzzy systems and knowledge discovery, China, pp 621–625
6. James RE, Davidlson LO (2001) Introduction to simulation and risk analysis, 2nd edn. Prentice Hall, Upper Saddle River

An Incremental Learning Approach for Updating Approximations in Rough Set Model Over Dual-Universes

Jie Hu, Tianrui Li and Anping Zeng

Abstract The rough set model over dual-universes (RSMDU) is a generalized model of the classical rough sets theory (RST) on the two universes. It is an effective way to use incremental updating approximations method in the dynamic environment to better support data mining-related tasks based on RST. In this paper, we propose an incremental learning approach for updating approximations in RSMDU when the objects of two universes vary with time. An illustration is employed to show the validation of the presented method.

Keywords Rough sets · Dual-universes · Approximations · Incremental approach

1 Introduction

Rough sets theory (RST) proposed by Pawlak is a powerful mathematical tool for processing incomplete, inexact, and inconsistent information systems [19]. It has been successfully applied to pattern recognition, knowledge discovery, decision analysis, and medical diagnosis.

J. Hu (✉) · T. Li · A. Zeng
School of Information Science and Technology, Southwest Jiaotong University,
Chengdu 610031, China
e-mail: jiehu@swjtu.edu.cn

T. Li
e-mail: tril@swjtu.edu.cn

A. Zeng
e-mail: zengap@126.com

A. Zeng
School of Computer and Information Engineering, Yibin University, Yibin 644007, China

Z. Wen and T. Li (eds.), *Knowledge Engineering and Management*,
Advances in Intelligent Systems and Computing 278,
DOI: 10.1007/978-3-642-54930-4_5, © Springer-Verlag Berlin Heidelberg 2014

The classical RST is mainly concerned with one universe U, where the equivalence relation R between the objects is included in one universe U, and the approximation space is (U, R). Due to requirements of real applications, RST has been generalized from three aspects. One is to relax the equivalence relation to other relations, such as similarity relation [24], tolerance relation [22], dominance relation [7], neighborhood relation [8], and general binary relation [30, 32] over the universe. The second aspect is the combination of RST with other theories, such as fuzzy set theory and probability theory. Rough fuzzy set models and fuzzy rough set models [6], probability rough set models [31], variable precision rough [36], and Bayesian decision rough set models [23] have been proposed. The third aspect is the universe extension of the classical RST. We can use both generalized approximation space [30] and interval structure [26] to explain the rough set model over two universes. A series of rough set model over dual-universes (RSMDU) have been proposed, and their corresponding properties have been discussed in the literature [11–13, 16–18, 20, 21, 25, 27–29, 33]. Although these models have different properties and calculation methods, they start with "almost the same" framework (two universes and a relation between them). It should also be noted that, in these models, the approximated sets and the approximating sets are always located at two different universes [12].

In our real-life applications, most practical problems are related to two of the domain situations. For example, commodity attribute set and customer characteristic set in the personalized marketing, problem set and solutions set in the diagnosis of business enterprises, disease symptoms set and drug set in the disease diagnosis, fault set and solution set in the mechanical fault diagnosis, and so on can be seen as two different universes. The characteristics of RSMDU determine it can be applied to these multi-attribute decision making and multi-objective decision problems. For example, in the personalized marketing, we can regard the product features and characteristics of consumers as two different universes, use RSMDU to analyze and reason, acquire marketing personnel rules, and provide a reference for the marketing personnel for customers personalized marketing.

Computation of approximations is a necessary step to induce rules in the knowledge discovery and data mining based on rough sets. In order to use RST in processing dynamical information systems, researchers have proposed incremental learning techniques. Some results have been generated in the incremental learning approaches based on rough set methodologies [1–5, 9, 10, 14, 15, 34, 35]. These incremental learning approaches based on RST can be divided into three kinds of cases: the variation of the object set [1, 4, 14, 15, 35], the variation of the attribute set [5, 9, 10, 34], and the variation of attribute values [2, 3]. The existing RSMDU were proposed based on static information systems. It is necessary to develop a feasible and effective incremental learning method for updating approximations in RSMDU to better support our decision making.

The remainder of this paper is organized as follows. Section 2 introduces the basic concepts of RSMDU. Section 3 discusses the principles for incrementally updating approximations of RSMDU when objects of two universes vary with time. Section 4 gives some illustrative examples. The paper ends with conclusions in Sect. 5.

2 Preliminaries

In this section, we briefly review some concepts, notations, and results of RSMDU [26, 28].

Let U and V be two nonempty finite sets, called universes. R is an arbitrary binary relation from U to V. The triple (U, V, R) is called a generalized approximation space.

Definition 1 For a generalized approximation space (U, V, R), $\forall x \in U, y \in V$, the neighborhood $R(x)$ of an element x in universe U is defined as follows:

$$R(x) = \{y \in V | (x, y) \in R\} \tag{1}$$

Definition 2 Given a generalized approximation space (U, V, R), $\forall Y \subseteq V$, the lower and upper approximations of Y in terms of the relation R are defined as:

$$
\begin{aligned}
\underline{R}(Y) &= \{x \in U | R(x) \subseteq Y \wedge R(x) \neq \emptyset\} \\
\overline{R}(Y) &= \{x \in U | R(x) \cap Y \neq \emptyset \vee R(x) = \emptyset\}.
\end{aligned}
\tag{2}
$$

If $\underline{R}(Y) \neq \overline{R}(Y)$, $R(Y) = (\underline{R}(Y), \overline{R}(Y))$ is called as a rough set of Y respect to universes U and V.

In order to discuss the relation R between U and V clearly, we use the matrix to describe the relation between U and V.

Definition 3 For a generalized approximation space (U, V, R), $\forall x \in U, y \in V$, the characteristic function of R is defined as:

$$
\chi_R(x, y) = \begin{cases} 1, & (x, y) \in R \\ 0, & (x, y) \notin R \end{cases}
\tag{3}
$$

Definition 4 The relation matrix of R between U and V is defined as follows:

$$
M^R_{m \times n} = (\phi_{ij})_{m \times n}, \text{where } \phi_{ij} = \begin{cases} 1, & \chi_R(x_i, y_j) = 1 \\ 0, & \chi_R(x_i, y_j) = 0 \end{cases}
\tag{4}
$$

3 An Incremental Method for Updating Approximations of RSMDU

In this section, we consider the problem of updating approximations of a subset Y of V when the objects of two universes vary with time in RSMDU.

For a generalized approximation space (U, V, R), the variation of objects varying with time contains two situations: the updating of the objects in the universe U (including adding objects and removing objects in U) and the updating of the objects in the universe V (including adding objects and removing objects in V).

To illustrate this method clearly, we first present several assumptions and the basic structure of the incremental approach.

3.1 Assumptions and Basic Structure for the Incremental Approach

We assume the incrementally learning process lasts from time t to time $t + 1$.

To describe a dynamic generalized approximation space, we denote a generalized approximation space at time t as (U, V, R), Y is an arbitrary subset of universe V. At time $t + 1$, some objects enter while some go out of the original two universes. Then U and V will be changed into U' and V'. The relation R will be changed into R'. Hence, the generalized approximation space (U, V, R) will be changed into (U', V', R'), and the set Y will be changed into Y', accordingly.

With these stipulations, we focus on the algorithms for updating approximations of the set Y. As the change of multi-objects can be considered as the cumulative change of single object, in this paper, we only discuss the case that single object changes. Let x', x_i, y', y_j denote the object which is added into U, deleted from U, added into V, deleted from V, respectively.

The change of single object can be divided in the following four cases:

(1) One object enters into the universe U at time $t + 1$ while the universe V keeping constant;
(2) One object gets out of the universe U at time $t + 1$ while the universe V keeping constant;
(3) One object enters into the universe V at time $t + 1$ while the universe U keeping constant;
(4) One object gets out of the universe V at time $t + 1$ while the universe U keeping constant.

3.2 Addition of one New Object into the Universe U

Proposition 1 *Now $U' = U \cup \{x'\}$, $V' = V$, $Y' = Y$, $R' \subseteq U' \times V'$.*
The following results hold:

a. *If $R'(x') \subseteq Y'$ and $R'(x') \neq \emptyset$, then $\underline{R'}(Y') = \underline{R}(Y) \cup \{x'\}$; Otherwise $\underline{R'}(Y') = \underline{R}(Y)$.*

b. *If* $R'(x') \cap Y' \neq \emptyset$ *or* $R'(x') = \emptyset$, *then* $\overline{R'}(Y') = \overline{R}(Y) \cup \{x'\}$; *Otherwise* $\overline{R'}(Y') = \overline{R}(Y)$.

Proof It follows directly from the Definition 2. □

3.3 Deletion of One Object from the Universe U

Proposition 2 *Now* $U' = U - \{x_i\}$, $V' = V$, $Y' = Y$, $R' \subseteq U' \times V'$ *The following results hold*:

a. *If* $x_i \in \underline{R}(Y)$, *then* $\underline{R'}(Y') = \underline{R}(Y) - \{x_i\}$; *Otherwise* $\underline{R'}(Y') = \underline{R}(Y)$.
b. *If* $x_i \in \overline{R}(Y)$, *then* $\overline{R'}(Y') = \overline{R}(Y) - \{x_i\}$; *Otherwise* $\overline{R'}(Y') = \overline{R}(Y)$.

3.4 Addition of One New Object into the Universe V

Now $U' = U$, $V' = V \cup \{y'\}$. It can be divided into the following two cases:

Case 1 $y' \notin Y'$.

Proposition 3 *Now* $Y' = Y$, *the following results hold*:

a. $\underline{R'}(Y') = \underline{R}(Y) - \{x\}$, *where* $x \in \underline{R}(Y)$, *and* $R'(x) = R(x) + \{y'\}$.
b. $\overline{R'}(Y') = \overline{R}(Y) - \{x\}$, *where* $\forall x \in (\overline{R}(Y) - \underline{R}(Y))$, *if* $R(x) = \emptyset$ *and* $R'(x) = \{y'\}$.

Case 2 $y' \in Y'$.

Proposition 4 *Now* $Y' = Y \cup \{y'\}$, *the following results hold*:

a. $\underline{R'}(Y') = \underline{R}(Y) \cup \{x\}$, *where* $\forall x \in (\overline{R}(Y) - \underline{R}(Y))$, *if* $R(x) = \emptyset$ *and* $R'(x) = \{y'\}$.
b. $\overline{R'}(Y') = \overline{R}(Y) \cup \{x\}$, *where* $\forall x \in (\overline{R}(Y))^C$ *and* $R'(x) = R(x) \cup \{y'\}$.

Remark 1 A^C denotes the complement of set A.

3.5 Deletion of One Object from the Universe V

Now $U' = U$, $V' = V - \{y_j\}$. It can be divided into two cases:

Case 1 $y_j \notin Y$.

Proposition 5 *Now* $Y' = Y$, *the following results hold*:

a. $\underline{R'}(Y') = \underline{R}(Y) \cup \{x\}$, *where* $\forall x \in (\overline{R}(Y) - \underline{R}(Y))$, $R(x) - Y = \{y_j\}$.
b. $\overline{R'}(Y') = \overline{R}(Y) \cup \{x\}$, *where* $\forall x \in (\overline{R}(Y))^C$, $R(x) = \{y_j\}$.

Case 2 $y_j \in Y$.

Proposition 6 *Now* $Y' = Y - \{y_j\}$, *the following results hold*:

a. $\underline{R'}(Y') = \underline{R}(Y) - \{x\}$, *where* $x \in \underline{R}(Y)$ *and* $R(x) = \{y_j\}$.
b. $\overline{R'}(Y') = \overline{R}(Y) - \{x\}$, *where* $x \in \overline{R}(Y)$, $R(x) \cap Y = \{y_j\}$ *and* $|R(x)| > 1$.

Remark 2 $|A|$ denotes the cardinality of set A.

4 An Illustration

Let (U, V, R) be a generalized approximation space at time t (shown in Table 1), U and V denote the two different universes of discourse, respectively, where $U = \{x_1, x_2, \ldots, x_7\}$, $V = \{y_1, y_2, \ldots, y_6\}$. R is an arbitrary binary relation from U to V. $R(x_1) = \emptyset$, $R(x_2) = \{y_4\}$, $R(x_3) = \{y_2, y_4\}$, $R(x_4) = \{y_4, y_5\}$, $R(x_5) = \{y_1, y_3, y_4, y_6\}$, $R(x_6) = \{y_1, y_2, y_5, y_6\}$, $R(x_7) = \{y_1\}$. For an arbitrary subset Y of the universe V, suppose $Y = \{y_2, y_3, y_4\}$ at time t. Then we can compute the approximations of Y: $\underline{R}(Y) = \{x_2, x_3\}$, $\overline{R}(Y) = \{x_1, x_2, x_3, x_4, x_5, x_6\}$.

Next, we consider the following four cases at time $t + 1$:

4.1 A New Object x_8 Enters into the Universe U

Suppose $R(x_8) = \{y_2\}$. Now $U' = U \cup \{x_8\}$, $V' = V$, $Y' = Y$, $R' \subseteq U' \times V'$.
Since $R'(x_8) \subseteq Y'$ and $R(x_8) = \{y_2\} \neq \emptyset$, then $\underline{R'}(Y') = \underline{R}(Y) \cup \{x_8\}$.
Since $R'(x_8) \cap Y' = \{y_2\} \neq \emptyset$, then $\overline{R'}(Y') = \overline{R}(Y) \cup \{x_8\}$.

4.2 An Object x_4 Gets Out from the Universe U

Now $U' = U - \{x_4\}$, $V' = V$, $Y' = Y$, $R' \subseteq U' \times V'$.
Since $x_4 \notin \underline{R}(Y)$, then $\underline{R'}(Y') = \underline{R}(Y)$.
Since $x_4 \in \overline{R}(Y)$, then $\overline{R'}(Y') = \overline{R}(Y) - \{x_4\}$.

Table 1 A generalized approximation space at time t

	y_1	y_2	y_3	y_4	y_5	y_6
x_1	0	0	0	0	0	0
x_2	0	0	0	1	0	0
x_3	0	1	0	1	0	0
x_4	0	0	0	1	1	0
x_5	1	0	1	1	0	1
x_6	1	1	0	0	1	1
x_7	1	0	0	0	0	0

4.3 A New Object y_7 Enters into the Universe V

Suppose $R(y_7) = \{x_1, x_2, x_5, x_7\}$. Now $U' = U, V' = V \cup \{y_7\}, R' \subseteq U' \times V'$.

(1) Suppose $y_7 \notin Y'$, that is to say $Y' = Y$.

Since $x_2 \in \underline{R}(Y)$, and $R'(x_2) = R(x_2) \cup \{y_7\}$, then $\underline{R'}(Y') = \underline{R}(Y) - \{x_2\}$.

Since $x_1 \in (\overline{R}(Y) - \underline{R}(Y))$, $R(x_1) = \emptyset$ and $R'(x_1) = \{y_7\}$, then $\overline{R'}(Y') = \overline{R}(Y) - \{x_1\}$.

(2) Suppose $y_7 \in Y'$, that is to say $Y' = Y \cup \{y_7\}$.

Since $x_1 \in (\overline{R}(Y) - \underline{R}(Y))$, $R(x_1) = \emptyset$ and $R'(x_1) = \{y_7\}$, then $\underline{R'}(Y') = \underline{R}(Y) \cup \{x_1\}$.

Since $x_7 \in (\overline{R}(Y))^C$ and $R'(x_7) = R(x_7) \cup \{y_7\}$, then $\overline{R'}(Y') = \overline{R}(Y) \cup \{x_7\}$.

4.4 An Object y_4 Gets Out from the Universe V

Now $U' = U, V' = V - \{y_4\}, Y' = Y - \{y_4\}, R' \subseteq U' \times V'$.

Since $y_4 \in Y$, $x_2 \in \underline{R}(Y)$ and $R(x_2) = \{y_4\}$, then $\underline{R'}(Y') = \underline{R}(Y) - \{x_2\}$.

Since $x_4 \in \overline{R}(Y)$, $R(x_4) \cap Y = \{y_4\}$ and $|R(x_4)| = |\{y_4, y_5\}| = 2 > 1$, then $\overline{R'}(Y') = \overline{R}(Y) - \{x_4\}$.

5 Conclusions

The incremental method for updating approximations is an efficient method to acquire knowledge in the dynamic information environment. RSMDU is an important and general model in information systems. In this paper, we proposed an incremental method to update lower and upper approximations of RSMDU when the objects evolve over time. We gave an example to illustrate the proposed approach. Our future work will focus on the development of algorithms and experiments to verify the effectiveness of the proposed method.

Acknowledgments This work is supported by the National Science Foundation of China (Nos. 60873108, 61175047, 61100117), the Fundamental Research Funds for the Central Universities (No. SWJTU12BR045), the Scientific Research Foundation of Sichuan Provincial Education Department (No. 13ZB0210), and the 2013 Doctoral Innovation Funds of Southwest Jiaotong University.

References

1. Bang WC, Bien Z (1999) New incremental inductive learning algorithm in the framework of rough set theory. Int J Fuzzy Syst 1(1):25–36
2. Chen H, Li T, Qiao S, Ruan D (2010) A rough set based dynamic maintenance approach for approximations in coarsening and refining attribute values. Int J Intell Syst 25(10):1005–1026. doi:10.1002/int.20436
3. Chen H, Li T, Ruan D (2012) Maintenance of approximations in incomplete ordered decision systems while attribute values coarsening or refining. Knowl-Based Syst 31:140–161. doi:10.1016/j.knosys.2012.03.001
4. Chen H, Li T, Ruan D, Lin J, Hu C (2013) A rough-set-based incremental approach for updating approximations under dynamic maintenance environments. IEEE Trans Knowl Data Eng 25(2):274–284. doi:10.1109/TKDE.2011.220
5. Cheng Y (2011) The incremental method for fast computing the rough fuzzy approximations. Data Knowl Eng 70(1):84–100
6. Dubois D, Prade H (1990) Rough fuzzy-sets and fuzzy rough sets. Int J Gen Syst 17(2–3):191–209. doi:10.1080/03081079008935107
7. Greco S, Matarazzo B, Slowinski R (2002) Rough approximation by dominance relations. Int J Intell Syst 17(2):153–171. doi:10.1002/int.10014
8. Hu Q, Yu D, Liu J, Wu C (2008) Neighborhood rough set based heterogeneous feature subset selection. Inf Sci 178(18):3577–3594. doi:10.1016/j.ins.2008.05.024
9. Li S, Li T, Liu D (2013) Incremental updating approximations in dominance-based rough sets approach under the variation of the attribute set. Knowl-Based Syst 40:17–26. doi:10.1016/j.knosys.2012.11.002
10. Li T, Ruan D, Geert W, Song J, Xu Y (2007) A rough sets based characteristic relation approach for dynamic attribute generalization in data mining. Knowl-Based Syst 20(5):485–494. doi:10.1016/j.knosys.2007.01.002
11. Li TJ (2008) Rough approximation operators on two universes of discourse and their fuzzy extensions. Fuzzy Sets Syst 159(22):3033–3050. doi:10.1016/j.fss.2008.04.008
12. Li TJ, Zhang WX (2008) Rough fuzzy approximations on two universes of discourse. Inf Sci 178(3):892–906. doi:10.1016/j.ins.2007.09.006
13. Liu C, Miao D, Zhang N (2012) Graded rough set model based on two universes and its properties. Knowl-Based Syst 33:65–72. doi:10.1016/j.knosys.2012.02.012
14. Liu D, Li T, Ruan D, Zhang J (2011) Incremental learning optimization on knowledge discovery in dynamic business intelligent systems. J Global Optim 51(2):325–344. doi:10.1007/s10898-010-9607-8
15. Liu D, Li T, Ruan D, Zou W (2009) An incremental approach for inducing knowledge from dynamic information systems. Fundamenta Informaticae 94(2):245–260. doi:10.3233/FI-2009-129
16. Liu G (2010) Rough set theory based on two universal sets and its applications. Knowl-Based Syst 23(2):110–115. doi:10.1016/j.knosys.2009.06.011

17. Ma W, Sun B (2012) On relationship between probabilistic rough set and bayesian risk decision over two universes. Int J Gen Syst 41(3):225–245. doi:10.1080/03081079.2011. 634067 2011, 634067
18. Ma W, Sun B (2012) Probabilistic rough set over two universes and rough entropy. Int J Approximate Reasoning 53(4):608–619. doi:10.1016/j.ijar.2011.12.010
19. Pawlak Z (1982) Rough sets. Int J Comput Inform Sci 11(5):341–356. doi:10.1007/BF01001956
20. Pei DW, Xu ZB (2004) Rough set models on two universes. Int J Gen Syst 33(5):569–581. doi:10.1080/0308107042000193561
21. Shen Y, Wang F (2011) Variable precision rough set model over two universes and its properties. Soft Comput 15(3):557–567. doi:10.1007/s00500-010-0562-6
22. Skowron A, Stepaniuk J (1996) Tolerance approximation spaces. Fundamenta Informaticae 27(2–3):245–253
23. Slezak D, Ziarko W (2002) Bayesian rough set model. Proc. of FDM, pp 131–135
24. Slowinski R, Vanderpooten D (2000) A generalized definition of rough approximations based on similarity. IEEE Trans Knowl Data Eng 12(2):331–336. doi:10.1109/69.842271
25. Sun B, Ma W (2011) Fuzzy rough set model on two different universes and its application. Appl Math Model 35(4):1798–1809. doi:10.1016/j.apm.2010.10.010
26. Wong SKM, Wang LS, Yao YY (1995) On modelling uncertainty with interval structures. Comput Intell 11(2):406–426
27. Yan R, Zheng J, Liu J, Qin C (2012) Rough set over dual-universes in fuzzy approximation space. Iran J Fuzzy Syst 9(3):79–91
28. Yan R, Zheng J, Liu J, Zhai Y (2010) Research on the model of rough set over dual-universes. Knowl-Based Syst 23(8):817–822. doi:10.1016/j.knosys.2010.05.006
29. Yang HL, Li SG, Wang S, Wang J (2012) Bipolar fuzzy rough set model on two different universes and its application. Knowl-Based Syst 35:94–101. doi:10.1016/j.knosys.2012.01.001
30. Yao Y (1998) Generalized rough set models. Rough Sets Knowl Disc 1:286–318
31. Yao Y (2008) Probabilistic rough set approximations. Int J Approximate Reasoning 49(2):255–271. doi:10.1016/j.ijar.2007.05.019
32. Yao Y, Lin T (1996) Generalization of rough sets using modal logic. Intell Autom Soft Comput 2(2):103–120
33. Zhang HY, Zhang WX, Wu WZ (2009) On characterization of generalized interval-valued fuzzy rough sets on two universes of discourse. Int J Approximate Reasoning 51(1):56–70. doi:10.1016/j.ijar.2009.07.002
34. Zhang J, Li T, Ruan D, Liu D (2012) Neighborhood rough sets for dynamic data mining. Int J Intell Syst 27(4):317–342. doi:10.1002/int.21523
35. Zhang J, Li T, Ruan D, Liu D (2012) Rough sets based matrix approaches with dynamic attribute variation in set-valued information systems. Int J Approximate Reasoning 53(4):620–635. doi:10.1016/j.ijar.2012.01.001
36. Ziarko W (1993) Variable precision rough set model. J Comput Syst Sci 46(1):39–59. doi:10.1016/0022-0000(93)90048-2

Research on Risk Assessment of Ship Repair Based on Case-Based Reasoning

Lu Yao, Zhi-Cheng Chen and Jian-Jun Yang

Abstract As it is difficult to describe the mechanism of ship repair risk, common methods are less credible for its assessment. By analyzing ship repair risk, this paper identifies the causes and consequences of risk, and utilizes frame-based representation to construct the case representation for ship repair risk. In addition, similarity functions are developed for enumerated attributes, numeric attributes, and fuzzy attributes, so as to perform reasoning of repair risk from the approach of K-Nearest Neighbor. Case analysis presents that the application of case-based reasoning in ship repair risk assessment is easy to understand and extend.

Keywords Ship repair · Risk assessment · Case-based reasoning · Frame-based representation · K-Nearest Neighbor

1 Introduction

The ship system has a complicated formation and structure with multiple functions, which contains a large number of electronic and mechanical equipment and software. As the technology of ship system becomes more and more complex, its repair is also increasingly difficult. Hence, a lot of risk sources exist in ship repair, easily leading to quality and safety problems as well as overspending and schedule delay. Therefore, the research on ship repair risk assessment has significant realistic meanings.

Nevertheless, there are few researches on this issue in the academic circles at present, and those existing researches are still superficial. For instance, Ref. [1] analyzes major safety risks in ship repair and puts forward some corresponding

L. Yao · Z.-C. Chen (✉) · J.-J. Yang
Department of Management Engineering, Naval University of Engineering, Wuhan, China
e-mail: chenzhicheng1979@163.com

Z. Wen and T. Li (eds.), *Knowledge Engineering and Management*,
Advances in Intelligent Systems and Computing 278,
DOI: 10.1007/978-3-642-54930-4_6, © Springer-Verlag Berlin Heidelberg 2014

coping measures, but overlooks how to assess risks. With regard to the special issue of risk assessment technology, there are also many different opinions in the academic world. Some scholars take probability distribution approach [2, 3] to study the historical data of similar projects, estimate the probability distribution of project risks, and then utilize distribution function to determine the risk probability of future projects. This method must depend on sufficient historical data, which is often difficult to realize in practice. Also, many scholars utilize subjective judgment methods that depend on experts' subjective judgment for risk assessment, which is represented by risk matrix method [4] and fuzzy comprehensive evaluation [5]. These methods are too subjective, so their assessment results are poorly reliable. At present, some popular research methods include Bayesian network [6] and system simulation [7], etc. However, these methods require extensively understanding the mechanical of risk, so they are not very useful to ship repair risk as it is difficult to gather the information on its rules. This paper assesses the risks of new repair task from the approach of case-based reasoning (CBR), which not only effectively utilizes the verified historical data, but also integrates the knowledge of experts, so that the assessment results are more credible and explainable than those from traditional approaches.

2 Overview of Case-Based Reasoning

Case-based reasoning is an emerging method for machine learning and reasoning and focuses on solving new problems based on the experience from problem solving in the past. Its process can be concluded as "4R" [8], namely Retrieve, Reuse, Revise, and Retain, as shown in Fig. 1.

1. Past cases are stored in a case database in terms of certain logic and structure, and a new problem is converted into a new case;
2. Retrieve one or several cases that are most relevant to the new case;
3. Reuse the retrieved case(s);
4. Revise the suitable cases based on actual situation and experts' opinions;
5. Apply revised cases in the current problem, and record the solved case in the base.

CBR inherits some characteristics from the process of human thinking and has many advantages:

- Describe a problem in natural language and only point out the key attributes of the problem;
- Acquire new knowledge through reasoning and store it in case base, so as to give some capability of learning to CBR system;
- Operate CBR system even with only a few cases since the number of cases keeps increasing in the case base;
- Provide the specific cases for user, which can be easily understood.

Fig. 1 Case-based reasoning process

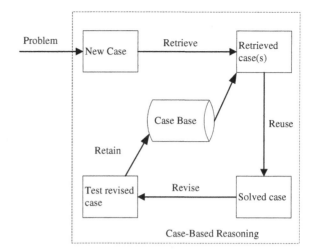

Therefore, CBR is highly advantageous to solving some less theoretical problems that are difficult to acquire knowledge and define rules, such as, ship repair risk assessment.

3 Ship Repair Risk Assessment Based on CBR

3.1 Case Representation of Repair Risk

Analyzing the existing ship repair risk event, it is discovered that repair risk factors mainly include:

- *Basic factors of repair work*: e.g., category of repair work, quantity of repair work (man-hour), technical difficulty of repair, accessibility of spare parts for repair, probability of involving other repair work, quantity of other repair work involved, etc.;
- *Factors of repairman*: e.g., technical level of repairman, repairman's sense of responsibility, repairman's thoughts, frequency of repairman change, etc.;
- *Management factors*: e.g., management level of management personnel, and completeness of rules and regulations, etc.;
- *Environmental factors*: e.g., adversity of repair environment, probability of force majeure (e.g. typhoon), and degree of impacts on other works, etc.

The consequences of repair risk include:

- Repair schedule delay
- Repair cost overspending
- Unacceptable repair quality
- System damage

Table 1 Repair risk case frame

Frame name: ship repair risk case	
Basic factors of repair work	Category of repair work (crw)
	Quantity of repair work (qrw)
	Technical difficulty of repair (tdr)
	Accessibility of spare parts for repair
	...
Factors of repairman	Technical level of repairman (tlr)
	Repairman's sense of responsibility
	...
Management factors	Management level of management personnel (mlmp)
	Completeness of rules and regulations
	...
Environmental factors	Adversity of repair environment (are)
	Probability of force majeure
	...
Consequences of repair risk	Repair schedule delay
	Repair cost overspending
	...

- Injury or death of people
- Others

Clearly, repair risk case is highly structural. Frame-based representation employs a frame consisting of several slots, and each slot can be divided into several profiles according to actual needs. A slot is used to describe the attributes of the object in an aspect, while a profile is employed to describe each aspect of an attribute, which is subject to specific constraints. This is suitable for representing structural knowledge. Moreover, this knowledge representation can be easily realized by means of relation database in computer. Therefore, we employ frame-based representation to present a repair risk case as follows (Table 1).

3.2 CBR Strategy

This paper employs a CBR strategy—K-Nearest Neighbor (KNN), which mainly calculates the similarity of cases. The calculation equation is as follows [9]:

$$S_{\mathrm{IR}} = \sum_{i=1}^{n} w_i \mathrm{sim}(f_i^{\mathrm{I}}, f_i^{\mathrm{R}}) \qquad (1)$$

in which, S_{IR} is the similarity between case I and case R, w_i stands for the weight of the ith attribute among the attributes of the case, sim() is the calculation function for the similarity of attributes, f_i^{I}, f_i^{R} stand for the value of the ith attribute in case I and case R, respectively, and n is the number of attributes in the case.

In Eq. (1), w_i can be obtained using Delphi method or analytic hierarchy process and consulting with experts in the field. Thus, the major difficulty in determining case similarity is to determine the similarity function of attributes. As repair risk case has various attributes, it is necessary to design the similarity function of different attributes in the case.

(1) Similarity function of enumerated attributes
    Enumerated categories of repair work, including electrical repair, mechanical repair and plumbing, etc., are essentially different from each other. Thus, the similarity function of these attributes is defined as follows:

$$\text{sim}\left(f_i^{\text{I}}, f_i^{\text{R}}\right) = \begin{cases} 1 & \left(f_i^{\text{I}} = f_i^{\text{R}}\right) \\ 0 & \left(f_i^{\text{I}} \neq f_i^{\text{R}}\right) \end{cases}. \tag{2}$$

(2) Similarity function of numeric attributes
    Quantity of repair work is an example of such attributes. When different attributes of repair work have larger differences between their values, they have poorer similarity. Therefore, the absolute value of the difference between attributes reflects the degree of similarity, but it is still necessary to normalize them, in order to unify measuring unit. Hence, the similarity function of these attributes is defined as follows:

$$\text{sim}\left(f_i^{\text{I}}, f_i^{\text{R}}\right) = 1 - \frac{\left|f_i^{\text{I}} - f_i^{\text{R}}\right|}{f_{i\max} - f_{i\min}}. \tag{3}$$

(3) Similarity function of fuzzy attributes
    Fuzzy attributes include technical difficulty of repair and technical level of repairman, etc. As it is difficult to determine the accurate values of these attributes, fuzzy levels are given instead, such as "very high," "high," "medium," "low," and "very low," etc. The similarity function of these attributes is defined as follows:

1. Calculate the membership of attribute f_i to each fuzzy level A: $\mu_A(f_i) = \frac{\text{Times}}{n}$, in which Times stands for the times that experts believe that the attribute belongs to A, and n is the number of experts;
2. Give a weight to each fuzzy level A in the monotone decreasing principle of weights, e.g.:

$$\omega_{\text{very high}} > \omega_{\text{high}} > \omega_{\text{medium}} > \omega_{\text{low}} > \omega_{\text{very low}}.$$

3. Calculate the similarity function:

$$\text{sim}\left(f_i^{\text{I}}, f_i^{\text{R}}\right) = 1 - \left|\sum_{A=1}^{m} \omega_A \left(\mu_A\left(f_i^{\text{I}}\right) - \mu_A\left(f_i^{\text{R}}\right)\right)\right|. \tag{4}$$

In which m stands for the number of fuzzy level.

Table 2 Risk case set

No.	Category of repair work	Quantity of repair work (man-hour)	Technical difficulty of repair	Technical level of repairman	Management level of management Personnel	Adversity of repair environment
1	Electrical repair	50	(0.5, 0.3, 0.2)	(0.2, 0.5, 0.3)	(0.3, 0.6, 0.1)	(0.7, 0.2, 0.1)
2	Electrical repair	60	(0.4, 0.5, 0.1)	(0.7, 0.3, 0)	(0.4, 0.5, 0.1)	(0.6, 0.2, 0.2)
3	Mechanical repair	53	(0.3, 0.4, 0.3)	(0.6, 0.2, 0.2)	(0.7, 0.3, 0)	(0.4, 0.5, 0.1)

4 Case Analysis

It is assumed that ship repair risk case database contains the following risk cases (only simplified attributes of factors) and the weights of attributes are

$$(w_{crw} = 0.25, \ w_{qrw} = 0.1, \ w_{tdr} = 0.2, \ w_{tlr} = 0.2, \ w_{mlmp} = 0.15, \ w_{are} = 0.1).$$

Fuzzy attributes are classified into three levels represented by membership degree. For instance, (0.5, 0.3, 0.2) are the membership degrees of levels (high, medium, low) respectively. Moreover, this defines (Table 2)

$$\omega_{high} = 0.5, \ \omega_{medium} = 0.3, \ \omega_{low} = 0.2.$$

Assuming that a new repairman is needed for risk assessment and case attributes are as follows:

- Category of repair work: Electrical repair
- Quantity of repair work: 55 man-hour
- Technical difficulty of repair: (0.6, 0.3, 0.1)
- Technical level of repairman: (0.3, 0.4, 0.3)
- Management level and management personnel: (0.4, 0.5, 0.1)
- Adversity of repair environment: (0.5, 0.4, 0.1)

As specified in Sect. 3, the similarities of this case to the cases in database are calculated as follows: $S_1 = 0.933$, $S_2 = 0.919$, $S_3 = 0.6595$. Obviously, this case is most similar to case 1. Therefore, the risk assessment of new repair work is conducted based on the consequences of repair risk in case 1. It is calculated that risk consequences will be 10 % in schedule delay and 15 % in overspending.

If technical level of repairman for new repair work is improved by means of technical training, its membership degree is changed to (0.6, 0.3, 0.1) and other factors remain unchanged; it is calculated that similarities are $S_1 = 0.917$, $S_2 = 0.935$, $S_3 = 0.6695$. Obviously, new repair work is most similar to case 2. Therefore, it is calculated based on case 2 that risk consequences will have no schedule delay and 10 % in overspending. As a result, corresponding measures should be taken to reduce repair risk.

Lastly, this new repair work can be revised based on actual condition and retained in the case database, in order to further improve the capability of case-based reasoning.

5 Conclusion

As shown in the case analysis, CBR can be used to conveniently assess ship repair risk, which has unknown mechanism. Along with the expansion of case database, the capability and accuracy of CBR will be improved gradually. Since reasoning depends on actual repair cases, its results will be more convincing than other methods. The future research should focus on improving the efficiency of CBR as much as possible, in order to enrich the results of risk assessment.

References

1. Qingming Z (2006) Safety risk analysis and control of ship-repair. China Ship Repair 19(3):5–8 (In Chinese)
2. Xinguang T, Zhiming Q, Miyi D (2008) Progress risk analysis of naval gun development based on multiple risk probability models. ACTA Armamentarii 29(5):521–525 (In Chinese)
3. Ge Zhihao X, Haojun LL et al (2007) Risk probability assessment method for complex system. J Beijing Univ Aeronaut Astronaut 33(9):1025–1028 (In Chinese)
4. Jian C, Zhongmin L, Shuchun T et al (2008) Risk assessment of purchasing weapon system based on improved risk matrix method. J Syst Eng Electron 30(10):1918–1923 (In Chinese)
5. Ping S, Weizhen M, Yuanfu T (2005) Application of fuzzy comprehensive evaluation method in project risk management. J Lanzhou Jiaotong Univ (Nat Sci Ed) 24(6):53–55 (In Chinese)
6. Hong Z, Yaju L, Tao S (2008) Project risk management based on Bayesian network. J Shenyang Univ Technol 1(3):239–243 (In Chinese)
7. Zhe X, Jinjin W, Ligang Y (2008) Quantitative appraisement model of technical risk for weapon system based on GERT simulation technique. J Syst Simul 20(7):1655–1660, 1664 (In Chinese)
8. Watoson I (1999) Case-based reasoning is a methodology not a technology. Knowl-Based Syst 12(5,6):303–308
9. Nakhaeizadeh G (1994) Learning prediction of time series. A theoretical and empirical comparison of CBR with some other approaches. In: Topics in Case-Based Reasoning. Springer, Berlin, pp 65–76

A Rule-Based Inference Method Using Dempster–Shafer Theory

Liuqian Jin and Yang Xu

Abstract The Dempster–Shafer theory of evidence for attribute aggregation provides a method to deal with uncertainty reasoning. In this paper, uncertainty reasoning method based on rule-base with certainty interval is investigated. First, knowledge representation with interval uncertainty is defined and the matching principle is given. Then, a rule-based inference method under interval numbers using Dempster–Shafer theory is derived. A numerical example is examined to show the implementation process of the proposed method.

Keywords Uncertainty reasoning · Dempster–Shafer theory · Knowledge representation · Rule-base · Interval number

1 Introduction

Uncertainty has been an important topic for decades. Uncertainty can occur because information is not clearly described, or only by partial and imprecise evidence, which is a result of loose concepts in observations or due to inaccuracy and poor reliability of instruments used to make the observations [1]. In our daily life, the information that is used for inference may be imprecise, incomplete, or unreliable, so uncertainty reasoning is very useful. There are two aspects for uncertainty reasoning research: (1) how to infer the consequent by antecedent and rule; (2) how to infer the uncertainty of consequent by the uncertainty of

L. Jin (✉)
Intelligent Control Development Center, Southwest Jiaotong University,
Chengdu 610031, China
e-mail: jinliuqian@163.com

Y. Xu
School of Economics and Management, Southwest Jiaotong University,
Chengdu 610031, China

Z. Wen and T. Li (eds.), *Knowledge Engineering and Management*,
Advances in Intelligent Systems and Computing 278,
DOI: 10.1007/978-3-642-54930-4_7, © Springer-Verlag Berlin Heidelberg 2014

antecedent and rule. Interval number (for e.g., $\alpha = [\alpha^L, \alpha^R]$) also named interval-valued fuzzy number, is important in various fields; in this paper interval numbers are used to express uncertainty.

Evidential theory was presented by Dempster in the 1960s, extended and progressed by Shafer, and then the Dempster–Shafer theory was established. D–S theory is available for uncertainty reasoning. D–S theory has some advantages, for eg., its powerful evidence combination rule and its reasonable requirement for the basic probability assignments that given a piece of evidence, the commitment of certainty in a hypothesis does not necessarily mean that the remaining certainty must be assigned to the complement of the hypothesis, but to the whole sample space [2, 3]. D–S theory is well suited for handling incomplete uncertainty. D–S theory can deal with incomplete uncertainty in a more rational way than other tools in that given a piece of evidence, the unassigned certainty in a hypothesis is just supposed to denote the unknown uncertainty, which instead of being necessarily assigned to the complement of the hypothesis, may eventually be assigned to any hypothesis in the sample space when more evidence is gathered, so we have chosen the D–S theory for uncertainty reasoning.

The paper is organized as follows. In Sect. 2, an inference rule method and D–S theory extended to combine interval evidence are given, in the preliminary part. In Sect. 3, knowledge representation and rule inference based on D–S theory under interval number are given; in this part, the matching method, the antecedents aggregation function, antecedent-consequent sequential reasoning method, and the same consequent parallel reasoning are given. In Sect. 4, using a numerical example, which is adapted from the literature [4] with different knowledge representation, to illustrate the application and detailed implementation process of the proposed approach. The paper is concluded in Sect. 5.

2 Preliminaries

The Dempster–Shafer theory of evidence for attribute aggregation is applied for uncertainty reasoning. The Dempster–Shafer theory is as follow [5]:

Suppose Θ is an identification space, $P(\Theta)$ is the power set of Θ, mass function m is a mapping from $P(\Theta)$ to $[0, 1]$, i.e. $m\colon P(\Theta) \to [0, 1]$, m satisfies the following conditions:

(1) $m\varnothing = 0$
(2) $\sum\limits_{E \subseteq \Theta} m(E) = 1$

where m is Basic Probability Assignment Function.

Belief factor is written as

$$\mathrm{Bel}(E) = \sum_{D \subseteq E} m(D), \quad \forall E \subseteq \Theta$$

where $\mathrm{Bel}(E)$ is the lower bound of the certainty degree of E.
    Plausibility factor is written as

$$Pl(E) = 1 - \mathrm{Bel}(\bar{E}) = \sum_{D \cap E \neq \varnothing} m(D), \quad \forall E \subseteq \Theta$$

where $Pl(E)$ is the upper bound of the certainty degree of E, $\bar{E}$ is the complement of E. m_1 and m_2 are basic probability assignment functions on Θ, the combination of m_1 and m_2 is represented as:

$$m(D) = m_1 \oplus m_2(D) = \begin{cases} \frac{1}{N} \sum_{E \cap F = D} m_1(E) m_2(F) & D \neq \varnothing \\ 0 & D = \varnothing \end{cases}$$

where $N = \sum_{E \cap F \neq \varnothing} m_1(E) m_2(F) > 0$.

3 Rule Inference

Knowledge representation is used to represent knowledge with formalization methods that can be identified and compiled by computers. In this paper, the knowledge representation method is an "If-then" rule with certainty factors.
    A rule-base is represented as

$$R = \langle (X, A), (Y, C), CF, F \rangle$$

where $X = \{X_i | i = 1, \ldots, M\}$ is the set of antecedent attributes; $A = \{A_i | i = 1, \ldots, M\}$ is a finite set, and $A_i = \{A_{i,I_i} | I_i = 1, \ldots, L_{A_i}\}$ is the set of attribute values for an antecedent attribute $X_i(i = 1, \ldots, M)$; $Y = \{Y_j | j = 1, \ldots, N\}$ is the set of consequent attributes; $C = \{C_j | j = 1, \ldots, N\}$ is a finite set, and $C_j = \{C_{j,Jj} | J_j = 1, \ldots, L_{C_j}\}$ is the set of attribute values for a consequent attribute $Y_j(j = 1, \ldots, N)$; CF is certainty factor, the degree of certainty; F is reflecting the set of the relation between antecedent and its associated consequent in rule.
    The certainty factor $CF(\Delta)$ is an interval number in $[0, 1]$, means the certainty factor of Δ. As $CF(\Delta)$ is an interval number, $CF(\Delta)$ also can be named as certainty interval, then remark $CF(\Delta)$ as

$$\mathrm{CF}(\Delta) = [\mathrm{Bel}(\Delta), \mathrm{Pl}(\Delta)]$$

where $0 \leq \mathrm{Bel}(\Delta) \leq \mathrm{Pl}(\Delta) \leq 1$, $\mathrm{Bel}(\Delta)$ is the lower bound of the certainty factor of Δ (belief factor), $\mathrm{Pl}(\Delta)$ is the upper bound of the certainty factor of Δ (plausibility factor); for different Δ, $\mathrm{Bel}(\Delta)$ and $\mathrm{Pl}(\Delta)$ are well-determined by different mass functions.

The kth rule in forms of a conjunctive "If-then" rule can be written as R^k:

$$\text{if } \left(X_1 = A_1^k, \ \mathrm{CF}^k(X_1 = A_1^k)\right) \wedge \cdots \wedge \left(X_M = A_M^k, \mathrm{CF}^k(X_M = A_M^k)\right)$$
$$\text{then } \left(Y_1 = C_1^k, \ \mathrm{CF}^k(Y_1 = C_1^k)\right) \wedge \cdots \wedge \left(Y_N = C_N^k, \ \mathrm{CF}^k(Y_N = C_N^k)\right)$$

with $\mathrm{CF}^k\left(R^k\right)$
where $A_i^k \in A_i$, $\quad i = 1, 2, \ldots, M$, $\qquad C_j^k \in C_j$, $\quad j = 1, 2, \ldots, N$.

The kth rule can be abbreviated as R^k

$$\text{if}\left(A_1^k, \ \alpha_1^k\right) \wedge \left(A_2^k, \ \alpha_2^k\right) \wedge \cdots \wedge \left(A_M^k, \ \alpha_M^k\right)$$
$$\text{then}\left(C_1^k, \ \beta_1^k\right) \wedge \left(C_2^k, \ \beta_2^k\right) \wedge \cdots \wedge \left(C_N^k, \ \beta_N^k\right) \text{ with } \gamma^k$$

where $\alpha_i^k = [\mathrm{Bel}(A_i^k), \mathrm{Pl}(A_i^k)]$, $\beta_j^k = [\mathrm{Bel}(C_j^k), \mathrm{Pl}(C_j^k)]$, $\gamma^k = [\mathrm{Bel}(R^k), \mathrm{Pl}(R^k)]$; α_i^k, β_j^k and γ^k are certainty intervals in $[0, 1]$.

Remark

$$A^k = \left\{A_i^k \big| A_i^k \in A_i, \quad i = 1, 2, \ldots, M \right\}; \quad \wedge A^k = A_1^k \wedge A_2^k \wedge \cdots \wedge A_M^k;$$
$$C^k = \left\{C_j^k \big| C_j^k \in C_j, \quad j = 1, 2, \ldots, N \right\}; \quad \wedge C^k = C_1^k \wedge C_2^k \wedge \cdots \wedge C_N^k;$$

then the kth rule can be remarked as

If $(\wedge A^k, \alpha^k)$ then $(\wedge C^k, \beta^k)$ with γ^k where α^k is the certainty factor of $\wedge A^k$, β^k is the certainty factor of $\wedge C^k$.

There is an actual input vector remarked as

$$\mathrm{Input}() = \{(a_1, \alpha_1), (a_2, \alpha_2), \ldots, (a_M, \alpha_M)\}$$

where a_i is in correspondence with X_i, α_i is the certainty factor of $a_i.\alpha_i$, $(i = 1, 2, \ldots, M)$, α^k, and β^k are certainty factors:

$$\alpha_i = [\mathrm{Bel}(a_i), \mathrm{Pl}(a_i)], \quad (i = 1, 2, \ldots, M);$$
$$\alpha^k = \left[\mathrm{Bel}(\wedge A^k), \mathrm{Pl}(\wedge A^k)\right]; \quad \beta^k = \left[\mathrm{Bel}(\wedge C^k), Pl(\wedge C^k)\right].$$

When all actual input values a_i, $(i = 1, 2, \ldots, M)$ are identifying with the ith antecedent attribute values A_i^k, $(i = 1, 2, \ldots, M)$, this rule is matched successful.

Suppose that rule R^k is matched successful with the actual input vector Input(). There is a similarity degree between the antecedent attribute value A_i^k and the actual input attribute value a_i, the similarity degree is as follows:

$$\mathrm{Sd}(\alpha_i^k, \alpha_i) = \left[\mathrm{Sdl}(\alpha_i^k, \alpha_i), \mathrm{Sdr}(\alpha_i^k, \alpha_i)\right] = \min\left\{1 - \alpha_i^k + \alpha_i, \ 1 - \alpha_i + \alpha_i^k\right\}. \quad (1)$$

Proposition For any interval number $\mu = [\mu^L, \mu^R]$, $(0 \leq \mu^L \leq \mu^R \leq 1)$ and $\eta = [\eta^L, \eta^R]$, $(0 \leq \eta^L \leq \eta^R \leq 1)$, Sd satisfies:

(1) (Reflexivity) $\mathrm{Sd}(\mu, \mu) = 1$;
(2) (Symmetry) $\mathrm{Sd}(\mu, \eta) = \mathrm{Sd}(\eta, \mu)$.

Proof

(1) $\mathrm{Sd}(\mu, \mu) = \min\{1 - \mu + \mu, 1 - \mu + \mu\} = 1$;
(2) Let us discuss the following situations:
(1) If $\mu < \eta$, then $0 < 1 - \eta + \mu < 1 < 1 - \mu + \eta$; as a result,

$$\mathrm{Sd}(\mu, \eta) = \min\{1 - \mu + \eta, 1 - \eta + \mu\} = 1 - \eta + \mu;$$
$$\mathrm{Sd}(\eta, \mu) = \min\{1 - \eta + \mu, 1 - \mu + \eta\} = 1 - \eta + \mu.$$

So $\mathrm{Sd}(\mu, \eta) = \mathrm{Sd}(\eta, \mu)$.
(2) If $\mu > \eta$, then $0 < 1 - \mu + \eta < 1 < 1 - \eta + \mu$; as a result,

$$\mathrm{Sd}(\mu, \eta) = \min\{1 - \mu + \eta, 1 - \eta + \mu\} = 1 - \mu + \eta;$$
$$\mathrm{Sd}(\eta, \mu) = \max\{1 - \eta + \mu, 1 - \mu + \eta\} = 1 - \mu + \eta.$$

So $\mathrm{Sd}(\mu, \eta) = \mathrm{Sd}(\eta, \mu)$.
(3) If $\mu = \eta$, then $\mathrm{Sd}(\mu, \eta) = \mathrm{Sd}(\eta, \mu) = 1$.
In conclusion, $\mathrm{Sd}(\mu, \eta) = \mathrm{Sd}(\eta, \mu)$.

Based on the order relation $\leq_{\mathrm{LR}}$ [6] and arithmetic operation of interval number [7], we have the following formulas:

$$\mathrm{Sdl}(\alpha_i^k, \alpha_i) = \min\{\min\{1 - \mathrm{Bel}(A_i^k) + \mathrm{Bel}(a_i), \ 1 - \mathrm{Bel}(a_i) + \mathrm{Bel}(A_i^k)\},$$
$$\min\{1 - \mathrm{Pl}(A_i^k) + \mathrm{Pl}(a_i), \ 1 - \mathrm{Pl}(a_i) + \mathrm{Pl}(A_i^k)\}\};$$
$$\mathrm{Sdr}(\alpha_i^k, \alpha_i) = \max\{\min\{1 - \mathrm{Bel}(A_i^k) + \mathrm{Bel}(a_i), \ 1 - \mathrm{Bel}(a_i) + \mathrm{Bel}(A_i^k)\},$$
$$\min\{1 - \mathrm{Pl}(A_i^k) + \mathrm{Pl}(a_i), \ 1 - \mathrm{Pl}(a_i) + \mathrm{Pl}(A_i^k)\}\},$$

Based on the order relation $\leq_{\mathrm{LR}}$ we have the following rule:

(1) If $\alpha_i^k < \alpha_i$, means the ith antecedent attribute is completely contented by the actual input vector, then the certainty factor α_i^k is updated to $\tilde{\alpha}_i^k = [1, 1]$.

(2) If $\alpha_i^k > \alpha_i$, then

$$1 - \mathrm{Bel}(A_i^k) + \mathrm{Bel}(a_i) < 1 - \mathrm{Bel}(a_i) + \mathrm{Bel}(A_i^k);$$
$$1 - \mathrm{Pl}(A_i^k) + \mathrm{Pl}(a_i) < 1 - \mathrm{Pl}(a_i) + \mathrm{Pl}(A_i^k);$$

as a result,

$$\mathrm{Sd}(\alpha_i^k, \alpha_i) = \left[\min\{1 - \mathrm{Bel}(A_i^k) + \mathrm{Bel}(a_i), \quad 1 - \mathrm{Pl}(A_i^k) + \mathrm{Pl}(a_i)\},\right.$$
$$\left.\max\{1 - \mathrm{Bel}(A_i^k) + \mathrm{Bel}(a_i), \quad 1 - \mathrm{Pl}(A_i^k) + \mathrm{Pl}(a_i)\}\right]$$

(3) If $\alpha_i^k = \alpha_i$, then $\mathrm{Sd}(\alpha_i^k, \alpha_i) = [1, 1]$.

With the similarity degree, the certainty factor is updated as:

$$\tilde{\alpha}_i^k = \left[\mathrm{Bel}(A_i^k, a_i), \mathrm{Pl}(A_i^k, a_i)\right]$$
$$= \left[\min\{[1,1], \quad \min\{1 - \mathrm{Bel}(A_i^k) + \mathrm{Bel}(a_i), \quad 1 - \mathrm{Pl}(A_i^k) + \mathrm{Pl}(a_i)\}\},\right.$$
$$\left.\min\{[1,1], \quad \max\{1 - \mathrm{Bel}(A_i^k) + \mathrm{Bel}(a_i), \quad 1 - \mathrm{Pl}(A_i^k) + \mathrm{Pl}(a_i)\}\}\right]$$

The antecedent of a rule is the combination of conjunction, based on Łukas-iewicz logic [8] and the order relation $\leq_{LR}$, the certainty factor of the kth rule's antecedent is as follows:

$$\tilde{\alpha}^k = \left[\mathrm{Bel}(\wedge\tilde{A}^k), \mathrm{Pl}(\wedge\tilde{A}^k)\right] = \min\{\tilde{\alpha}_1^k, \tilde{\alpha}_2^k, \ldots, \tilde{\alpha}_M^k\}$$
$$= \left[\min\{\mathrm{Bel}(A_1^k, a_1), \ldots, \mathrm{Bel}(A_M^k, a_M)\}, \min\{\mathrm{Pl}(A_1^k, a_1), \ldots, \mathrm{Pl}(A_M^k, a_M)\}\right]$$

$$(2)$$

where $\wedge\tilde{A}^k$ is the antecedent with new certainty factor after the rule is matched successful with the actual input vector.

The rule for the certainty factor of consequent under the actual input vector can be remarked as R^k:

$$\text{If } (A_1^k, \tilde{\alpha}_1^k) \wedge (A_2^k, \tilde{\alpha}_2^k) \wedge \cdots \wedge (A_M^k, \tilde{\alpha}_M^k)$$
$$\text{then } (C_1^k, \tilde{\beta}_1^k) \wedge (C_2^k, \tilde{\beta}_2^k) \wedge \cdots \wedge (C_N^k, \tilde{\beta}_N^k) \text{ with } \gamma^k$$

abbreviated as:

$$(\wedge\tilde{A}^k, \tilde{\alpha}^k) \text{ then } (\wedge C^k, \tilde{\beta}^k) \text{ with } \gamma^k$$

Based on Dempster–Shafer theory and Łukasiewicz logic, the certainty factor of consequent can be given as the following:

If $\gamma^k = [1,1]$ is said to be the kth rule complete certainty, then the result of inference is $\tilde{\beta}^k = \tilde{\alpha}^k$.

If $\gamma^k \neq [1,1]$, then the inference procedure is given as follows.

Remark

$$A^k \times C^k = \left\{ \left(A_{i,I_i}, C_{j,J_j}\right) \middle| A_{i,I_i} \in A^k \text{ and } C_{j,J_j} \in C^k \right\},$$
$$A \times C = \left\{ \left(A_{i,I_i}, C_{j,J_j}\right) \middle| A_{i,I_i} \in A \text{ and } C_{j,J_j} \in C \right\},$$
$$A^k + C^k = \left(A^k \times C\right) \cup \left(A \times C^k\right) = \left\{ \left(A_{i,I_i}, C_{j,J_j}\right) \middle| A_{i,I_i} \in A^k \text{ or } C_{j,J_j} \in C^k \right\},$$
$$A^k \rightarrow C^k = \bar{A}^k + C^k = \left\{ \left(A_{i,I_i}, C_{j,J_j}\right) \middle| A_{i,I_i} \notin A^k \text{ or } C_{j,J_j} \in C^k \right\};$$

As a result

$$\text{Bel}\left(R^k\right) = \text{Bel}\left(\bar{A}^k + C^k\right); \; \text{Pl}\left(R^k\right) = \text{Pl}\left(\bar{A}^k + C^k\right).$$

$\text{Bel}(R^k)$ and $\text{Pl}(R^k)$ are the belief factor and plausibility factor on $A \times C$. The mass function on $A \times C$ is as follows:

$$m_{R^k}\left(\bar{A}^k + C^k\right) = \text{Bel}\left(\bar{A}^k + C^k\right) = \text{Bel}\left(R^k\right),$$
$$m_{R^k}\left(\overline{\bar{A}^k + C^k}\right) = 1 - \text{Pl}\left(\bar{A}^k + C^k\right) = 1 - \text{Pl}\left(R^k\right),$$
$$m_{R^k}(A \times C) = \text{Pl}\left(R^k\right) - \text{Bel}\left(R^k\right).$$

As we know, $\tilde{\alpha}^k = [\text{Bel}(\tilde{A}^k), Pl(\tilde{A}^k)]$ is the certainty factor of A^k, so $\text{Bel}(\tilde{A}^k)$ and $\text{Pl}(\tilde{A}^k)$ are the belief factor and plausibility factor on A. The mass function on A is as follows:

$$m_{F^k}\left(A^k\right) = \text{Bel}\left(\tilde{A}^k\right), \; m_{F^k}\left(\bar{A}^k\right) = 1 - \text{Pl}\left(\tilde{A}^k\right),$$
$$m_{F^k}(A) = \text{Pl}\left(\tilde{A}^k\right) - \text{Bel}\left(\tilde{A}^k\right).$$

Extend the mass function m_{F^k} from A to $A \times C$:

$$m_{F^k}(D) = \begin{cases} m_{F^k}(B), & D = B \times C \\ 0, & \text{others} \end{cases} = \begin{cases} \text{Bel}\left(\tilde{A}^k\right), & D = A^k \times C \\ 1 - \text{Pl}\left(\tilde{A}^k\right), & D = \bar{A}^k \times C \\ \text{Pl}\left(\tilde{A}^k\right) - \text{Bel}\left(\tilde{A}^k\right), & D = A \times C \\ 0, & \text{others} \end{cases}$$

with Dempster–Shafer theory,

$$N^k = 1 - \left(1 - \text{Pl}\left(R^k\right)\right)\left(1 - \text{Pl}\left(\tilde{A}^k\right)\right),$$
$$m_k\left(A^k \times C^k\right) = \text{Bel}\left(\tilde{A}^k\right)\text{Bel}\left(R^k\right)/N^k, \quad m_k\left(A^k \times \bar{C}^k\right) = \text{Pl}\left(\tilde{A}^k\right)\left[1 - \text{Pl}\left(R^k\right)\right]/N^k,$$
$$m_k\left(A^k \times C\right) = \text{Bel}\left(\tilde{A}^k\right)\left[\text{Pl}\left(R^k\right) - \text{Bel}\left(R^k\right)\right]/N^k,$$
$$m_k\left(\bar{A}^k \times C\right) = \text{Pl}\left(R^k\right)\left[1 - \text{Pl}\left(\tilde{A}^k\right)\right]/N^k,$$
$$m_k\left(\bar{A}^k + C^k\right) = \text{Bel}\left(R^k\right)\left[\text{Pl}\left(\tilde{A}^k\right) - \text{Bel}\left(\tilde{A}^k\right)\right]/N^k,$$
$$m_k(A \times C) = \left[\text{Pl}\left(\tilde{A}^k\right) - \text{Bel}\left(\tilde{A}^k\right)\right]\left[\text{Pl}\left(R^k\right) - \text{Bel}\left(R^k\right)\right]/N^k,$$

then

$$\text{Bel}\left(C^k\right) = \sum_{D \subseteq A \times C^k} m_k(D) = \text{Bel}\left(\tilde{A}^k\right)\text{Bel}\left(R^k\right)/N^k,$$
$$\text{Bel}\left(\bar{C}^k\right) = \sum_{D \subseteq A \times \bar{C}^k} m_k(D) = \text{Pl}\left(\tilde{A}^k\right)\left[1 - \text{Pl}\left(R^k\right)\right]/N^k,$$
$$\text{Pl}\left(C^k\right) = 1 - \text{Bel}\left(\bar{C}^k\right) = 1 - \text{Pl}\left(\tilde{A}^k\right)\left[1 - \text{Pl}\left(R^k\right)\right]/N^k,$$
$$\text{Pl}\left(\bar{C}^k\right) = 1 - \text{Bel}\left(C^k\right) = 1 - \text{Bel}\left(\tilde{A}^k\right)\text{Bel}\left(R^k\right)/N^k,$$

where $\bar{\beta}^k = [\text{Bel}(\bar{C}^k), \text{Pl}(\bar{C}^k)]$ is the certainty factor of $\bar{C}^k$.

So the certainty factor of consequent can be represented as:

$$\tilde{\beta}^k = \left[\text{Bel}\left(\tilde{A}^k\right)\text{Bel}\left(R^k\right)/N^k, 1 - \text{Pl}\left(\tilde{A}^k\right)\left[1 - \text{Pl}\left(R^k\right)\right]/N^k\right], \tag{3}$$

then the certainty factors of each consequent attributes are:

$$\tilde{\beta}_j^k = \left(1 - \beta_j^k + \tilde{\beta}^k\right)\beta_j^k, \quad (j = 1, 2, \ldots, N). \tag{4}$$

If the consequents of two rules are C^k, with $\text{CF}_1(C^k) = [\text{Bel}_1(C^k), \text{Pl}_1(C^k)]$ and $\text{CF}_2(C^k) = [\text{Bel}_2(C^k), \text{Pl}_2(C^k)]$. With the definitions of belief factor and plausibility factor, the mass functions can be represented as:

$$m_1\left(C^k\right) = \text{Bel}_1\left(C^k\right), \quad m_1\left(\bar{C}^k\right) = 1 - \text{Pl}_1\left(C^k\right), \quad m_1(C) = \text{Pl}_1\left(C^k\right) - \text{Bel}_1\left(C^k\right),$$
$$m_2\left(C^k\right) = \text{Bel}_2\left(C^k\right), \quad m_2\left(\bar{C}^k\right) = 1 - \text{Pl}_2\left(C^k\right), \quad m_2(C) = \text{Pl}_2\left(C^k\right) - \text{Bel}_2\left(C^k\right),$$

with Dempster–Shafer theory,

$$N_{C^k} = 1 - \text{Bel}_1\left(C^k\right)\left[1 - \text{Pl}_2\left(C^k\right)\right] - \text{Bel}_2\left(C^k\right)\left[1 - \text{Pl}_1\left(C^k\right)\right],$$

$$m^k\left(C^k\right) = \left\{\text{Bel}_1\left(C^k\right)\text{Pl}_2\left(C^k\right) + \text{Bel}_2\left(C^k\right)\text{Pl}_1\left(C^k\right) - \text{Bel}_1\left(C^k\right)\text{Bel}_2\left(C^k\right)\right\}/N_{C^k},$$

$$m^k\left(\bar{C}^k\right) = \left\{ \begin{array}{l} \left(1 - \text{Bel}_1\left(C^k\right) - \text{Bel}_2\left(C^k\right) + \text{Bel}_1\left(C^k\right)\text{Pl}_2\left(C^k\right)\right. \\ \left. + \text{Bel}_2\left(C^k\right)\text{Pl}_1\left(C^k\right) - \text{Pl}_1\left(C^k\right)\text{Pl}_2\left(C^k\right)\right) \end{array} \right\} \Big/ N_{C^k},$$

$$m^k\left(C\right) = \left[\text{Pl}_1\left(C^k\right) - \text{Bel}_1\left(C^k\right)_1\right]\left[\text{Pl}_2\left(C^k\right) - \text{Bel}_2\left(C^k\right)\right]/N_{C^k},$$

then

$$\text{Bel}\left(C^k\right) = \left\{\text{Bel}_1\left(C^k\right)\text{Pl}_2\left(C^k\right) + \text{Bel}_2\left(C^k\right)\text{Pl}_1\left(C^k\right) - \text{Bel}_1\left(C^k\right)\text{Bel}_2\left(C^k\right)\right\}/N_{C^k},$$

$$\text{Bel}\left(C^k\right) = \left\{ \begin{array}{l} 1 - \text{Bel}_1\left(C^k\right) - \text{Bel}_2\left(C^k\right) + \text{Bel}_1\left(C^k\right)\text{Pl}_2\left(C^k\right) \\ + \text{Bel}_2\left(C^k\right)\text{Pl}_1\left(C^k\right) - \text{Pl}_1\left(C^k\right)\text{Pl}_2\left(C^k\right) \end{array} \right\} \Big/ N_{C^k},$$

$$\text{Pl}\left(C^k\right) = 1 - \left\{ \begin{array}{l} 1 - \text{Bel}_1\left(C^k\right) - \text{Bel}_2\left(C^k\right) + \text{Bel}_1\left(C^k\right)\text{Pl}_2\left(C^k\right) \\ + \text{Bel}_2\left(C^k\right)\text{Pl}_1\left(C^k\right) - \text{Pl}_1\left(C^k\right)\text{Pl}_2\left(C^k\right) \end{array} \right\} \Big/ N_{C^k},$$

$$\text{Pl}\left(C^k\right) = 1 - \left[\text{Pl}_1\left(C^k\right) - \text{Bel}_1\left(C^k\right)_1\right]\left[\text{Pl}_2\left(C^k\right) - \text{Bel}_2\left(C^k\right)\right]/N_{C^k}$$

So the certainty factor can be represented as:

$$\text{CF}\left(C^k\right) = \left[\frac{\begin{array}{l}\left(\text{Bel}_1\left(C^k\right)\text{Pl}_2\left(C^k\right) + \text{Bel}_2\left(C^k\right)\text{Pl}_1\left(C^k\right)\right.\\ \left. - \text{Bel}_1\left(C^k\right)\text{Bel}_2\left(C^k\right)\right)\end{array}}{N_{C^k}}, \quad 1 - \frac{\begin{array}{l}\left(1 - \text{Bel}_1\left(C^k\right) - \text{Bel}_2\left(C^k\right) + \text{Bel}_1\left(C^k\right)\text{Pl}_2\left(C^k\right)\right.\\ \left. + \text{Bel}_2\left(C^k\right)\text{Pl}_1\left(C^k\right) - \text{Pl}_1\left(C^k\right)\text{Pl}_2\left(C^k\right)\right)\end{array}}{N_{C^k}} \right].$$

$$(5)$$

4 Example

Each rule used in [4] only one consequent attribute with certainty factor in $[0, 1]$, the certainty factors of antecedent attributes are 1 and without the certainty factor of rule, which is a special case of the knowledge representation in Sect. 3. In this section, the rule base is given in Table 1 with the interval certainty factor belonging to the closed interval $[0, 1]$ of antecedent, consequent, and rule.

There is an actual input vector:

$$\text{Input}() = \{(A_{1,2}, [0.3, 0.4]), (A_{2,3}, [0.4, 0.6]), (A_{3,3}, [0.7, 0.8]), (\varnothing, 0),$$
$$(A_{5,2}, [0.5, 0.5]), (A_{6,1}, [0.6, 0.7]), (A_{7,2}, [0.8, 0.9]), (A_{8,3}, [0.2, 0.3])\}.$$

The matched rules are R^2, R^3, R^4, and R^{17}; the certainty factors are given in Table 2.

The consequents of R^3 and R^4 are $A_{4,3}$, so the aggregation certainty factor using (5) is $[0.84, 0.9]$. Then there are three new facts as antecedent attributes:

Table 1 Rule base

Number R^k	Certainty factor γ^k	Antecedent and α_i^k	Consequent and β_j^k
R^1	[1, 1]	$(x_1 = A_{1,1}, [0.9, 1])$	$(x_2 = A_{2,1}, [0.6, 0.7])$
R^2	[0.9, 1]	$(x_1 = A_{1,2}, [0.4, 0.5])$	$(x_2 = A_{2,2}, [0.8, 0.8])$
R^3	[0.5, 0.6]	$(x_2 = A_{2,3}, [0.7, 0.8])$	$(x_4 = A_{4,3}, [0.9, 1])$
R^4	[1, 1]	$(x_3 = A_{3,3}, [0.9, 1])$	$(x_4 = A_{4,3}, [0.8, 0.9])$
R^5	[0.6, 0.7]	$(x_2 = A_{2,1}, [0.9, 0.9]) \wedge (x_3 = A_{3,2}, [0.9, 1])$	$(x_4 = A_{4,2}, [0.7, 0.8])$
R^6	[0.5, 0.7]	$(x_2 = A_{2,2}, [0.5, 0.5]) \wedge (x_3 = A_{3,1}, [0.6, 07])$	$(x_4 = A_{4,2}, [0.5, 0.5])$
R^7	[1, 1]	$(x_2 = A_{2,3}, [0.6, 07]) \wedge (x_3 = A_{3,2}, [0.5, 0.5])$	$(x_4 = A_{4,2}, [0.6, 0.7])$
R^8	[1, 1]	$(x_5 = A_{5,1}, [0.6, 0.8]) \wedge (x_4 = A_{4,1}, [0.6, 0.7])$	$(x_6 = A_{6,1}, [0.4, 0.5])$
R^9	[0.6, 0.8]	$(x_5 = A_{5,1}, [0.4, 0.5]) \wedge (x_4 = A_{4,2}, [0.5, 0.5])$	$(x_6 = A_{6,2}, [0.4, 0.5])$
R^{10}	[0.8, 0.9]	$(x_5 = A_{5,2}, [0.6, 0.7]) \wedge (x_4 = A_{4,1}, [0.8, 0.9])$	$(x_6 = A_{6,2}, [0.5, 0.6])$
R^{11}	[1, 1]	$(x_5 = A_{5,2}, [0.9, 1]) \wedge (x_4 = A_{4,2}, [0.6, 0.7])$	$(x_6 = A_{6,2}, [0.8, 0.9])$
R^{12}	[0.7, 0.8]	$(x_5 = A_{5,2}, [0.6, 0.7]) \wedge (x_4 = A_{4,3}, [0.8, 0.8])$	$(x_6 = A_{6,3}, [0.5, 0.6) \wedge (x_8 = A_{8,1}, [0.6, 0.7])$
R^{13}	[0.5, 0.6]	$(x_5 = A_{5,3}, [0.5, 0.6]) \wedge (x_4 = A_{4,1}, [0.7, 0.7])$	$(x_6 = A_{6,3}, [0.6, 0.7])$
R^{14}	[0.8, 1]	$(x_5 = A_{5,3}, [0.8, 0.9]) \wedge (x_4 = A_{4,2}, [1, 1])$	$(x_6 = A_{6,3}, [0.8, 0.8])$
R^{15}	[0.6, 0.7]	$(x_7 = A_{7,1}, [0.5, 0.6]) \wedge (x_1 = A_{1,2}, [0.7, 0.8])$	$(x_8 = A_{8,2}, [0.5, 0.6])$
R^{16}	[1, 1]	$(x_7 = A_{7,2}, [0.8, 0.9]) \wedge (x_1 = A_{1,1}, [0.6, 0.7])$	$(x_8 = A_{8,1}, [0.7, 0.8])$
R^{17}	[0.9, 1]	$(x_7 = A_{7,2}, [0.6, 0.7]) \wedge (x_1 = A_{1,2}, [0.3, 0.4])$	$(x_8 = A_{8,2}, [0.3, 0.4]) \wedge (x_9 = A_{9,2}, [0.6, 0.7])$
R^{18}	[0.8, 0.8]	$(x_6 = A_{6,1}, [0.5, 0.6]) \wedge (x_8 = A_{8,2}, [0.6, 0.7])$	$(x_9 = A_{9,2}, [0.4, 0.5])$
R^{19}	[0.9, 1]	$(x_6 = A_{6,2}, [0.7, 0.8]) \wedge (x_8 = A_{8,3}, [0.5, 0.7])$	$(x_9 = A_{9,3}, [0.6, 0.7])$
R^{20}	[1, 1]	$(x_6 = A_{6,3}, [0.8, 0.9]) \wedge (x_8 = A_{8,1}, [0.6, 0.7])$	$(x_9 = A_{9,3}, [0.8, 0.8])$

Table 2 Certainty factors I

Number R^k	Certainty factor γ^k	Certainty factor $\tilde{\alpha}^k$	Certainty factor $\tilde{\beta}^k$	Consequent attributes and certainty factor
R^2	[0. 9,1]	[0.9, 0.9]	[0.81, 1]	$(x_2 = A_{2,2}, [0.808, 0.96])$
R^3	[0.5, 0.6]	[0.7, 0.8]	[0.38, 0.65]	$(x_4 = A_{4,3}, [0.432, 0.65])$
R^4	[1, 1]	[0.8, 0.8]	[0.8, 1]	$(x_4 = A_{4,3}, [0.8, 0.99])$
R^{17}	[0.9, 1]	[1, 1]	[0.9, 1]	$(x_8 = A_{8,2}, [0.48, 0.64]) \wedge (x_9 = A_{9,2}, [0.78, 0.91])$

Table 3 Certainty factors II

Number R^k	Certainty factor γ^k	Certainty factor $\tilde{\alpha}^k$	Certainty factor $\tilde{\beta}^k$	Consequent attributes and certainty factor
R^{12}	[0.7, 0.8]	[0.8, 0.9]	[0.7, 0.81]	$(x_6 = A_{6,3}, [0.535, 0.726]) \wedge (x_8 = A_{8,1}, [0.66, 0.777])$
R^{18}	[0.8, 0.8]	[0.88, 0.94]	[0.713, 0.81]	$(x_9 = A_{9,2}, [0.666, 0.779])$

$(A_{2,2}, [0.808, 0.96])$ $(A_{4,3}, [0.84, 0.9])$ $(A_{8,2}, [0.48, 0.64])$.

The matched rules are R^{12} and R^{18}, the certainty factors are given in Table 3. The consequents of R^{17} and R^{18} are $A_{9,2}$, so the aggregation certainty factor using (5) is [0.91, 0.923].

5 Conclusion

In this paper, the knowledge representation and inference methods are different from those of the past. Based on Dempster–Shafer and fuzzy set theory, a rule-base was designed for uncertainty reasoning. The inference process of such a rule-base was characterized by interval numbers and was implemented using the Dempster–Shafer theory. The methodology was further extended to uncertainty reasoning. A numerical example was used to illustrate the application of this method.

Acknowledgment This work is supported by the National Science Foundation of China (Grant No. 61175055), Sichuan Key Technology Research and Development Program (Grant No. 2011FZ0051), Radio Administration Bureau of MIIT of China (Grant No. [2011] 146), China Institution of Communications (Grant No. [2011] 051).

References

1. Chen W, Chen C (2010) Knowledge engineering and knowledge management. Tsinghua University Press, Beijing
2. Buchanan BG, Shortliffe EH (1984) Rule-based expert systems reading. Addison-Wesley, Boston

3. Lopez de Mantaras R (1990) Approximate reasoning models. Ellis Horwood Limited, Chichester
4. Yang J, Liu J, Wang J et al (2006) Belief rule-base inference methodology using the evidential reasoning approach–RIMER. IEEE Trans Syst Man Cybern Part A Syst Hum 36(2):266–285
5. Yager RR, Liu L (eds) (2008) Classic works of the dempster-shafer theory of belief functions. Springer, Berlin
6. Ishihuchi H, Tanaka H (1990) Multiobjective programming in optimization of the interval objective function. Eur J Oper Res 48:219–225
7. Tran L, Duckstein L (2002) Comparison of fuzzy numbers using a fuzzy distance measure. Fuzzy Sets Syst 130:331–341
8. Wang G, Zhou H (2009) Introduction to mathematical logic and resolution principle 2nd edn. Science Press, Beijing
9. Calzada A, Liu J, Wang H et al (2011) An intelligent decision support tool based on belief rule-based inference methodology. In: IEEE international conference on fuzzy systems, Taipei, 27–30 June 2011, 2638–2643
10. Yang G, Li X, Wang W et al (2012) Customer satisfaction survey based on evidential reasoning approach with belief intervals. J Ind Eng Manag 26(1):27–34
11. Couso I, Garrido L, Sánchez L (2013) Similarity and dissimilarity measures between fuzzy sets: a formal relational study. Inf Sci 229:122–141
12. Yang J, Singh MG (1994) An evidential reasoning approach for multiple-attribute decision making with uncertainty. IEEE Trans Syst Man Cybern 24(1):1–18
13. Zadeh LA (1975) Calculus of fuzzy restriction, In: Zadeh LA et al (eds) Fuzzy sets and their applications to cognitive and decision processes. Academic Press, New York
14. Zhang Q, Xu H (2012) Research on early-warning expert system for security of grain storage based on uncertain inference. Comput Dig Eng 40(2):79–82
15. Zhang W, Liang Y, Xu P (2007) Uncertainty reasoning based on inclusion degree. Tsinghua University Press, Beijing, p 3
16. Luo X, Cai J, Qiu Y (1994) A new interval-based uncertainty reasoning model. J Southwest China Normal Univ 19(6):591–600
17. Hu Z, Shen T, Li G et al (2009) Interference finding expert system based on case reasoning and rule reasoning. Comput Eng 35(18):185–190

Blog Topic Diffusion Prediction Model Based on Link Information Flow

Dazhen Lin and Donglin Cao

Abstract How to predict the topic diffusion is a challenging research work in social media data mining. The classical research works in Twitter and Micorblog mainly focus on diffusion links that ignore the importance of diffusion content. In this paper, we propose a Link Information Flow-based topic diffusion prediction model, which combines the link view and content view in diffusion. The experiment results show that our model achieves good performance in topic diffusion prediction.

Keywords Topic diffusion prediction · Link information flow · Author behavior

1 Introduction

Although more and more research focuses on information diffusion in Twitter and Microblog, the problem of information fragmentation in Twitter and Microblog leads to the difficulty of analyzing the content of information diffusion. This is because each message in Twitter and Microblog contains less than 140 words, and it is hard to reveal the deep content information in each message. Unlike Twitter and Microblog, Blog is a content sharing platform that encourages authors to share more meaningful information without any limitations. Therefore, Twitter and Microblog as an application focus on information diffusion, and Blog is an application that focuses on deep content information sharing. Although the

D. Lin · D. Cao (✉)
Cognitive Science Department, Xiamen University, Xiamen, China
e-mail: another@xmu.edu.cn

D. Lin
e-mail: dzlin@xmu.edu.cn

Z. Wen and T. Li (eds.), *Knowledge Engineering and Management,*
Advances in Intelligent Systems and Computing 278,
DOI: 10.1007/978-3-642-54930-4_8, © Springer-Verlag Berlin Heidelberg 2014

diffusion speed of Blog is slower than Twitter and Microblog, it is valuable in revealing the diffusion feature of the deep content of information.

The classical diffusion models in Twitter and Microblog consider two kinds of author behaviors: reading behavior and content copy behavior. Through the comment link and forward link, it is easy to model the reading behavior and content copy behavior. However, there exist two problems in classical diffusion models. First, the content is excursive with topic diffusion and forward links do not ensure that the new content is in accordance with the original topic. Second, classical diffusion models did not give the relation between authors and their articles. The root cause of these two problems is that classical diffusion models did not effectively combine the deep content and link information in diffusion. To solve this problem, we propose a Link Information Flow-based topic diffusion prediction model. The advantage of this model is that we combine the content view and link view through the information flow.

The rest of the paper is organized as follows. Section 2 describes the related work. Details of our Link Information Flow approach are elaborated in Sect. 3. Experimental results that show the performance of our algorithm are given in Sects. 4 and 5 concludes the paper.

2 Related Work

Many information diffusion researches focus on how to find the diffusion law in the social network. These researches can be categorized into two kinds. The first kind studies use of some statistics methods to model the diffusion process in the social media [1–12]. Gruhl et al. [1] first studied information diffusion in blog. They used diffusion probability to model the diffusion law; the diffusion probability is computed by the probability of reading the topic and the probability of copying the content of the topic. Qamra et al. [2] proposed a Content-Community-Time model to mine the topic event from the connection of content, time, and blog group. Yan et al. [3] proposed a real-time information diffusion model used in vehicle ad hoc networks. This model chooses time and distance as the most important information to ensure the safety of a moving vehicle. Kuo et al. [4] designed a learning-based framework to predict the diffusion of a new topic. The second kind studies some diffusion law in the social media [13–18]. Kwon et al. [13] analyze the feature of information diffusion in blog. The most important contribution of this research work is that they found that almost 85 % of information diffusion is processed in the nonexplicit way. This indicates that most of the information diffusion in blog is not performed among friends in blog. Wang et al. [14] found that the popularity of Microblog satisfies the stretched exponential model. Furthermore, they used multiplicative cascade model to model the popularity of Microblog. Romero et al. [15] found that there exists complex contagion in Twitter. In other words, users accept the topic in condition of receiving various different information sources.

Tsung-Ting Kuo's work is similar to ours. In their prediction framework, topic information, user information, user-topic information, and global information are used in the prediction process. To improve the performance, Latent Dirichlet Allocation (LDA) method is used to represent the semantic meaning. The main difference between Tsung-Ting Kuo's work and ours is that we construct a unified model to represent the information flow in diffusion.

3 Link Information Flow-Based Topic Diffusion Prediction Model

3.1 Link View and Content View in Diffusion

From the content view, topic diffusion is performed according to the process shown in Fig. 1. In the first step, author A writes an article. In the second step, author B reads the article written by author A. In the third step, author B writes an article or comment about the same topic. In the fourth step, author B copies some words from author A's article.

From the link view, topic diffusion is performed according to the process shown in Fig. 2. In the first step, because of the authority of author A, author A obtains an information. In the second step, because of the relation between authors A and B, author B receives the information from author A. In the third step, author B spreads the information to his friends.

The content view considers the diffusion information and the link view considers the diffusion path. To combine the advantage of two views, we proposed a Link Information Flow based topic diffusion prediction model which is shown in Fig. 3. This model is a two-level model. In the lower level, this model uses the link view to model the diffusion path. In the upper level, this model uses the content view to model the diffusion information.

3.2 Link Information Flow-Based Model

To formulate our model, we denote two authors as A and B. In the diffusion process, author A writes an article d_1, and author B writes an article d_2, where $T_{d_1} < T_{d_2}$ which means that author A writes d_1 first. Using the above denotation, probability of Link Information Flow from author A to B is formulated as follows:

$$P_{\text{diffusion}}(A \rightarrow B) = P(A)P(d_1|A)P(B|A)P(d_2|B, d_1) \tag{1}$$

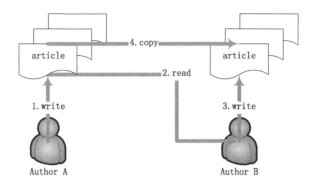

Fig. 1 Content diffusion

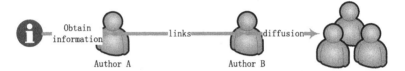

Fig. 2 Link diffusion

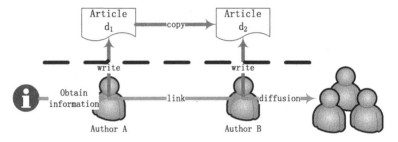

Fig. 3 Link information flow-based topic diffusion

where $P(A)$ is the influence of author A, $P(d_1|A)$ is the probability that author A writes article d_1, $P(B|A)$ is the diffusion probability from A to B, $P(d_2|B, d_1)$ is the probability that author B write articles d_2 after reading d_1.

To simplify the model, we assume that the writing process and citing (copying) process are independent. Therefore, Eq. (1) is rewritten as follows:

$$P_{\text{diffusion}}(A \rightarrow B) = P(A)P(d_1|A)P(B|A)P(d_2|B)P(d_2|d_1). \qquad (2)$$

According to Fig. 3, Eq. (2) is transformed into a graph model shown in Fig. 4. There are two kinds of nodes in this graph model: author nodes and article nodes. Each information message is spread according to the information diffusion flow in the diffusion graph.

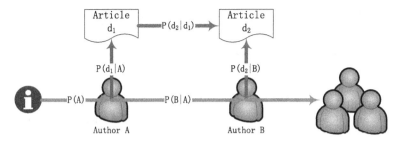

Fig. 4 Link information flow based graph model

By using the Link Information Flow graph, the topic diffusion prediction model is described as follows:

Input: topic T

Output: information diffusion path L

1. Retrieval all documents $\{d_1, d_2, \cdots, d_n\}$ which is relevant to the topic T according to the classical topic tracking algorithm.
2. Obtain the publish time of each document. The time sequence is denoted as $\{t_1, t_2, \ldots, t_n\}$.
3. Obtain author of each document. The authors are denoted as $\{a_1, a_2, \ldots, a_n\}$. Because each author can write more than one article, there exist $a_i = a_j$ and $i \neq j$.
4. Compute the average Link Information Flow probability of every two authors by using the friend links. The average Link Information Flow probability is denoted as $\text{avg}(P_{\text{diffusion\_friends}})$.
5. For any two authors a_i and a_j who publish the document d_i and d_j, respectively, compute the Link Information Flow probability $P_{\text{diffusion}}$. If $P_{\text{diffusion}} \geq \text{avg}(P_{\text{diffusion\_friends}})$, then there exists diffusion relation between a_i and a_j, this relation is denoted as (a_i, a_j). Update the diffusion path by $L = L + \{(a_i, a_j)\}$.
6. Output the diffusion path L.

4 Experiments and Analysis

4.1 Experiment Setup

We crawled 444,951 bloggers from Sina blog (http://blog.sina.com.cn), totally 4,212,354 articles. The total compressed size of dataset is 50.3 G. The publish time of article ranged from January 2006 to April 2009. To test the performance of our model, we chose some hot topics in 2008 and 2009, including sports, entertainment, economy, science, etc., totally 50 topics. The measurement of

Table 1 Prediction performance

Method	Average precision	Average recall	Average F-measure
Friend links-based model (FL)	0.00596912	0.00858066	0.007041
Diffusion probability-based model (DP)	0.264995	0.373489	0.310024
Diffusion probability and copy probability-based model (DPCP)	0.279717	0.435129	0.340529
Link information flow-based model (LIF)	**0.317169**	**0.886249**	**0.467154**

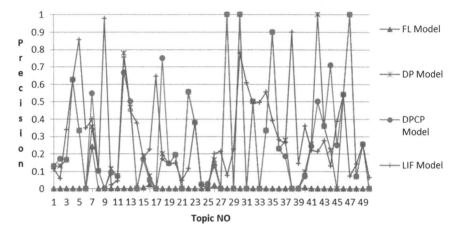

Fig. 5 Comparison in precision

experiments is precision, recall, and F-measure. Because the experiments contain 50 topics, we use average precision, average recall, and average F-measure as the final measurements of diffusion path prediction.

$$\text{Precision} = \frac{\#\text{correct diffusion path in prediction}}{\#\text{diffusion path in prediction}}$$

$$\text{Recall} = \frac{\#\text{correct diffusion path in prediction}}{\#\text{all diffusion path}}$$

$$\text{F} - \text{measure} = \frac{2 \times \text{precision} \times \text{recall}}{\text{precision} + \text{recall}}.$$

To illustrate the performance of our model, we compared our model with other three methods. The first method is based on the friend links which only uses the friend links as the diffusion path. The second method is based on the diffusion probability $P(B|A)$. The third method is based on the diffusion probability $P(B|A)$ and copy probability $P(d_2|d_1)$.

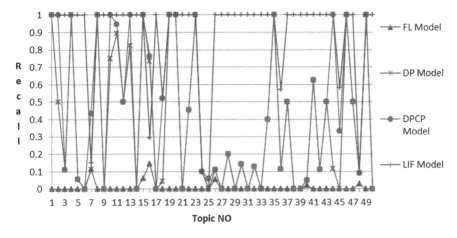

Fig. 6 Comparison in recall

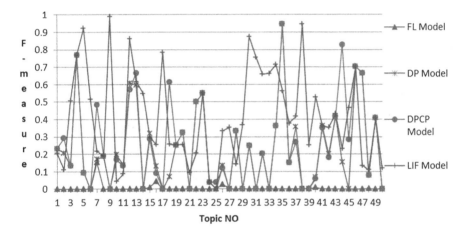

Fig. 7 Comparison in f-measure

4.2 Experiment Analysis

The experiment results are shown in Table 1 where bold indicates the best result of each measurement. It indicates that the topic diffusion prediction model based on Link Information Flow outperforms the other three methods in all three measurements. It also shows that friend links-based prediction model is the worst. The reason is that topic diffusion in friends is a small part of topic diffusion, and most of the authors who join the diffusion are not close friends in the social media.

For different topics, the comparisons are shown in Figs. 5, 6 and 7. The horizontal axis is the number of topics and the vertical axis is one of measurement. It indicates that the Link Information Flow-based topic diffusion prediction model outperforms the other three models in recall and f-measure.

5 Conclusions

With social media obtaining more and more importance, how to predict topic diffusion in social media is a valuable research field. To achieve better prediction performance, we propose a Link Information Flow-based topic diffusion prediction model that combines the link view and content view in diffusion. The experiment results show that our model achieves a high recall in topic diffusion prediction.

Acknowledgments This work is supported by the Nature Science Foundation of China (No. 61202143 and No. 61305061), the Natural Science Foundation of Fujian Province (No. 2013J05100, No. 2010J01345 and No. 2011J01367), the Key Projects Fund of Science and Technology in Xiamen (No. 3502Z20123017), the Fundamental Research Funds for the Central Universities (No. 2013121026 and No. 2011121052), the Research Fund for the Doctoral Program of Higher Education of China (No. 201101211120024), the Special Fund for Developing Shenzhen's Strategic Emerging Industries (No. JCYJ20120614164600201), the Hunan Provincial Natural Science Foundation (12JJ2040), and the Hunan Province Research Foundation of Education Committee(09A046).

References

1. Gruhl D, Liben-Nowell D et al (2004) Information diffusion through blogspace. SIGKDD Explor Newsl 6(2):43–52
2. Qamra A, Tseng B et al (2006) Mining blog stories using community-based and temporal clustering. In: Proceedings of the 15th ACM international conference on information and knowledge management. ACM, Arlington, pp 58–67
3. Yan G, Rizvi S et al (2010) A time-critical information diffusion model in vehicle ad hoc networks. In: Proceedings of the 8th international conference on advances in mobile computing and multimedia. ACM, Paris, pp 102–108
4. Kuo T-T, Hung S-C, Lin W-S, Peng N, Lin S-D, Lin W-F (2012) Exploiting latent information to predict diffusions of novel topics on social networks. In: Proceedings of the 50th annual meeting of the association for computational linguistics, pp 344–348
5. Gaussier E (2011) Models of information diffusion in social networks. In: Proceedings of the second symposium on information and communication technology. ACM, Hanoi, pp 2–2
6. Ishikawa M, Geczy P et al (2007) Information diffusion approach to cold-start problem. In: Proceedings of the 2007 IEEE/WIC/ACM international conferences on web intelligence and intelligent agent technology—workshops. IEEE Computer Society, pp 129–132
7. Lee C, Kwak H et al (2010) Finding influentials based on the temporal order of information adoption in twitter. In: Proceedings of the 19th international conference on world wide web. ACM, Raleigh, pp 1137–1138
8. Leskovec J (2011) Social media analytics: tracking, modeling and predicting the flow of information through networks. In: Proceedings of the 20th international conference companion on world wide web. ACM, Hyderabad, pp 277–278
9. Lim S-H, Kim S-W et al (2011) Construction of a blog network based on information diffusion. In: Proceedings of the 2011 ACM symposium on applied computing. ACM, TaiChung, pp 937–941
10. Ratkiewicz J, Conover M et al (2011) Truthy: mapping the spread of astroturf in microblog streams. In: Proceedings of the 20th international conference companion on world wide web. ACM, Hyderabad, pp 249–252

11. Sambasivan N, Cutrell E et al (2010) ViralVCD: tracing information-diffusion paths with low cost media in developing communities. In: Proceedings of the 28th international conference on human factors in computing systems. ACM, Atlanta, pp 2607–2610

12. Zinoviev D, Duong V (2011) A game theoretical approach to broadcast information diffusion in social networks. In: Proceedings of the 44th annual simulation symposium. Society for Computer Simulation International, Boston, pp 47–52

13. Kwon Y-S, Kim S-W et al (2009) The information diffusion model in the blog world. In: Proceedings of the 3rd workshop on social network mining and analysis. ACM, Paris, pp 1–9

14. Wang D, Li Z et al (2011) The pattern of information diffusion in microblog. In: Proceedings of the ACM CoNEXT student workshop. ACM, Tokyo, pp 1–2

15. Romero DM, Meeder B et al (2011) Differences in the mechanics of information diffusion across topics: idioms, political hashtags, and complex contagion on twitter. In: Proceedings of the 20th international conference on world wide web. ACM, Hyderabad, pp 695–704

16. Agrawal D, Budak C et al (2011) Information diffusion in social networks: observing and affecting what society cares about. In: Proceedings of the 20th ACM international conference on information and knowledge management. ACM, Glasgow, pp 2609–2610

17. Kim S-W, Faloutsos C et al (2011) Blogcast effect on information diffusion in a blogosphere. In: Proceedings of the 34th international ACM SIGIR conference on research and development in information retrieval. ACM, Beijing, pp 1149–1150

18. Kwon Y-S, Kim S-W et al (2009) An analysis of information diffusion in the blog world. In: Proceedings of the 1st ACM international workshop on complex networks meet information knowledge management. ACM, Hong Kong, pp 27–30

Construction of Multidimensional Dynamic Knowledge Map Based on Knowledge Requirements and Knowledge Connection

Yanjie Lv, Gang Zhao, Pu Miao and Yujie Guan

Abstract In the purpose to solve the problem of difficult construction, poor effect, complex update, and poorly comprehensive presentation, the paper proposes a method for the knowledge map construct model based on knowledge requirements and knowledge connection. According to the task context, product structure, and the personal knowledge structure, this method can get the knowledge requirements of the users to generate the knowledge retrieval expressions to obtain the knowledge points. Then it can construct the multi-dimensional dynamic local knowledge map through the analysis of multiple dimensions and using the design of aircraft landing gear as an example to verify the feasibility of this method.

Keywords Knowledge requirement · Knowledge connection · Multi-dimension · Knowledge map

1 Introduction

The development of complex products is a collaborative process, which needs the collaboration of design teams of multiple areas to accomplish work under users' requirement and enterprise resource constraints, based on knowledge sharing and reuse. As intangible resources and production factors [1], the quick discovery and reuse of knowledge is an important way to speed up product development, avoid repetitive error, and improve business efficiency. Therefore, in order to make good

Y. Lv (✉) · G. Zhao · P. Miao · Y. Guan
School of Mechanical Engineering and Automation, Beijing University of Aeronautics and Astronautics, Beijing 100191, China
e-mail: lvyanjie@126.com

G. Zhao
e-mail: zhaog@buaa.edu.cn

Z. Wen and T. Li (eds.), *Knowledge Engineering and Management*,
Advances in Intelligent Systems and Computing 278,
DOI: 10.1007/978-3-642-54930-4_9, © Springer-Verlag Berlin Heidelberg 2014

use of knowledge in the process of complex product development, a large number of domestic and foreign scholars have conducted a lot of research on the implementation and application of knowledge management from different aspects. For the lack of appropriate tools to reveal and recall the knowledge, experience, and methods, the phenomenon of knowledge silos, knowledge flooding, and knowledge trek is very serious. Knowledge map, as a form of knowledge management, has the function of knowledge navigation and tacit knowledge acquisition [2−4]. It is playing an important role in the knowledge management system to express the knowledge hierarchy and knowledge connection.

Literature [5] has proposed a method that uses knowledge map as a tool for knowledge acquisition and reuse. According to the knowledge requirement of mechanical product design, literature [6] has conducted some research on the classification and function of the knowledge map. Literature [7] has proposed a method to construct the knowledge map for product development based on meta-knowledge and XML visualization. Literature [8] has conducted some research on the principle of knowledge map construction. Literature [9] has proposed a frame model of the knowledge map. The shortcoming of literature [5–9] is that: ① The structure of complex product is very complex, which involves experience of many areas, and it is very difficult to construct a full set of knowledge map. ② The structure of knowledge map is too simple. Different user gets knowledge from different aspects. So it is unreasonable to construct a knowledge map of a single dimension. ③ Static knowledge map affects the results of knowledge map. Knowledge is updating and the users' requirement is changing, so the dynamics local knowledge map connected to users can improve the efficiency using of knowledge map.

To solve these deficiencies, this paper proposes a set of multi-dimensional dynamic partial knowledge map construction method, integrating people, knowledge, and scenarios. By analyzing the knowledge requirement, this method can select the appropriate knowledge for the right user, build knowledge maps from different application scenarios, and finally demonstrate dynamics local knowledge map from multiple aspects by 3D visualization techniques.

2 Knowledge Map Construction Model Based on Knowledge Requirement and Connection

To solve the problem of complex product resulting from too many related fields and a big quality of knowledge, building knowledge map is difficult and its update is complex with poor application effects. This paper proposes a knowledge map construction model based on knowledge requirement and connection, for the building process: ① Getting aware of the activities context information of current user which concludes the information of task, product structure, and personal information, and then analyzing the knowledge requirements to generate the

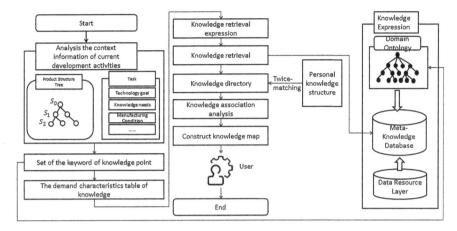

Fig. 1 Knowledge map construction model based on knowledge requirement and connection

demand characteristics of knowledge, constructing knowledge retrieval expression. ② Computing the similarity between the meta-knowledge and knowledge retrieval expression,getting the knowledge directory. ③ Analyzing the relationship between the knowledge objects of the knowledge map from different dimension. ④ Using the knowledge visualization technology to construct a multi-dimension knowledge map according to knowledge elements and the context dimension choose by the user.

Figure 1 shows the key technology of the knowledge map construct model based on knowledge requirements and knowledge connection: ① Constructing knowledge retrieval expression based on knowledge requirement. ② Context similarity-based knowledge retrieval. ③ Knowledge correlation analysis and multi-dimension knowledge visualization technology.

3 Knowledge Retrieval Expressions Constructed Based on Knowledge Requirements

3.1 Knowledge Requirement

The purpose of knowledge management is to improve the efficiency of product development. It contains two meanings: one is to assign task to the worker according to his knowledge background, another is to choose the right knowledge at the right and push it to the user. Figure 2 shows the process of complex product development based on knowledge.

The task is decomposed according to the project's needs and then assigned to the user according to the knowledge requirement and the knowledge structure of

Fig. 2 Complex product development process diagram based on the knowledge

the user to finish it. After accepting the task, the user gets help from the enterprise's knowledge database and using their skills and experiences to finish it. Knowledge as the main intellectual resource runs through the whole process in the product development.

The application of knowledge is based on the analysis of knowledge requirements. The nature of knowledge requirements analysis is based on the development context information of the product to determine the characteristics of the current design activities to generate the knowledge retrieval expressions. Figure 2 shows that the knowledge requirement in the process is depending on three factors: task requirements, product structure demand and the user's experience, and cognitive structure. Therefore, this paper defines knowledge requirement as follows:

$$KR = \{TaskR, \ ProductR, \ PersonR\} \tag{1}$$

Knowledge requirement consists of a three tuple: TaskR, ProductR, and PersonR which are the context requirement information connected to the current task, product structure, and user's knowledge structure which can be formalized as: $TaskR = \{t1, t2, \ldots ti \ldots, tn\}$, $ProductR = \{p1, p2, \ldots pi \ldots, pn\}$, $PersonR = \{u1, u2, \ldots ui \ldots, un\}$, ti and pi are the knowledge points required by task and product structure, ui is the knowledge point master by the user.

3.2 Constructing Knowledge Retrieval Expression

The process of the knowledge retrieval expression is constructed according to the knowledge requirements of the product development. Through the aware of the context information about the current development activities to get a collection of topics in each dimension and then constructing the demand characteristics table of the knowledge, generating knowledge retrieval expression.

The context elements in the process of product development involve many aspects. The paper mainly considers context information of three dimensions: task, product, and user. In order to realize the marching between context and knowledge, this paper defines context awareness of current activities:

Fig. 3 Obtain the task
knowledge requirements

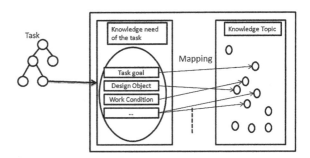

Fig. 4 Knowledge
requirements of product
structure

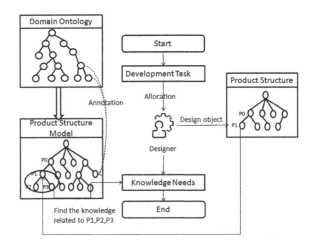

$$K\text{Content} = \{\text{Task, PD, Person}\} \tag{2}$$

Getting the context information, and then transferring it to the knowledge requirements. The basic requirement of product development is to meet the structure requirement and the function requirement. The transfer process is shown in Fig. 3 and the knowledge demand task is as follows:

$$\text{TR}_m = \{\ W_{t1}\quad W_{t2}\quad \ldots\quad W_{ti}\quad \ldots\quad W_{tm}\ \} \tag{3}$$

$W_{ti} = (w_{ti}, f_{ti})$ is the knowledge point, w_{ti} is the topic, f_{ti} is the weight that is obtained by mapping from task requirements to knowledge topics, and m is the number of knowledge topics.

The product structure and components is the results of task execution. At first, the product structure has only one node and then gradually added on in the design processing. Therefore, in the view of the current design activities, users have no idea of the substructure of the current product. To solve the problem, a product structure model is constructed by referring to the previous product structure and identified the knowledge needs of each node as the reference information and avenues to knowledge needs, as shown in Fig. 4.

$$SR_n = \{ W_{s1} \quad W_{s2} \quad \ldots \quad W_{si} \quad \ldots \quad W_{sn} \} \tag{4}$$

$W_{si} = (w_{si}, f_{si})$ is the topic and its weight that obtained by the mapping from product structure requirements to knowledge topics, and m is the number of knowledge topics.

The demand characteristics of knowledge can be constructed according to the task and the product structure. Specific expression is as follows:

$$KR = \left\{ \begin{matrix} TR_m \\ SR_n \end{matrix} \right\} = \left\{ \begin{matrix} W_{t1} & W_{t2} & \ldots & W_{ti} & \ldots & W_{tr} \\ W_{s1} & W_{s2} & \ldots & W_{si} & \ldots & W_{sr} \end{matrix} \right\} \tag{5}$$

If the number of knowledge topics of TR_m and SR_n is different, then makes r equal to the bigger one between M and N. Inadequate topics are added by null value. Knowledge retrieval expression is constructed according to the demand characteristics.

$$L = (W_{t1} \cup W_{t2} \ldots \cup W_{tr}) \cap (W_{s1} \cup W_{s2} \ldots \cup W_{sr}) \tag{6}$$

4 Knowledge Retrieval Based on Multiple Context Similarity

This paper carried on the analysis to the knowledge requirements from different context dimensions, getting the topics of knowledge requirements from different context dimensions and then constructing the knowledge retrieval expressions. The relationship between the requirement and knowledge is built through the topics. So that the knowledge can be obtained by comparing the similarity between the topics of knowledge requirements and knowledge points, as the multiple dimensions of knowledge requirement, the paper proposed a method of knowledge retrieval based on multiple context similarity which mainly considered the context dimensions of task and product structure and then sort the results according to the knowledge structure of the product development user.

In order to calculate the similarity to express the topics related to a knowledge point as follows:

$$U_r = \left\{ \begin{matrix} W'_{t1} & W'_{t2} & \ldots & W'_{ti} & \ldots & W'_{tr} \\ W'_{s1} & W'_{s2} & \ldots & W'_{si} & \ldots & W'_{sr} \end{matrix} \right\} \tag{7}$$

$W'_{ti} = (w_{ti}, f'_{ti})$, $W'_{si} = (w_{si}, f'_{si})$, f'_{ti} and f'_{si} are the weight of the topic. The following is the formula of calculating the similarity between the knowledge points and knowledge requirements.

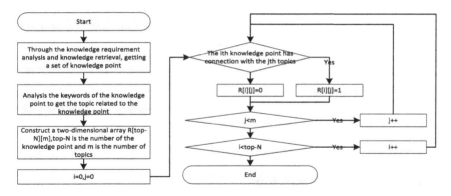

Fig. 5 Flow chart to build the association chain of topic-knowledge point-topic

$$\text{Sim}(U_r, L) = \alpha \frac{\sum_{i=1}^{r} f_{ti}^* f_{ti}'}{\sqrt{\sum_{i=1}^{r} f_{ti}^2} \sqrt{\sum_{i=1}^{r} f_{ti}'^2}} + \beta \frac{\sum_{i=1}^{r} f_{si}^* f_{si}'}{\sqrt{\sum_{i=1}^{r} f_{si}^2} \sqrt{\sum_{i=1}^{r} f_{si}'^2}} \tag{8}$$

In the expression, α and β are the weights of the knowledge requirements of task and product structure. The greater of $\text{Sim}(U_r, L)$ is, the more in line with the current context requirements of the knowledge point.

5 Multi-dimensional Knowledge Map Construction Based on Knowledge Connection

5.1 Knowledge Connection

Contiguity theory from psychology shows that the objects once felt together are usually related together, so that at the thought of one of them, the others will be recalled that once they occurred at the same time. Based on the theory, this paper analyzes the knowledge connections from the perspective of the co-occurrence analysis of thesauruses, knowledge points, and their owner to mine the implicit relations among the knowledge points, and then constructing the multi-dimensions knowledge maps based on knowledge connections. The study will resolve the problem that traditional knowledge maps are of single structure and can only display knowledge from a single dimension.

Based on the co-occurrence analysis, knowledge connections are divided into four types, which are thesaurus-knowledge point-thesaurus connection chain, thesaurus-knowledge owner-thesaurus connection chain, knowledge point-knowledge owner-knowledge point connection chain, and knowledge point-knowledge node reference connection chain. Taking the first type as an example, the flow of building connection chains is as shown in Fig. 5.

In order to combine knowledge map with the requirements in the product development process, so as to improve work efficiency, this paper adopts the object-oriented approach and metadata technology to model and store the knowledge objects and constructs domain ontology to label them. Computers can analyze users' requirements according to their current contextual information, search the meta-knowledge base, extract the related knowledge nodes, and draw the objects and their relations from multiple dimensions using the drawing engine.

5.2 Two Multi-dimensions Map Model Based on Knowledge Connection

Knowledge map occurred as a kind of cognitive map initially, proposed by Brooks, an English information scientist, who believed that according to the intrinsic connections, the knowledge units in the subject areas can be joined into subject cognitive map. Afterward, cognitive map gradually evolved into a kind of knowledge management tool by which domain knowledge units and their relations are organized and described in the form of visual nodes. For better support to resolve the problems arising in the product development process, this paper constructs knowledge map from the knowledge connections under different scenario dimensions.

Multi-dimensions knowledge map based on knowledge connections is divided into four tiers: presentation tier, requirements analysis tier, meta-knowledge tier, and resource tier. As shown in Fig. 6, in resource tier, various knowledge documents, CAD models, standards, and so on are stored. Meta-knowledge tier is a description of the data resources, in which the data resources are labeled via metadata. In requirements analysis tier, through perceiving users' tasks and the related contextual information such as product structure to build search expressions by analyzing requirements, the needed knowledge points are acquired and returned along with their relations through connection analysis. The presentation tier displays the knowledge points and their relations from different dimensions using visualization technology. Knowledge map is made up of knowledge points, knowledge connections, and knowledge links.

① *Knowledge node* It represents the knowledge objects extracted from the various resources and the related context elements or attribute information, such as key words, product structure nodes, or task nodes.
② *Knowledge connection* It refers to the various kinds of relations existing among the nodes, by which users can find one node from another, or discover and reveal the implicit relations.
③ *Knowledge link* It provides the mapping between the nodes and its detailed information or carrier, through which the detailed information, the source or the provider can be found.

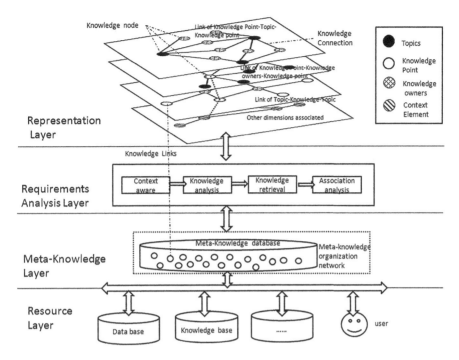

Fig. 6 Layered model of knowledge map

6 Examples

In the process of aircraft design, designers need to acquire knowledge from knowledge database, communicate with others to exchange and share knowledge. For designer's knowledge requirement is changing according to different context. There is a need to analyze the designer's requirement to obtain the knowledge points, so to keep the work going on smoothly.

This paper utilizes the technology system of Java 2 platform Enterprise Edition, development environment of MyEclipse 8.6, database of Oracle 9.0 to developed a knowledge management prototype system, and established a module of knowledge map. Using the ontology modeling tools of Protégé developed by Stanford University to develop the plane design domain ontology, ontology analytical tools Jena developed by HP Labs to operate the concepts in the ontology. Knowledge documents are annotated in the prototype system to form the knowledge directory, as shown in Fig. 7, and taking the design of landing gear as an example to construct the multi-dimensional dynamic local knowledge map of the landing gear.

Figure 8a is the task tree and Fig. 8b is the product structure tree, the knowledge requirement is the landing gear, they are the current context information. The

---

Parameter Design of Landing Gear Shock Absorber for Flexible Airplane

Topic : aircraft structure design; landing gear; parameter design; flexible airplane; ......

Creator : *Shi Youjin*

Create Time : 2012-03-06

Reference : ①XXXXX;
                    ②XXXXX;
                    ....

Referenced : ①XXXXX;
                    ....

Verifier : Liu Yuming
....

Abstract : For settling problems in aircraft landing gear shock absorber design without considering the fuselage flexibility, parameters of the shock absorber are calculated by using a three-mass equivalent aircraft system and a landing energy ......

Link address : V226----Parameter Design of Landing Gear Shock Absorber for Flexible Airplane.pdf

---

Fig. 7 Knowledge annotation

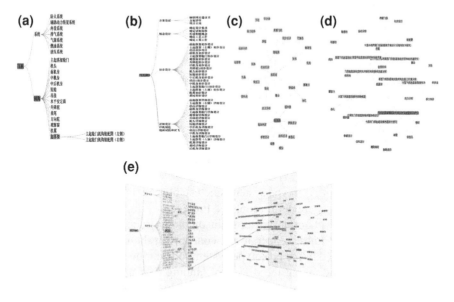

Fig. 8 Knowledge map of landing gear

Table 1 Set of knowledge points

Name of Knowledge Points	Topics	Knowledge owner
某型飞机起落架的收放与转动机构	起落架 机构可靠性 敏感性	朱伟
大柔性飞机起落架缓冲器设计研究	起落架 减震 着陆响应 参数设计	史友进
飞机起落架着陆动力学分析及减震技术研究	起落架 模拟 减震 动力分析 降落区	史友进
某型飞机起落架收放机构及舱门开度分析	起落架 收放机构 动力分析 耦合	张振
大型水陆两栖飞机起落架方案设计及相关技术研究	两栖飞机 起落架 初步设计 收放机构	魏晓辉
基于仿真计算的某型飞机起落架收放机构的仿真研究	起落架 仿真 收放机构 连接臂	朱林
某型号起落架收放机构性能仿真	收放机构 仿真 起落架 运动分析	朱林
飞机起落架收放机构与舱门机构的集成仿真分析	仿真 起落架 舱门机构 收放机构	李田园
大型飞机起落架制造技术	起落架 支柱 钛合金 承力构件	李铭
大柔性飞机起落架缓冲参数设计	起落架 结构设计 参数设计 着陆响应	史友进
大型飞机起落架缓冲数模型	缓冲器 摩擦力 油孔 起落架	陈福成

prototype system constructs the knowledge retrieval expression by analyzing the context information, and gets the set of knowledge points as shown in Table 1. Figure 8c is the knowledge organization network diagram based on the knowledge points and the relationship between the knowledge points and topics, Fig. 8d is based on the relationship between the knowledge owners and topics. From different dimensions to analyze the relationships among the knowledge points and different dimensions, it gets the multi-dimensional dynamic local knowledge map of the landing gear as shown in Fig. 8e.

7 Conclusion

For the purpose of solving the problems of knowledge overflow and knowledge treks, this paper proposed a method to construct the knowledge map of complex product development and verified it through an instance. This method can solve the difficult construction to complete a whole knowledge map and change as updating the knowledge. It can make users focus on the knowledge which is useful to the current task to improve the work efficiency.

A further research can be done in the following aspects about this paper: ① the paper analyzes the knowledge needs from context of the task, product structure and user scenarios, the future mining and expansion about the context information can be done; ② the knowledge retrieval expression is constructed based on topics, in order to get better knowledge points, the method of knowledge retrieval can be improved.

References

1. Shi M, Wang T, Chen Y et al (2011) Knowledge push system based on business process and knowledge need. Comput Integr Manuf Syst 17(4):882–887
2. Wei JD, Chen JY, Tung TY et al (2009) Integration of a concept map generator and a knowledge-portal-based e-learning system. Computer science and information engineering, WRI World Congress, Washington DC, 2009, pp 356–360
3. Mike U, Michael G (1996) Ontologies: principles, methods and applications. Knowl Eng Rev 11(2):93–136
4. Liao S-H (2003) Knowledge management technologies and applications-literature review from 1995–2002. Expert Syst Appl 25(2):155–164
5. Kim S, Suh E, Hwang H (2003) Building the knowledge map: an industrial case study. J Knowl Manage 7(2):34–45
6. Zhang S, Yang C, Liu Z (2011) Applied research on hierarchical design structure matrix in product modular design. Modular Mach Tool Autom Manuf Tech 1:18–25
7. Su H, Jiang Z, Wu H (2005) Building knowledge map for product development. J Shanghai Jiaotong Univ 39(12):2034–2039
8. Yang T, Xiao T, Zhang L (2004) Context-centered design knowledge management. Comput Integr Manuf Syst 10(12):1541–1545
9. Yang Y, Yu S, Chen D et al (2013) Complex product virtual maintenance training method based on knowledge map. Comput Integr Manuf Syst 21(2):209–214

An Evolution System for Traditional Chinese Medicine Prescription

Liang Yao, Yin Zhang and Baogang Wei

Abstract In Traditional Chinese Medicine (TCM), the prescription is the crystallization of clinical experience of doctors, which is the main way to cure diseases in China for thousands of years. The relationship between prescriptions is the important research field of the pharmacology of traditional Chinese medical formulae. In this paper, we present a system which mines the relationship between TCM prescriptions from prescription literatures, including the composition and function of prescriptions. The relationship of prescription composition is established by Trie-based tree, and the relationship of prescription function is established by the topic model. The evolution of prescriptions is presented to users in a visualization way. Finally, the experiment validates the effectiveness of our method.

Keywords Traditional Chinese Medicine · Prescription Trie tree · Topic model · Visualization

1 Introduction

In Traditional Chinese Medicine (TCM), a prescription is a set of medicines. In the long Chinese history, a large number of prescriptions have been invented to cure diseases, but the relationship between prescriptions is not very clear. Although

L. Yao · Y. Zhang (✉) · B. Wei
College of Computer Science and Technology, Zhejiang University,
Hangzhou 310027, China
e-mail: yinzh@zju.edu.cn

L. Yao
e-mail: yaoliang@zju.edu.cn

B. Wei
e-mail: wbg@zju.edu.cn

Z. Wen and T. Li (eds.), *Knowledge Engineering and Management*,
Advances in Intelligent Systems and Computing 278,
DOI: 10.1007/978-3-642-54930-4_10, © Springer-Verlag Berlin Heidelberg 2014

some TCM literatures have classified some prescriptions into different classes [1], the amount of prescriptions which have been classified is few (several thousands), far less than the prescriptions we have already known (nearly 100 thousand), and their composition relationship (with more medicines or less medicines or common medicines) is implicit. To present the relationship between prescriptions in a direct way, we develop a system which can find the composition and the function relationship between prescriptions, and present the relationship in a visualization way.

In Sect. 2, we briefly introduce some related works. In Sect. 3, we first give the framework of the prescription evolution system, then describe the key technologies of our method, especially the Trie tree and the LDA model. In Sect. 4, we show the experiment of the method proposed in Sect. 3 and do further analysis. The experiment result shows our method could correctly find the composition relationships, and label most of prescriptions in a convincing way. Meantime, we give a visualization example of the evolution result. In Sect. 5, we discuss some works should be done in the next step.

2 Related Work

As mentioned above, most of prescriptions have not been classified yet, and their composition relationship is implicit. It will be time consuming for TCM experts to find the relationship manually.

Ren [2] proposed an analysis method for TCM prescriptions. This method can explain several relations existing in prescriptions, such as relationship between dosage and function, relationship between herb and functions. Jing and Dong-qing [3] presented a reduction method of high dimensional data based on attribute similarity algorithm, which could be applied in automatic induction of Chinese traditional medicine prescription. With the reduction method, the complexity of relationship finding will be reduced. Gao et al. [4] developed algorithms of classifying herbs, used for intelligent mining of Chinese medicine prescriptions efficacies, which can help to achieve automatically implementation of classified herbs in practice. The classification method is similar to a decision tree and gives us a possible solution to the prescription classification.

Although previous works have made a lot of analysis on prescriptions, these works mainly focus on analysis on single prescription, but could not present relationship among prescriptions explicitly. Motivated by this, our method mainly cares about the relationship among prescriptions.

3 Our Evolution System of TCM Prescription

Here, we briefly describe the framework of our evolution system of TCM prescription, as shown in Fig. 1.

Firstly, the system preprocesses the data collected from the digital TCM prescription books. It uses Chinese word segmentation tools to segment sentences into words, then gets the prescription composition list and the prescription function corpus by dosages filter and domain dictionary filter.

After the preprocessing, the system mines the composition relationship of prescription via Trie-based tree. Based on the composition relationship and the prescription function corpus, the system mines the function evolution with topic model. Finally, the relationship of prescriptions is presented to users. Users can query the evolution of their input prescriptions online, and see the result in an interactive, visualization way.

3.1 Data Preprocessing

We collected prescriptions from the "Dictionary of Chinese medicine prescriptions" [5], which contains about 100 thousand TCM prescriptions. Each prescription (as shown in Fig. 2) is described by four items, i.e., its source book, composition, usage, and function. Here, we mainly use its composition and function to get the evolution relationship of prescriptions.

Take the prescription "gonghan decoction" in Fig. 2 for example, its composition string is "rhizoma alpiniae officinarum and cortex cinnamomi 50 g, radix glycyrrhize 150 g" (the second directory), from which we can see the prescription are made of three TCM medicines separated by spaces. We can extract medicines by separating the string according to spaces and filtering the dosages. The function string is "regain yang qi, expel evil cold and counteract malaria epidemic" (the fourth directory). We can get medicines by separating the string by spaces and filtering the dosages. To utilize the prescription function description, we first use word segmentation tools to segment the sentence into separating words, then remove stop words and filter some irrelative words by TCM domain dictionary.

3.2 Trie-Based Prescription Composition Evolution

The first step is to derive the composition of the prescriptions. For example, if a user wants to know whether there is a prescription A when he adds a medicine to the prescription "gonghan decoction." If there is, whether there is a prescription when he adds a medicine to A in the previous step? Thus, the system should iteratively generate such relationship among prescriptions based on the Trie tree.

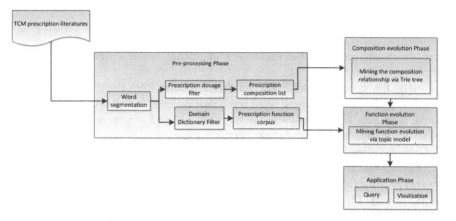

Fig. 1 Framework of TCM prescription evolution system

Fig. 2 Prescription
description in "Dictionary of
Chinese medicine
prescriptions"

38970 **攻寒汤**

【方源】《普济方》卷一九九。

【组成】高良姜　桂心各一两　甘草三两

【用法】将姜、桂碎锉，以清油半两煎，不
住搅，候焦褐色，取出，旋放冷，三味同为
末。空心沸汤入盐点服。

【功用】复阳气，逐寒邪，辟瘴疫。

Trie [6] is a variant of a tree, also called dictionary tree, which allows us to share prefixes that are common among strings, and reduce the complexity of query operations. For instance, if we have strings "abcd," "abce," "abcf," the Trie tree should be like Fig. 3, "abc" is their common prefix.

In standard Trie tree, a node stores a character. Here, we use string type to represent a medicine. Notice that a prescription is a set of medicines, which have no order. To make the medicines in order, we use Algorithm 1 to iteratively build a Trie tree.

Algorithm 1: Build Compostion evolution Trie tree

Require: $inputprescription, prescriptionlist$
    $function\ derive(inputprescription, prescriptionlist)$
    while a $prescription$ in $prescriptionlist$ **do**
        if $prescription.containsAll(inputprescription)$ and $prescription.size=$ $inputprescription.size + 1$ **then**
            $temp \leftarrow inputprescription$
            $suffix \leftarrow prescription.removeAll(inputprescription)$
            $temp \leftarrow temp + suffix$
            $Trietree.add(temp)$ {add to the Trie tree,the node type is string}
            $derive(temp, prescriptionlist)$ {iteration}
        end if
    end while

Fig. 3 Standard Trie tree

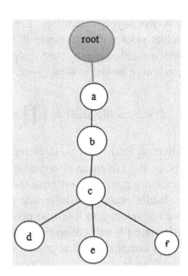

Algorithm 1 takes a prescription entered by a user and all the prescriptions in the Dictionary of Chinese medicine prescriptions as the input. A prescription is a list of medicines (strings), the algorithm traverses all the prescriptions to check whether each prescription (a string's list, label it A) contains the input list (label it B), and the size of A is equal to the size of B plus 1, where containsAll and removeAll are methods in Java's ArrayList class. If the procedure finds an eligible A, the procedure sorts medicines in prescription A by putting the medicine not in B to the end of list A, then adds A to a Trie tree with string node type. Then the procedure searches next prescription in a depth-first search manner, namely iteratively call the procedure, takes A and all the prescriptions as input. After the initial call ends, the composition evolution Trie tree is built.

Then the algorithm uses getWithPrefix method of Trie class to attain the prescriptions with additional medicines.

3.3 Function Evolution Based on Topic Model

Topic modeling provides methods for automatically organizing, understanding and summarizing large electronic archives. In topic models, latent Dirichlet allocation (LDA) [7] is a typical model.

We now briefly review LDA. In LDA, it is assumed that observed words in each document are generated by a document-specific mixture of corpus-wide latent topics. We define the input corpus of length N with the word vector $w = w_1...w_N$. At corpus position i, the element d_i in $d = d_1...d_N$ denotes the document containing observed token w_i. Likewise, the vector $z = z_1...z_N$ is the hidden topic assignments of each observed token. The number of latent topics is fixed to some T, and each topic is $t = 1...T$ is associated with a topic-word multinomial ϕ_t over a W-word-type vocabulary. Each multinomial ϕ is generated by a conjugate Dirichlet prior with parameter β. Similarly, each document $j = 1...D$ is associated with a multinomial θ_j over T topics, with another Dirichlet prior parameter α. The generative model is then given by

$$P(w, z, \varphi, \theta | \alpha, \beta, d) \propto \left(\prod_t^T p(\varphi_t | \beta) \right) \left(\prod_j^D p(\theta_j | \alpha) \right) \left(\prod_i^M \varphi_{z_i}(w_i) \theta_{d_i}(z_i) \right) \quad (1)$$

where $\phi_{z_i}(w_i)$ is the w_i-th element in vector ϕ_{z_i}, and $\theta_{d_i}(z_i)$ is the z_i-th element in vector θ_{d_i}. Given an observed corpus and hyper-parameters (α, β), one important modeling goal is to estimate the latent variables (z, ϕ, θ).

While exact LDA inference is very hard, a kind of approximate schemes have been developed to infer the posterior. Two main inference methods are variation inference [7] and collapsed Gibbs sampling [8, 9]. In this work we use collapsed Gibbs sampling. The approach iteratively re-samples a new value for each latent topic assignment z_i conditioned on the current values of all other z values. After running the Markov chain for a fixed number of iterations, we infer the topic-word multinomial ϕ and the document-topic mixture weights θ from the final z sample, using the means of their posteriors given by

$$\phi_t(w) \propto n_{tw} + \beta \quad (2)$$

$$\theta_j(t) \propto n_{jt} + \alpha \quad (3)$$

where n_{tw} is the number of times word w is assigned to topic t, and n_{jt} is the number of times topic t is used in document j, with both counts being taken with respect to the final sample z. The topic-word multinomials ϕ_t for each topic t are learned topics and each document-topic multinomial θ_d represents the proportion of topics within document d [10].

We choose the most probable topic in θ_d to represent a prescription function, and recommend some relative words in that topic by following expression

$$T = \arg \max_t \theta_d(t) \quad (4)$$

$$W_t = k - \arg \max_w \varphi_t(w) \quad (5)$$

where $k = 10$, which represents the top ten of most probably words.

In an alternative view, we can treat a prescription as a document, and a medicine as a word, a class of function as a topic. So the LDA model of prescription can be written as follows:

$$P(m, f, \varphi, \theta | \alpha, \beta, p) \propto \left(\prod_t^T p(\varphi_t | \beta) \right) \left(\prod_j^P p(\theta_j | \alpha) \right) \left(\prod_i^M \varphi_{f_i}(m_i) \theta_{p_i}(f_i) \right) \quad (6)$$

where m denotes medicines, f denotes function, p denotes prescriptions, and others are the same as (1).

Likewise, we could recommend some medicines under a certain topic (function) by following expression

$$F = \arg\max_{f} \theta_d(f) \tag{7}$$

$$W_m = k - \arg\max_{m} \varphi_t(m) \tag{8}$$

where F denotes the selected function with max probability, f denotes a class of functions, m denotes the medicines, and W_m denotes the top k medicines with max probabilities.

4 Experiment

Now we have developed the evolution system based on prescription function corpus (including 98,334 prescriptions) and TCM domain dictionary which uses IKAnayzer to segment and filter words, LDA to train the corpus, and D3 [11] to visualize the Trie tree of one prescription evolution. In our system, after each filtered word is indexed, the corpus is transformed to a two-dimensional integer array. We set the topic numbers of diseases and syndromes to 20 and 11 respectively [12], and run Gibbs sampling for 10,000 iterations with hyper-parameters $\alpha = 2$, $\beta = 0.5$.

Here to illustrate the feasibility and validity of our system method, let us see one experimental example in which we use the prescription named "heilong mini-pills" as an input and the output is its visualized Trie-based evolution result.

As shown in Fig. 4, the composition evolution result of "heilong mini-pills" actually is a Trie tree.

When a user inputs a prescription named "heilong mini-pills," which consists of sulfur and croton seed, the system returns prescriptions "zhentou pie," "sulfur bag," "daoshui pie," "shenxian bulao mini-pills" with one additional medicine, "huanjing powder," "bixia mini-pills" with two additional medicines, and "mercury pill," "bixia pill" with three additional medicines. The output Trie tree correctly finds the composition relationship between prescriptions.

Table 1 lists nine prescriptions resulting from the input prescription "heilong mini-pills" and their topic labels (syndrome or disease class) with corresponding probability (when the topic number is 20, if a document has no preference to any topic, all of 20 probabilities should be 0.05, and when the topic number is 11, all of 11 probabilities should be 0.909 if a document has no preference to any topic).

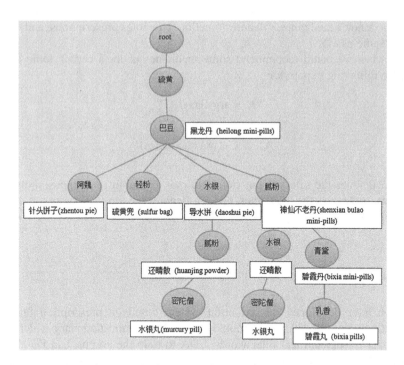

Fig. 4 Result Trie tree based on the input prescription

Table 2 lists the result of topic labeling when treating a medicine as a word. Other conditions are the same to the Table 1. Note that the topic labels are not identical to labels in Table 1, because they are output of two different Gibbs sampling experiments.

From Table 1 we can see that most of the topic labels are correct and could be explained, except the functions "swell" and "tumidness" are assigned with different topics.

Also, Table 2 shows that most of the prescriptions are assigned with the same topic due to the fact that most of their medicines are the same, but the topic labeling result could not correctly reflect their function. Because in LDA, it is assumed that all words are independent, while medicines are not. Medicines have the relationship like "monarch, minister." So in the future work we consider incorporating the TCM domain knowledge into the LDA model in the future work.

At last, our system uses the popular visualization library D3 to represent the evolution tree, as shown in Fig. 5.

The preceding output Trie tree is presented in a web browser after the user inputs the prescription "heilong mini-pills."

Table 1 Latent topic assignment of prescription functions

Prescription	Heilong mini-pills	Zhentou pie	Sulfur bag	Daoshui pie	Shenxian bulao mini-pills	Huanjing powder	Bixia mini-pills	Mercury pill	Bixia pill
Topic (syndrome class)	4 (warm disease syndrome)	6 (typhoid fever syndrome)	6 (typhoid fever syndrome)	0 (systemic syndrome)	2 (viscera syndrome)	9 (skin syndrome)	7 (children syndrome)	6 (typhoid fever syndrome)	7 (children syndrome)
Probability	0.1357	0.1572	0.1285	0.1046	0.156	0.1939	0.171	0.1346	0.194
Topic (disease class)	12 (epidemic, poisoning and other diseases)	18 (infections diseases and parasitic diseases)	9 (ulcers diseases)	14 (liver diseases)	4 (kidney diseases)	16 (skin diseases)	8 (children diseases)	8 (children diseases)	8 (children diseases)
Probability	0.0723	0.0873	0.0702	0.0535	0.0865	0.1078	0.1014	0.0829	0.1111
Function description	Typhoid fever with yin yang toxicity	Cholera, vomiting and diarrhea, bellyache	Swell	Tumidness	Ten kinds of water pathogen	Lumbodorsal carbuncle and all kinds of abscess	1. In Shenghui: children acute infantile convulsions. 2. In Puji Fang: children with diaphragm excess and saliva is filled	Children with nodules of breast, hard lateral thorax, breathing heavily, and sometimes aching	Infantile surprised

Table 2 Latent topic assignment of prescription compositions

Prescription	Heilong mini-pills	Zhentou pie	Sulfur bag	Daoshui pie	Shenxian bulao mini-pills	Huanjing powder	Bixia mini-pills	Mercury pill	Bixia pill
Topic (11 in total)	4	4	4	4	4	4	10	4	4
Probability	0.151	0.1654	0.1908	0.1774	0.148	0.662	0.157	0.201	0.175
Topic (20 in total)	10	17	10	10	10	10	10	10	10
Probability	0.0732	0.0758	0.0924	0.09305	0.0738	0.0912	0.0969	0.1044	0.0945

Fig. 5 Output evolution tree presented in the browser by our system

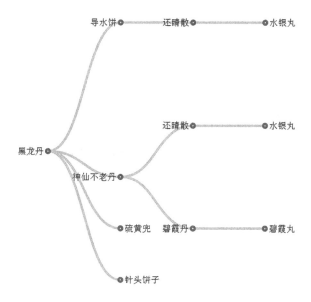

5 Conclusion and Future work

As we can see in the experiment result, the composition evolution is correct but the topic labels of the prescriptions are not very precise, the main reason for which is that some of the texts is relatively short, and the vocabulary size is too large, so the function description could not represent the topic (function) very well.

In the next step, we consider supplying the prescription function by combining the function's detail explanation. In addition, as mentioned before, we consider incorporating the TCM domain knowledge into the LDA model when treating a medicine as a word.

Because the medicine dosage of a prescription plays an important role in the prescription function, our system will show the change of the prescription function when the dosage changes. Moreover, the temporal information of a prescription is also an important attribute, so we will present the timeline of prescriptions across the Chinese history in our system in the future work.

Acknowledgment This research was supported by Chinese Knowledge Center of Engineering Science and Technology (CKCEST) and China Academic Digital Associative Library (CADAL).

References

1. Chinese medicine Thesaurus. http://books.google.com.hk/books?id=iS6gQwAACAAJ
2. Tingge R (2008) An Intelligence Processing Method related to Efficacy Messages of Traditional Chinese Medicine Prescriptions. In: ICICIC' 08
3. Jing P, Dong-qing Y (2007) A novel reduction method based on attributes similarity Chinese traditional medicine prescriptions. In: PACRIM' 07

4. Gao Q (2009) A classified herbs method and searching algorithm of classification tree used for mathematical measurement of effect. In: ICICIC'09
5. Dictionary of Chinese medicine prescriptions, vol 5. http://books.google.com.hk/books?id= 0hfBngEACAAJ
6. http://en.wikipedia.org/wiki/Trie
7. Blei D, Ng A, Jordan M (2003) Latent Dirichlet allocation. J Mach Learn Res 3:1107–1135
8. Griffiths TL, Steyvers M (2004) Finding scientific topics. PNAS 101:5228–5235
9. Mark S (2007) Probabilistic topic model. University of California, Ivrine
10. David A (2011) Latent topic feedback for information retrieval. In: KDD 2011
11. http://d3js.org/
12. Differential diagnosis of TCM Syndrome, 2nd edn. http://books.google.com.hk/books?id= TQghAgAACAAJ

Evaluating Side Effects to Hide Sensitive Itemsets Through Transaction Deletion

Chun-Wei Lin, Tzung-Pei Hong and Hung-Chuan Hsu

Abstract In this paper, a novel *hiding-missing-artificial utility* (HMAU) algorithm is proposed to hide sensitive itemsets through transaction deletion. It considers three side effects of hiding failure, missing itemsets, and artificial itemsets for evaluating whether the processed transactions are required to be deleted or not, in sanitization process. Experiments show that the proposed HMAU algorithm has better performance whether in the execution times, the number of deleted transactions, and the number of side effects.

Keywords Privacy preserving data mining · Side effects · Information hiding · Data sanitization · Security

1 Introduction

With the rapid growth of data mining technologies in recent years, the derived knowledge from those of techniques can be simply classified into association rules [1–4], sequential patterns [5, 6], classification [7], clustering [8, 9], and utility

C.-W. Lin
Shenzhen Key Laboratory of Internet Information Collaboration,
School of Computer Science and Technology, Innoviative Information Industry Research
Center, Harbin Institute of Technology Shenzhen Graduate School, Shenzhen, China

T.-P. Hong (✉) · H.-C. Hsu
Department of Computer Science and Information Engineering,
National University of Kaohsiung, Kaohsiung 811, Taiwan, ROC
e-mail: tphong@nuk.edu.tw

T.-P. Hong
Department of Computer Science and Engineering, National Sun Yat-sen University,
Kaohsiung, Taiwan, ROC

Z. Wen and T. Li (eds.), *Knowledge Engineering and Management*,
Advances in Intelligent Systems and Computing 278,
DOI: 10.1007/978-3-642-54930-4_11, © Springer-Verlag Berlin Heidelberg 2014

mining [10–13], among others. Among them, association-rule mining is the most common way to determine the relationships of purchased items in large data sets.

In the past, privacy preserving data mining (PPDM) [14] was thus proposed to reduce the risk of privacy threats by hiding sensitive information but still has the capability to mine the required information from the provided databases. In this study, a novel *hiding-missing-artificial utility* (HMAU) algorithm is thus proposed for evaluating whether the processed transactions are required to be deleted or not for hiding the sensitive itemsets. It considers three designed dimensions of hiding failure dimension (HFD), missing itemset dimension (MID), and artificial itemset dimension (AID) as the evaluation criteria. Each dimension can be given by an adjustable weight according to users' preferences. It is also obvious to see that the proposed HMAU algorithm has good performance than the aggregation approach [15] in the experiments.

2 Review of Related Works

In this section, the data mining techniques and PPDM techniques are, respectively, reviewed.

2.1 Data Mining Techniques

Traditional data mining is used to extract useful itemsets or rules from a binary database. The most common approach is to generate association rules from a transactional database such that the presence of certain items in a transaction implies the presence of some other items. Agrawal and Srikant [5] proposed Apriori algorithm for mining association rules from a set of transactions level by level. The downward closure property is used to prune unpromising candidate itemsets, thus improving the efficiency for discovering frequent itemsets. Many algorithms have been proposed for efficiently discovering the desired association rules [1, 3, 4], and the other algorithms are still developed in progress.

2.2 Privacy Preserving Data Mining

Due to the rapid growth of the Internet and computer techniques, different kinds of information can be efficiently derived by various data mining techniques in different applications [12, 16, 17]. The misuse of the techniques, however, may lead to privacy concerns and produce security problems. PPDM has thus become a critical issue for hiding the private, confidential, or secure information [14, 15, 18–21].

In data sanitization of PPDM, it is an intuitive way to directly delete the sensitive data for hiding sensitive information. Leary [22] mentioned that data mining techniques can be concerned as threats for security and privacy issues. Atallah et al. [18] stated that the optimal sanitization is an NP-hard problem. Amiri [15] has proposed aggregate approach, disaggregate approach, and hybrid approach to determine whether the items or transactions are required to be deleted for hiding sensitive information. Divanis and Verykios [19] protected sensitive information by hiding the sensitive itemsets instead of hiding the rules.

3 Proposed Algorithm Through Transactions Deletion

In this section, the HMAU algorithm is described step-by-step as follows.
Proposed HMAU algorithm

INPUT: An original database D, a minimum support threshold ratio λ, a risky bound μ, a set of large (frequent) itemsets FI = $\{fi_1, fi_2, ..., fi_p\}$, a set of small (non-frequent) 1-itemsets SI^1 = $\{si_1, si_2, ..., si_q\}$, and a set of sensitive itemsets to be hidden HS = $\{hs_1, hs_2, ..., hs_r\}$.

OUTPUT: A sanitized database D^* with no sensitive information mined out.

STEP 1 Select the transactions to form a projected database D', each transaction T_k in D' consists of any sensitive itemsets hs_i within it, where $1 \le i \le r$.

STEP 2 Process each frequent itemset fi_j in the set of FI to determine whether its frequency satisfies the condition as $freq(fi_j) \le \lceil \lceil |D| \times \lambda \rceil \times (1 + \mu) \rceil$, where $|D|$ is the number of transactions in the original database D, $freq(fi_j)$ is the occurrence frequency of the large itemset fi_j. Put the unsatisfied fi_j into the set of FI_{tmp}.

STEP 3 Process each small 1-itemset si_a in the set of SI^1 to determine whether its frequency satisfies the condition as $freq(si_a) \ge \lfloor \lceil |D| \times \lambda \rceil \times (1 - \mu) \rfloor$, where $freq(si_a)$ is the occurrence frequency of the small 1-itemset si_a. Put the unsatisfied si_a into the set of SI^1_{tmp}.

STEP 4 Calculate the *maximal count* (MAX_{HS}) of the sensitive itemsets hs_i in the set of HS as:

$$MAX_{HS} = \max\{freq(hs_i), \forall hs_i, 1 \le i \le r\},$$

where $freq(hs_i)$ is the occurrence frequency of the sensitive itemset hs_i in the set of HS.

STEP 5 Calculate the HFD of each transaction T_k. Do the substeps as follows:

Substep 5-1: Calculate the HFD of each sensitive itemsets hs_i within T_k as:

$$\mathrm{HFD}^k(\mathrm{hs}_i) = \frac{\mathrm{MAX_{HS}} - \mathrm{freq(hs}_i) + 1}{\mathrm{MAX_{HS}} - \lceil |D| \times \lambda \rceil + 1}.$$

Substep 5-2: Sum the HFD of each sensitive itemsets hs_i within T_k as:

$$\mathrm{HFD}^k = \frac{1}{\displaystyle\sum_{i=1}^{r} \mathrm{HFD}^k(\mathrm{hs}_i) + 1}.$$

Substep 5-3: Normalize the HFD^k for all transactions T_k in D'.

STEP 6 Calculate the *maximal count* ($\mathrm{MAX_{FI}}$) of the large itemsets fi_j in the set of FI as:

$$\mathrm{MAX_{FI}} = \max\{\mathrm{freq(fi}_j), \forall \mathrm{fi}_j, 1 \le j \le p\}.$$

STEP 7 Calculate the MID of each transaction T_k. Do the substeps as follows:

Substep 7-1: Calculate the MID of each large itemsets within T_k as:

$$\mathrm{MID}^k(\mathrm{fi}_j) = \frac{\mathrm{MAX_{FI}} - \mathrm{freq(fi}_j) + 1}{\mathrm{MAX_{FI}} - \lceil |D| \times \lambda \rceil + 1}.$$

Substep 7-2: Sum the MID of each large itemsets fi_j within T_k as:

$$\mathrm{MID}^k = \sum_{j=1}^{p} \mathrm{MID}^k(\mathrm{fi}_j).$$

Substep 7-3: Normalize the MID^k for all transactions T_k in D'.

STEP 8 Calculate the *minimal count* ($\mathrm{MIN_{SI^1}}$) of the small 1-itemsets si_a in the set of SI^1 as:

$$\mathrm{MIN_{SI^1}} = \min\{\mathrm{freq(si}_a), \forall \mathrm{si}_a, 1 \le a \le q\}.$$

STEP 9 Calculate the AID of each transaction T_k. Do the substeps as follows:

Substep 9-1: Calculate the AID of each small 1-itemset within T_k as:

$$\mathrm{AID}^k(\mathrm{si}_a) = \frac{\mathrm{freq(si}_a) - \mathrm{MIN_{SI^1}} + 1}{\lceil |D| \times \lambda \rceil - \mathrm{MIN_{SI^1}}}.$$

Substep 9-2: Sum the AID of each small 1-itemset si_a within T_k as:

Table 1 An original database

TID	Item	TID	Item
T_1	a, b, c, e	T_6	b, c, e
T_2	e	T_7	a, b, c, d, e
T_3	b, c, e, f	T_8	a, b, e
T_4	d, f	T_9	c, e
T_5	a, b, d	T_{10}	a, b, c, e

Table 2 Large itemsets

Large 1-itemset	Count	Large 2-itemset	Count	Large 3-itemset	Count
a	5	ab	5	abe	4
b	7	ae	4	bce	5
c	6	bc	5	–	–
e	8	be	6	–	–
–	–	ce	6	–	–

Table 3 Small 1-itemsets

Small 1-itemset	Count
d	3
f	2

$$\mathrm{AID}^k = \frac{1}{\sum\limits_{a=1}^{q} \mathrm{AID}^k(\mathrm{si}_a) + 1}.$$

Substep 9-3: Normalize the AID^k for all transactions T_k in D'.

STEP 10 Calculate the HMAU for three dimensions (HFD, MID and AID) of each transaction T_k as:

$$\mathrm{HMAU}^k = w_1 \times \mathrm{HFD}^k + w_2 \times \mathrm{MID}^k + w_3 \times \mathrm{AID}^k,$$

where w_1, w_2, and w_3 are the pre-defined weights by users' preferences.

STEP 11 Remove transaction T_k with $\min\{\mathrm{HMAU}^k, \forall T_k, 1 \le k \le |D'|\}$ value.

STEP 12 Update the minimum count ($=\lceil |D| \times \lambda \rceil$) of sanitized database.

STEP 13 Update the occurrence frequencies of all sensitive itemsets in the set of HS and HS_{tmp}. Put the hs_i into the set of HS_{tmp} if $\mathrm{freq}(\mathrm{hs}_i) <$ minimum count ($=\lceil |D| \times \lambda \rceil$), and put the hs_i into the set of HS, otherwise.

STEP 14 Update the occurrence frequencies of all large itemsets in the set of FI and FI_{tmp}. Put the fi_j into the set of FI_{tmp} if $\mathrm{freq}(\mathrm{fi}_j) <$ minimum count ($=\lceil |D| \times \lambda \rceil$), and put the fi_j into the set of FI, otherwise.

Table 4 Three dimensions of each transaction in projected database

TID	HFD	MID	AID	HMAU
T_1	0.57	1	1	0.785
T_3	1	0.33	1	0.733
T_6	1	0.33	1	0.733
T_7	0.57	1	0.5	0.735
T_8	0.57	0.67	1	0.652
T_{10}	0.57	1	1	0.785

STEP 15 Update the occurrence frequencies of all small 1-itemsets in the set of SI^I and SI^1_{tmp}. Put the si_a into the set of SI^1_{tmp} if freq(si_a) $\geq$ minimum count ($=\lceil |D| \times \lambda \rceil$), and put the si_a into the set of SI^I, otherwise.

STEP 16 Repeat STEP 4–15 until the set of HS is empty ($|HS| = 0$).

4 An Illustrative Example

Assume that a database has 10 transactions (tuples) with 6 items (denoted as a to f) shown in Table 1.

Each transaction can be concerned as a set of purchased items in a trade. The minimum support threshold is initially set at 40 %, and the risky bound is set at 10 %. A set of sensitive itemsets is considered to be hidden by the sanitization process, which is HS = {be:6, abe:4}. Based on the Apriori-like approach [2], the large (frequent) itemsets and small 1-itemsets are mined out and are, respectively, shown in Tables 2 and 3.

The transactions in Table 1 are then selected within any of the sensitive itemsets in HS. In this example, transactions 1, 3, 6, 7, 8, and 10 are then selected to form another database.

The frequent itemsets in FI are then processed to check whether the condition is satisfied or not according to the defined equation in STEP 2. In this example, the satisfied itemsets {a, ab, ae, bc, bce} still remain in the set of FI, but the unsatisfied itemsets {b, c, e, ce} are then put into the set of FI_{tmp}.

The infrequent 1-itemsets in SI^I are then processed to check whether the condition is satisfied or not according to the defined equation in STEP 3. In this example, the satisfied itemset {d} still remains in the set of SI^I, but the unsatisfied itemset {f} is then put into the set of SI^1_{tmp}.

The maximal count (MAX_{HS}) among the sensitive itemsets in the set of HS is then calculated. In this example, the maximal count of the sensitive itemsets {be} and {abe} is calculated as $MAX_{HS} = \max\{6, 4\} = 6$.

The HFD of each transaction is thus calculated to evaluate the side effects of hiding failure. After that, the HFDs for all transactions are normalized. The maximal count (MAX_{FI}) among the large itemsets in the set of FI is then

Table 5 Sanitized database

TID	Item	TID	Item
T_2	e	T_7	a, b, c, d, e
T_4	d, f	T_9	c, e
T_5	a, b, d	T_{10}	a, b, c, e

Fig. 1 The comparisons of the execution times for two algorithms

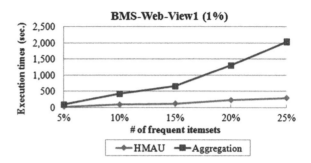

calculated as $\text{MAX}_{\text{FI}} = \max\{5, 5, 4, 5, 5\}$ ($= 5$). The MID of each transaction is thus calculated to evaluate the side effects of missing itemset. After that, the MIDs for all transactions are normalized. The minimal count (MIN_{SI^1}) among the small 1-itemsets in the set of SI^1 is then calculated as $\text{MIN}_{SI^1} = \min\{3\} = 3$. The AID of each transaction is thus calculated to evaluate the side effects of artificial itemset. After that, the AIDs for all transactions are normalized. The three dimensions for evaluating the selected transactions are organized in Table 4. The weights of hiding failure, missing itemset, and artificial itemset are then, respectively, defined as 0.5, 0.4, and 0.1. For example, the HMAU of transaction 7 is calculated as $\text{HMAU}^7 = 0.5 \times 0.57 + 0.4 \times 1 + 0.1 \times 0.5$ ($= 0.735$). The other transactions are processed in the same way and the results are shown in the last column in Table 4.

In this example, transaction 8 has the minimal one and will be directly removed from Table 4. The minimum count is then updated as $\lceil |10 - 1| \times 0.4 \rceil$ ($= 4$).

The occurrence frequencies of all sensitive itemsets in the set of HS and HS_{tmp} are updated. After the updating process, the itemset $\{abe\}$ is put into the set of HS_{tmp} since its count is below the minimum count ($3 < 4$). The occurrence frequencies of all large itemsets in the set of FI and FI_{tmp} are, respectively, updated. After the updating process, the itemset $\{ae\}$ is put into the set of FI_{tmp} since its count is below the minimum count ($3 < 4$). The occurrence frequencies of all small 1-itemsets in the set of SI^1 and SI^1_{tmp} are, respectively, updated, and nothing is done in this step in this example. After processing the above steps, the sensitive itemset $\{abe\}$ is already hidden, but the occurrence frequency of sensitive itemset $\{be\}$ is still larger than the minimum count. The same procedure is repeated from

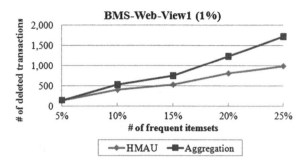

Fig. 2 The comparisons of the number of deleted transactions

STEPs 4–15 until the set of sensitive itemsets HS is empty ($|HS| = 0$). After all STEPs are processed, the sanitized database is as shown in Table 5.

5 Experimental Results

In this section, the experiments are made to show the performance of the proposed HMAU algorithm compared to the aggregation algorithm [15] for hiding the sensitive itemsets through transaction deletion. A real dataset BMS-WebView-1 [23] is used in the experiments. In the experiments, the weights of HFD, MID, and AID of the proposed algorithm are, respectively, set at 0.5, 0.4, and 0.1. The minimum support threshold is set at 1 %, and the ratios of sensitive itemsets are sequentially set at 5–25 % of the number of frequent itemsets in 5 % increments each time. The risky bound in the experiments is initially set at 10 %. Figure 1 shows the execution times of the obtained two algorithms at various sensitive percentages of the frequent itemsets.

Experiments are also made to evaluate the number of deleted transactions, and the results are shown in Fig. 2.

To summarize the performance of the proposed HMAU algorithm, the execution times and the number of deleted transactions achieve higher performance than the aggregation algorithm [15] through transaction deletion.

6 Conclusions

In this paper, a novel HMAU algorithm is proposed to hide sensitive itemsets in data sanitation by minimizing the side effects of hiding failure, missing itemsets, and artificial itemsets through transaction deletion. It considers three designed dimension for evaluating whether the processed transactions are required to be deleted or not in the sanitization process. Three flexible weights of dimensions can be set by users' preferences. Experimental results also show that the proposed

HMAU algorithm has better performance of execution times, the number of deleted transactions, and the number of side effects compared to the aggregation algorithm for transaction deletion.

References

1. Agrawal R, Imielinski T, Sawmi A (1993) Mining association rules between sets of items in large databases. In: ACM SIGMOD international conference on management of data, vol 22, pp 207–216
2. Agrawal R, Srikant R (1994) Fast algorithms for mining association rules in large databases. In: The international conference on very large data bases, pp 487–499
3. Hong TP, Lin CW, Wu YL (2008) Incrementally fast updated frequent pattern trees. Expert Syst Appl 34:2424–2435
4. Lin CW, Hong TP, Lu WH (2009) The pre-fufp algorithm for incremental mining. Expert Syst Appl 36:9498–9505
5. Agrawal R, Srikant R (1995) Mining sequential patterns. In: The international conference on data engineering, pp 3–14
6. Srikant R, Agrawal R (1996) Mining sequential patterns: generalizations and performance improvements. In: The international conference on extending database technology: advances in database technology, pp 3–17
7. Quinlan JR (1993) C4.5: Programs for machine learning, vol 16, pp 235–240
8. Berkhin P (2006) A survey of clustering data mining techniques. Grouping Multidimensional Data, pp 25–71
9. Jarvis RA, Patrick EA (1973) Clustering using a similarity measure based on shared near neighbours. IEEE Trans Comput 22:1025–1034
10. Liu Y, Liao WK, Choudhary A (2005) A two-phase algorithm for fast discovery of high utility itemsets. Lect Notes Comput Sci 3518:689–695
11. Lin CW, Hong TP, Lu WH (2011) An effective tree structure for mining high utility itemsets. Expert Syst Appl 38:7419–7424
12. Lan GC, Hong TP, Tseng VS (2011) Discovery of high utility itemsets from on-shelf time periods of products. Expert Syst Appl 38:5851–5857
13. Lin CW, Lan GC, Hong TP (2012) An incremental mining algorithm for high utility itemsets. Expert Syst Appl 39:7173–7180
14. Agrawal R, Srikant R (2000) Privacy-preserving data mining. In: ACM SIGMOD international conference on management of data, pp 439–450
15. Amiri A (2007) Dare to share: Protecting sensitive knowledge with data sanitization. Decis Support Syst 43:181–191
16. Lin CW, Hong TP, Chang CC, Wang SL (2013) A greedy-based approach for hiding sensitive itemsets by transaction insertion. J Inf Hiding Multimedia Signal Process 4:201–227
17. Lin CW, Hong TP (2013) A survey of fuzzy web mining. Wiley Interdisc Rev Data Min Knowl Discovery 3:190–199
18. Atallah M, Bertino E, Elmagarmid A, Ibrahim M, Verykios V (1999) Disclosure limitation of sensitive rules. In: The workshop on knowledge and data engineering exchange, pp 45–52
19. Gkoulalas-Divanis A, Verykios VS (2006) An integer programming approach for frequent itemset hiding. In: ACM international conference on information and knowledge management, pp 748–757
20. Oliveira SRM, Zaiane OR (2003) Protecting sensitive knowledge by data sanitization. In: IEEE international conference on data mining, pp 613–616
21. Wu YH, Chiang CM, Chen Arbee LP (2007) Hiding sensitive association rules with limited side effects. IEEE Trans Knowl Data Eng 19:29–42

22. Leary DEO (1991) Knowledge discovery as a threat to database security. In: International conference on knowledge discovery and databases, pp 507–516
23. Zheng Z, Kohavi R, Mason L (2001) Real world performance of association rule algorithms. In: ACM SIGKDD international conference on knowledge discovery and data mining, pp 401–406

Computing Concept Relatedness Based on Ontology

Yanping Lu, Xingwei Hao and Shaocun Tian

Abstract Concept relatedness is widely used in information retrieval, text classification, semantic extension, and other fields. So measuring the concept relatedness efficiently is an important task. Previous studies rarely distinguish between relatedness and similarity; they usually use a common formula. We suggest that concept relatedness consists of similarity and relevance, which should be computed differently. In this paper, we first give a similarity measure based on path length, taxonomy depth, and different relations between concepts. Then we propose a method to measure the specific association relation besides basic relations. Finally, incorporating both similarity and specific relevance, we get an overall formula of computing concept relatedness. Compared to existing methods, our measure of concept relatedness is more consistent with human judgment.

Keywords Domain ontology · Relevance · Similarity · Concept relatedness

1 Introduction

Relations between words are very complex in natural language. One is related to another is only a simplified summary. Actually semantic relatedness represents the degree of how words are related; it can be quantified by some general measure.

Semantic relatedness is widely used in information retrieval, text classification, semantic extension, and other fields. In particular, finding the semantic relatedness between two words has been one central problem in information retrieval for many

Y. Lu (✉) · X. Hao · S. Tian
School of Computer Science and Technology, Shandong University, Jinan, China
e-mail: luyanping1994@163.com

Z. Wen and T. Li (eds.), *Knowledge Engineering and Management*,
Advances in Intelligent Systems and Computing 278,
DOI: 10.1007/978-3-642-54930-4_12, © Springer-Verlag Berlin Heidelberg 2014

years. By extending the query with closely related words, performance of information retrieval system can be significantly improved [1]. The computation of concept relatedness is very important for many NLP applications. Most of the previous studies calculate similarity as the relatedness between the two concepts, while Resnik [2] has given an example to explain the difference between them. He points out that cars are dependent on the gasoline to move, while cars and bicycles are both vehicles and have some same components, they also share the common attribute of transportation. If we compute the relatedness by models considering only similarity, the relatedness of cars and bicycles is certainly greater than that of cars and gasoline, but from our knowledge of the real word, we know that cars and gasoline are more closely related. Therefore, Resnik [2] points out similarity was a special kind of relatedness; similar concepts are related to each other. In addition to similarity, there are other kinds of relations between words. We consider those special relations as semantic relevance. Then semantic relatedness includes both similarity and relevance.

In this paper, we propose a method of computing the relatedness between two concepts. Our method is based on measurement of similarity and relevance. For the part of similarity, a number of factors such as different semantic relations, shortest path length, etc., are considered. For the relevance part, we propose a computing method based on distance.

2 Related Works

There are two kinds of model for computing semantic relatedness. One is based on word co-occurrence of real corpus. It requires large-scale data for statistical analysis to get convergent results [3, 4]. The other is based on linguistic knowledge and taxonomy system. Usually, a common formula is used to calculate the semantic relatedness ignoring the difference between similarity and relatedness. When the relation is *is-a*, we get a measure of similarity, otherwise we get a measure of relatedness [5].

According to the corpus-based model, more times two words co-occur, more closely they are related. To some extent, this method reflects the degree of relatedness, but it can't further explain the particular semantic relations between words, and semantic relatedness is more about concepts than words, which makes it a less satisfying method to measure semantic relatedness.

As we have already mentioned above, relatedness and similarity computation share same calculation formulas in many previous models. There have been a number of algorithms proposed. For example, Liu et al. [6] proposed an algorithm based on HowNet, considering the distance of concepts. The simplest algorithm [7] only utilizes the shortest path among the possible paths between concepts. Short distance means high similarity. In spite of its simplicity, it has been applied to multiple constraints medical semantic web [8] and gives a rather good result. Leacock and Chodorow [9] extend the idea by scaling the path. Their method shows some improvements. But all methods above have the common flaw that same distance results in same relatedness, whatever their depths are. Wu and Palmer [10] not only consider the distance between concepts, but also take common parent nodes of two concepts into consideration, as is shown in the following formula:

$$\text{Sim}(X, Y) = \frac{2 * \text{depth}(\text{msc}(X, Y))}{\text{len}(X, Y) + 2 * \text{depth}(\text{msc}(X, Y))} \tag{1}$$

msc(X, Y) denotes the parent concept of concept X and Y. Their algorithm has better results compared to the previous two methods. Lin [11] defines similarity in term of information content besides the factors of length and depth. More common parent nodes two concepts share, they are more related, and otherwise less related.

Duan [12] proposes a new method which has better results than previous algorithms. The method is a nonlinear combination of path length, concept intersection, the union set of concepts, and the depth level. The formula is as follows:

$$\text{Sim}(X, Y) = \begin{cases} 1 & X = Y \\ \frac{\alpha \times \beta \times |N\text{Set}(X) \cap N\text{Set}(Y)|}{(\text{Dist}(X,Y) + \alpha) \times |N\text{Set}(X) \cup N\text{Set}(Y)| \times (\gamma |d(X) - d(Y)| + 1)} & X \neq Y \end{cases} \tag{2}$$

Although the above models have considered many factors, they have their own scope of application. For example, the target application of Liu' algorithm is machine translation. It considers the structure and the interpretation of the word, but does not consider cases where words have low similarity but high relevance. For example, by the algorithm, the similarity between "孔子" (Confucius) and "孟子" (Mencius) is 1, while the similarity between "孔子" (Confucius) and "论语" (The Analects) is 0.130233.

3 Concept Relatedness

3.1 Ontology and Conceptual Relation

Domain Ontology [13] is an abstraction of domain knowledge, including concepts of the discipline, attributes of concept and relations between concepts and attributes. The relatedness is the quantification of the relationship of the ontology. In the domain ontology, relation between concepts contains the basic relation and the associated relation. The basic relation contains *is-a, part-of, attribute-of, made-of* [8, 14]. The associated relation is defined by experts in particular field who are familiar with domain knowledge. This particular relation determines the relevance between concepts. A simple ontology graph of virus knowledge is given in Fig. 1. In the graph, solid lines represent the basic relation, while dotted lines represent the associated relation.

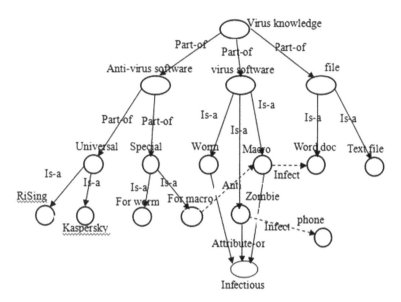

Fig. 1 A fragment of virus ontology

Though several basic relationships are marked in Fig. 1, it is not complete. Concepts usually have many attributes and many other components which the figure doesn't show.

3.2 Concept Similarity

In general, given an ontology graph, factors that affect the similarity between concepts are as follows: the shortest path length of the concept, the hierarchy depth of the concept, the density of concept, and the maximum common ancestor set [15]. In this paper, we give the following definition.

Definition 3.1 The length of relational edge, it refers to the weight of different relations between two concepts in the ontology. Because different relations have different contributions to the similarity, so we assign different weights to different relational edges. We define that $d(Is\text{-}a) = a_1$, $d(Part\text{-}of) = a_2$, $d(Is\ made\ of) = a_3$, d (an attribute of) $= a_4$. If it is required to define new basic relation, the length of which is $\max\{a_1, a_2, a_3, a_4\}$. For all i, we have $a_i \geq 1$.

Definition 3.2 The shortest path distance, it refers to the weighted sum of edge length in the shortest path between two concepts X and Y. We denote it as dist(X, Y). When the two nodes are not connected, dist(X, Y) $= \infty$.

Definition 3.3 The depth. In the ontology, the depth of root node is defined to be 1. The depth of any concept X except root node is calculated as:
depth(X) = depth(parent(X)) + 1

Definition 3.4 The sum of depth, it refers to the recursive sum of the depth of node X and its parent nodes. Here we use the symbol Sumdepth(X), then by definition, Sumdepth$(X) = \sum_{i=1}^{\text{depth}(X)} i$.

Definition 3.5 The upper set of concepts, the set of nodes in the shortest path from concept X to the Root node. It is denoted as US(X).

It is clear that the contribution to similarity of node from different levels is different. Deeper level represents finer concept granularity, accordingly, hence the contribution to similarity is larger. On the contrary, the contribution will be smaller. Similarity calculation is divided into two parts. The first part is determined by the upper set of concepts, the depth and the sum of depth. The second part is calculated based on the shortest distance. Then values of these two parts are combined, as:

$$
\text{Sim}(X, Y) = \begin{cases} 1 & X = Y \\ \alpha \dfrac{\sum\limits_{z \in \{US(X) \cap US(Y)\}} \text{depth}(z)}{\max\{\text{Sumdepth}(x), \text{Sumdepth}(Y)\}} + \beta \dfrac{\lambda}{\lambda + \text{disc}(X,Y)} & X \neq Y \end{cases}
\tag{3}
$$

α and β are parameters that act as weights of the two factors (the upper set of concepts and the shortest distance) in the integrated semantic similarity. The only constraint is $0 \leq \alpha, \beta \leq 1$ and $\alpha + \beta = 1$, but the specific values depend on specific application. The interval of the similarity is [0, 1]. Equation 3 clearly shows that nodes in different level have different weights. For a parent node Z, the depth of Z is depth(Z), the weight of Z is $\dfrac{\text{depth}(z)}{\max\{\text{Sumdepth}(x), \text{Sumdepth}(Y)\}}$. We can know that for the upper set of concepts, the deeper the node is, the greater the weight is.

And Eq. 3 satisfies the following conditions:

1. If the distance of the two concepts is 0, the similarity of them is 1;
2. The value of the similarity ranges from 0 to 1.
3. The greater the distance of two concepts is, the smaller the similarity is. The smaller the distance is, the greater the similarity is.
4. If the distance is infinite, the similarity is 0;
5. The more nodes the intersection of two concepts' upper sets has, the greater the similarity is.

3.3 Concept Relevance

The associated relation is defined by experts of specific area. These relations determine the relevance between concepts. For example, personalization is a relatively new word in the field of computer science. With the development of user-centered Web2.0, personalized search has become an important concept.

While generally personalization is not similar to search, in the field of computer science, we see a strong relatedness between these two concepts. Then an expert may define a special associated relation between them, which in turn will facilitate our calculation of relatedness.

The relevance is based on distance. We define that two concepts X and Y are relevant if only there is path between them that contains edges of associated relation. Since associated relation is less transitive than basic relations, every appearance of it will cause significant decrease in relevance.

$$\operatorname{Re} l(X, Y) = \frac{\gamma}{\gamma + \prod_{d \in \text{Allpath}} d} \tag{4}$$

where, Allpath is an aggregation of all the edges from concept X to Y, d is the length of the edge. The product in the denominator guarantees that the relevance will greatly decrease as the path becomes long. γ is a parameter controlling the maximum value of relevance.

3.4 Concept Relatedness

The semantic relatedness of concepts X and Y is the integration of similarity and relevance of concepts X and Y. It is calculated as:

$$\operatorname{Sim\_Re} l(X, Y) = \operatorname{Sim}(X, Y) + \operatorname{Re} l(X, Y) - \operatorname{Sim}(X, Y) * \operatorname{Re} l(X, Y) \tag{5}$$

The upper bound of relatedness is 1.

4 Experiments and Result

In this section, we give the experiment result of our method based on Fig. 1. We choose different pairs of words to show the influence of depth and other factors. The calculation follows the description in Sect. 3. In this paper, we set parameters as follows:

$$a_1 = 1.5, a_2 = a_3 = a_4 = 3, a_5 = 2;$$
$$\alpha = 0.5, \beta = 0.5, \lambda = 4, \gamma = 6;$$

In addition, we give a detailed comparison with other classic methods. They are introduced respectively by Wu and Palmer [10] (Eq. 1) and Duan [12] (Eq. 2). Results are summarized in Table 1. Column $R2$ shows the result of Wu and Palmer. Column $R2$ shows the result of Duan, we set parameters as the original paper $\alpha = 5, \beta = 1, \gamma = 0.2$. The last column is the result of our method.

Table 1 Results of the relatedness computation

Concept pairs		$R1$	$R2$	Col
Virus knowledge	Virus software	0.66667	0.34722	0.45238
Virus knowledge	Universal	0.50000	0.17007	0.28333
Virus knowledge	Worm	0.5	0.17007	0.31863
Anti-virus	Universal	0.80000	0.46296	0.53571
Anti-virus	For worm	0.66667	0.25510	0.38529
Universal	Rising	0.85714	0.52083	0.66364
Virus software	Zombie	0.80000	0.46296	0.61364
Zombie	Phone	0.85714	0.52083	0.75
Virus software	Phone	0.66667	0.25510	0.66667
Macro	Word doc	0.66667	0.52083	0.80929
For Macro	Word doc	0.5	0.34014	0.670
Special	Word doc	0.40000	0.19531	0.61063
Warm	Infectious	0.85714	0.52083	0.58571
Rising	Kaspersky	0.75	0.42857	0.58571
Warm	Macro	0.66667	0.35714	0.53571

From Table 1, it is clearly seen that Wu and Palmer [10] only considers the semantic distance and common nodes, it does not consider the concept granularity, which usually has impacts on relatedness. Duan's method [12] considers more comprehensive factors, thus the result is more reasonable, but it doesn't distinguish the different semantic relations. For example, the relatedness of Virus knowledge and Universal is equal to that of Virus and Worm.

In our method, edges of different relations of instance, part, properties, and composition have different length. In addition, relevance decreases rapidly as the number of associated edges included in the path between two concepts increases. Just as Table 1 shows, the relatedness of Macro and Word doc is much larger than that of For Macro and Word doc. Besides, for paths that contain different relationship between concepts, the results are different. The relatedness of Virus software and phone is 0.66667. The path between them contains two edges: one is the relationship "is-a", another is associated relationship. While the relatedness of For Macro and Word doc is 0.670. The path between these two concepts also contains two edges, but they are all the associated relationship. The result is more consistent with people's intuition.

5 Conclusions and Future Work

In this paper we proposed a novel algorithm of measuring the semantic relatedness between concepts. The model is based on a weighted graph for some domain ontology. Different relations have different weights. According to the experiment result, our algorithm is better than the previous approaches. And it can be further applied in data mining and information retrieval, etc.

The method in this paper is based on domain ontology, which largely depends on experts to define the semantic relations and semantic distance. The relationship between concepts must be well defined to achieve a better result. In future work, we will focus more on this challenging problem.

References

1. Li Y, Bandar ZA, McLean D (2003) An approach for measuring semantic similarity between words using multiple information sources. IEEE Trans Knowl Data Eng 15(4):871–882
2. Resnik P (1995) Using information content to evaluate semantic similarity in a taxonomy. arXiv preprint cmp-lg/9511007
3. Zazo ÁF, Figuerola CG, Berrocal JLA, Rodriguez E (2005) Reformulation of queries using similarity thesauri. Inf Process Manage 41(5):1163–1173
4. Dagan I, Lee L, Pereira FC (1999) Similarity-based models of word cooccurrence probabilities. Mach Learn 34(1–3):43–69
5. Liu H, Xu D (2012) Ontology based semantic similarity and relatedness measure review. Comput Sci 39(2):8–13
6. Liu Q, Li S (2002) Based on the "Text" lexical semantic similarity computation. Chin Comput Linguist 7(2):59–76
7. Rada R, Mili H, Bicknell E, Blettner M (1989) Development and application of a metric on semantic nets. IEEE Trans Syst Man Cybern 19(1):17–30
8. Kamps J, Marx M, Mokken RJ, De Rijke M (2004) Using WordNet to measure semantic orientations of adjectives
9. Leacock C, Chodorow M (1998) Combining local context and WordNet similarity for word sense identification. WordNet: An electronic lexical database. 49(2):265–283
10. Wu Z, Palmer M (1994) Verbs semantics and lexical selection. In: Proceedings of the 32nd annual meeting on association for computational linguistics, pp 133–138. Association for Computational Linguistics
11. Lin D (1998) An information-theoretic definition of similarity. In: ICML, vol 98, pp 296–304
12. Duan J, Yang Z, Gan J (2009) The comprehensive quantification research of semantic similarity and relevance based on domain ontology. Comput Sci Technol 11:011
13. Ruotsalo T, Hyvönen E (2007) A method for determining ontology-based semantic relevance. In: Proceedings of database and expert systems applications, pp 680–688. Springer, Berlin, Heidelberg
14. Alvarez MA, Lim S (2007) A graph modelling of semantic similarity between words. In: International conference on semantic computing 2007, ICSC 2007, pp 355–362. IEEE
15. Song L, Ma J, Lian L, Chen Z (2006) Fuzzy similarity from conceptual relations. In: IEEE Asia-Pacific conference on services computing 2006, APSCC'06, pp 3–10. IEEE

Extracting Academic Activity Transaction in Chinese Documents

Fang Huang, Shanmei Tang and Charles X. Ling

Abstract As academic relationship networks are implied in various academic activities which relate to study and work experiences of scholars, extracting the transaction information of academic activities is important for mining academic relations networks. For the characteristics of long-distance dependencies in Chinese sentences of academic activities, this paper proposes to extract the transaction information from Chinese documents by using sequence forecast based on Conditional Random Fields (CRF). Often, in research project applications, the resumes of applicant and team members include many complex Chinese sentences about academic activities. We design novel methods to analyze special sentence patterns in those resumes. More specifically, we focus on the design of feature templates according to the sentences characteristics of academic activities, and employ the regular matching method to deal with inaccurate words segmentation, especially for academic-specific words. Through evaluating tests, we choose the optimum feature templates and input to CRF++ model to label trunk words of the sentences. The transaction information extraction of academic activities is implemented. Experimental results show the effectiveness of the proposed approach.

Keywords Academic activity transaction information · Regular matching · Conditional random fields · Sequence labeling · Feature templates

F. Huang (✉) · S. Tang
School of Information Science and Engineering, Central South University,
Changsha 410083, China
e-mail: hfang@mail.csu.edu.cn

C. X. Ling
Department of Computer Science, University of Western Ontario, London, ON N6A 5B7,
Canada
e-mail: cling@csd.uwo.ca

Z. Wen and T. Li (eds.), *Knowledge Engineering and Management*,
Advances in Intelligent Systems and Computing 278,
DOI: 10.1007/978-3-642-54930-4_13, © Springer-Verlag Berlin Heidelberg 2014

1 Introduction

During the long-term learning, collaborative research, academic exchanges, and other academic activities, an academic-specific relationship is established between researchers, which formed academic relation networks. It is significant to the cultivation of high-level scientific research groups and evaluation of science technology to mine and analyze the special social networks. Since academic networks are contained in the abundant academic activity information, it has become a concerned topic by domestic and foreign scholars how to extract the activities information from a mass of Chinese text and mine potential academic relationship. Generally, academic relationships consist of teacher–students, classmates, cooperation projects, co-authors, etc., which are usually embedded in experiences of learning, education, research, project cooperation, co-published with monographs and papers, participation in academic community and conferences. And there is a concentrated reflection in resumes of applicant and team members in project applications, which provides powerful supports for building academic network. So we propose a scheme of text sequence labeling by using Conditional Random Fields (CRF) to extract academic activity information from research project applications in Chinese.

Text sequence labeling is a significant topic of Chinese information extraction. There are three frequently used approaches, Hidden Markov Model (HMM), Maximum Entropy Model (MEM), and CRF [1]. Compared with HMM and MEM, CRF is to calculate joint probability of the whole label sequence, then it effectively solved the problem of long-distance state labeling. In 2006, Hong and Zhang et al. proposed a Chinese parts of speech labeling based on CRF with context information [2]. Zhou et al. presented a hierarchical CRFs model to improve the efficiency and accuracy of automatic recognition Chinese organization names [3]. In 2010, Zhou and Li et al. raised an extraction algorithm for temporal relationships by CRF [4] to obtain time attribute from Chinese medical record. In the proposed scheme, the resumes of project applications are first divided into Chinese words, and then produce a mixture sequence of words and part of speech label. Since the long-distance dependencies of trunk words and the position of time attribute relevant to the academic activities occurred in Chinese sentences, the sequence forecast model with the whole sequence labeling, CRFs, is introduced into the information extraction.

2 Extraction Process of Academic Activity Transaction Information

There are three main tasks in the extraction process of academic activity information for project applications, which are text preprocessing, text sequence labeling of academic activity, and establishing transaction record database.

First, the resume sections are separated from applications after converting Word documents of project applications into TXT files. Then, we use Institute of Computing Technology, Chinese Lexical Analysis System (ICTCLAS) to divide Chinese words and label part of speeches, and employ the regular matching to merge some inaccurate divided specific words so that the segmentation can reflect the features of academic activity transaction information, and thereby improve the accuracy of next labeling.

In the text sequence labeling, the sequences with Chinese words and part of speech label of resume sections are divided into two parts, one part as training data, another as testing data. The CRF model is trained to lean by the input training data and feature templates. The trained CRF can forecast and label test data by using learning rules, and acquire sequence labels of academic activity transaction information.

Finally, we can pick up the trunk words of academic activities and time attribute from the labeled sequence to form transaction records of academic activities and establish transaction database.

3 Text Sequence Preprocessing with Regular Matching

3.1 Characteristics of Sequences with Chinese Words and Part of Speech Labels

Although the word dividing and part of speech labeling are achieved by words segmentation system, there are some professional words and organization names still incorrectly divided, and it will influence the accuracy of labeling of academic activity information. In order to properly segment these words, it is needed to add more specific words into ICTCLAS dictionary, such as academic organization names and major names, etc. Due to more special words added manually, the integrity of dictionary cannot be guaranteed, which will result in the inaccuracy of labeling. A sample of the text sequence of words and part of speech label is shown in Fig. 1.

Here, "nr" represents as a noun of name, "wd" as a comma. "t" indicates the time, "v" is a verb. "p" expresses preposition, "org" as a name of organization. "n" is a common noun, "vn" as noun verb, and "wj" as a period. "nsf" indicates a transliteration of place name, "vg" as verbal morpheme, and "ng" as noun morpheme. With the addition of institutions name in ICTCLAS dictionary, "湖南医科大学" is accurately divided, and marked a label of "org". However, since "美国匹兹堡大学" is not added in the dictionary, it is divided into "美国/nsf 匹/vg 茨/ng 堡/ng 大学/n", which is not as an institution name to identify. "医学/n检验/vn专业/n" also is inaccurate dividing, because the major category is various, and classification standard is not unified. It is very difficult to add all major names into the dictionary manually. So the segmentation for these specific words needs to be merged as a remedy. As the complexity of mixing sequence including words

李四/nr,　/wd　1962年/t　1月/t　生/v,　/wd　1999年/t　毕业/v　于/p 湖南医科大学/org 医学/n 检验/vn 专业/n 。/wj 现在/t 美国/nsf 匹/vg 茨/ng 堡/ng 大学/n 医学/n 中心 /n 遗传/vn 药理/n 研究/vn 中心/n教授/n 。/wj

Fig. 1 Text sequence of words and part of speech labels

segmentation and part of speech label, the pattern of specific words in sequence is complex. Fortunately, the regular matching can be used to locate phrases to implement the merger duo to its good fuzzy matching.

3.2 Context Words Classification in Regular Matching

In regular matching, a contextual pattern corresponding to specific words is converted into a regular expression to match and locate the text sequence to be merged. There are two kinds of matching words in the process, major name and institution nouns. The institution nouns include university, college, laboratory, company, research institute, which are five categories.

For example, the "医学/n 检验/vn专业/n" in the text sequence will be combined into "医学检验专业/n". The regular expression is "\b(org).{0, 24} (专业)\b". Here, "\b" is a start and end of the matching word between 0 and 24 characters. When found "org", "." can match any character excepting a line break, and the searching will be ended until finding "专业", it is a positioning keyword, which typically include "大学", "公司", etc. The context words are auxiliary words which are able to guide search for keywords, such as "攻读", "在", "进入", and so on. Merger is to remove superfluous characters between the context words and keywords, and remain a complete major or organization name. The length of characters between context words and keywords is adjusted by experiments. In this paper, the keyword set is divided into two categories of major name K_1 and institutional name K_2. The context words set C are classified into six groups related with K_1, as shown in Table 1. There are seven groups of context words related with K_2, and shown in Table 2.

With removing superfluous characters between the matched context keywords and keywords, "医学检验专业" and "美国匹茨堡大学" be merged into a complete major and organization words.

4 Academic Activity Information Extraction by Using CRF

4.1 The Principle of Conditional Random Fields

In the forecast model of CRF, let observed input sequence is $X = \{x_1, x_2, ..., x_n\}$, and the output state sequence is $Y = \{y_1, y_2, ..., y_n\}$. To gain the state of label sequence, $P(y|x)$ is a conditional probability distribution of random variables

Table 1 Context words related with K_1

Context words	Examples
系	计算机 系 n 软件工程专业 n
org	大学 org 生物医学专业 n
攻读	攻读 v 电子与通信工程专业 n
学院	学院 org 化学专业 n
任	任 v 微电子学专业 n
主修	主修 v 计算机科学与技术专业 n

Table 2 Context words related with K_2

Context words	Examples
在	在……大学|学院|研究所|公司
进入	进入……大学|学院|研究所|公司
于	于……大学|学院|研究所|公司
兼任	兼任……大学|学院|研究所|公司
获	获……大学|学院|研究所|公司
为	为……大学|学院|研究所|公司
从	从……大学|学院|研究所|公司

Y. The line chain CRF with parameter $R = \{\lambda_1, \lambda_2,..., \lambda_n\}$ come into being through training R to cause a maximum conditional probability $P(y|x)$ [5].

In Eq. (1), $f_i(y_{j-1}, y_j, x, j)$ is a two-valued feature function, which can be as state feature functions and transfer feature functions. The state function means an alone node feature, and the transfer function represents the sequence relationship feature. λ_i is the weight of feature function by training data to learn in order to obtain the maximum value $P_\lambda(y|x)$. There are two problems in establishing CRF model.

$$P_\lambda(y|x) = \frac{1}{Z_\lambda(x)} \exp\left(\sum_{j=1}^{n} \sum_{i=1}^{m} \lambda_i f_i(y_{j-1}, y_j, x, j)\right) \tag{1}$$

(1) **Parameter estimation** It is to solve $R = \{\lambda_1, \lambda_2,..., \lambda_n\}$ with learning from the training dataset for every weight parameters λi, which are acquired by the maximum log-likelihood estimation.

(2) **Feature function selection** It is to choose an appropriate feature function set from training dataset by the trained forecast model with inputted feature templates.

4.2 Labeling Academic Activity Transaction Information

We use CRF++ toolkit [6] to implement the text sequence labeling of academic activity information. In CRF++, the forecast model is trained by the labeled training data and designed feature templates, and the training will produce a model

Fig. 2 Training data samples	李四	nr	N
	,	wd	O
	1992年	t	T
	毕业	v	R
	于	p	O
	湖南医科大学	org	J
	医疗	n	O
	系	n	O
	,	wd	O
	获	v	R
	医学	n	I
	学士学位	n	I
	。	Wj	O

file as output including estimated parameters and selected functions, which are taken as input into the forecast model again to label the test data. For the academic activity information labeling, the process is divided into the following three tasks: preparation of training data and test data, designing the feature templates of sentences structure for academic activities, and sequence labeling of academic transaction information.

4.2.1 Training Data and Test Date

First of all, the sentences set of words and part of speech labels for personal resumes are divided into two parts, one for the training data, another part for the test data. There are three columns in training data text, as shown in Fig. 2. The first column is input sequence for words segmentation, the second is the part of speech labels, and the third is manually labels sequence in states of academic activity information. The test data only occupy the first and second columns, using CRF++ to predict, the states will be labeled and filled in the last column of test data text, which formed labeled state sequence.

4.2.2 The Labeling State of Academic Activity Information

Since the resume mainly related to academic activities occurred time, place, people and entity, etc., we collected six types of state which is shown in Table 3.

In particular with Table 3, "T" and "M" are taken as the start and end of time, respectively. The various expressions of time are collected, such as the period or point time of event occurred, and as shown in Table 4.

For the period time, between the start and end of time may be filled with other Chinese characters or connection symbols, whatever, these will be labeled by "O", which represents unrelated words.

Table 3 State labels of academic activity information

Labeling types	Labeling state (y_j)	Examples
Name	$N(y_1)$	张三, 李四
Time	$T, M(y_2)$	2001年
Academic relationship words	$R(y_3)$	毕业, 获得
Organization	$J(y_4)$	大学, 研究所
Academic noun	$I(y_5)$	学位, 学者, 专业
Unrelated words	$O(y_6)$	我们, 的, 并且

Table 4 Various expressions of time

Types of time	Examples
Point time	1991年; 1991年6月; 1991年9月1日
Period time	1992年–1998年; 2000年12月至2001年12月; 2000年9月1日至2001年6月30日
	1985–1989年; 1991年至今; 1991年 ~ 现在

4.3 Designing Templates with Characteristics of Academic Activity Sentences

In general, the target of feature template design is as many as including contextual information patterns. But, if the span of context information is too long, it can cause over fitting result in the accuracy of the sequence labeling decreased. We randomly picked up 100 resumes from project applications, and analyzed the syntactic structure of Chinese.

4.3.1 Syntactic Structure of Academic Activity Transaction Sentences

In the paragraph of resume, most sentences have a key verb of participating in academic activities, which is called academic relation words. According to the meaning of academic relationship in the special sentence, the academic relationship words are divided into seven categories, as shown in Table 5.

According to the grammatical structure, academic relationship word is a key trunk word as the support for the whole sentence, and its current position is set to 0, and then the span between the academic relationship word and other trunk words are show in Table 6. There are three kinds of position relations to be expressed, including the beginning, middle, and end of sentence. For example, in the first row, "毕业" is located in the beginning of the sentence, the current position is 0, and then the organization name "湖南医科大学" is 2 which is followed by the second position in the sentence.

Table 5 Classification of academic relationship words

Categories	Examples
Learning experiences	毕业, 攻读, 就读, 进入, 获, 作为, 获得, 参加, 赴, 研修, 取得, 学习, 考入, 读, 派往, 保送, 师从
Work experiences	进入, 任教, 担任, 参加, 指导, 任, 受聘 留校, 晋升, 入选, 调入, 从事 等
Research experiences	作为, 参加, 承担, 主持
Publish paper	合著, 发表, 出版
Others	生, 出生, 回国, 到, 评为, 剖析

Table 6 Sentence structure and span of academic activities

Position of academic relation word	Labeling state	Span value
Beginning of sentence	毕业R 于O 湖南医科大学J 医学检验专业I	[0, 2, 3]
Middle of sentence	在O 中南大学J 应用化学专业I 攻读R 博士学位I	[−2, −1, 0, 1]
End of sentence	1984年T 武汉大学J 生物系O 生物化学专业I 毕业 R	[−4, −3, −2, −1, 0]

4.3.2 Principle of Template Design

The labeled states of Chinese sentences will be utilized to identify the sequence relationships between the trunk words. According to the sentences structure, the principle of designing template is mainly following three aspects. At first, the state of context vicinity determines the labeling state of the current word. Although the context words with longer distance far away from the current words contribute to the labeling results smaller, its influence is actual existence. So, the context span is an important factor in the template design. Then, since the part of speech of context words can also impact on the state of current word, it is another important factor. The third, in order to improve the accuracy, multiple inputs information including context span and part of speech can reflect in the feature template simultaneously to form multidimensional compound feature templates.

In Fig. 3, the line "清华大学" is the current token in the input text sequence. The corresponding feature template is designed as shown in Fig. 4. As template U00:% X [−1, 0], "U00" is a template name, "−1" is a previous row relative to the current line, "0" is first column of the input text. U00, U01, and U02 described the span of the first column, U03, U04, and U05 expressed the span of second column, U06 and U07 are multidimensional feature templates. For example, U07:%x[0, 0]/ %x[−1, 1]/%x[1, 1] is a three-dimensional template, %x[0,0] corresponding to the "清华大学" in Fig. 4, %x[−1, 1] and %x[1, 1] locate to "t" and "n", respectively, previous and next row of current line in the second column, which effect on the state of current token to be labeled.

张三	nr（name）	N
,	wd（comma）	O
1991年	t（time）	T
清华大学	**org (organization)**	**J>>current token**
计算机科学与技术专业	n (noun)	I
硕士学位	n (noun)	I
毕业	v (verb)	R
。	wj (period)	O

Fig. 3 The current token of input sequence

Fig. 4 The sample of feature
templates

```
# Unigram1
U00:%x[-1, 0]
U01:%x[0, 0]
U02:%x[1, 0]
U03:%x[-1, 1]
U04:%x[0, 1]
U05:%x[1, 1]
U06:%x[-1, 0]/%x[0, 0]
U07: %x[0, 0]/%x[ -1, 1]/%x[1, 1]
```

5 Experiment and Analysis

To test and prove effect of the context span, part of speech, and multidimensional compound template to the academic activity information labeling, the comparative tests are carried out by three sets of templates, respectively. The control variable method is used to verify the merits of feature templates, and each time only to change one property of the template. We selected 300 project applications for the experiment, and use the ten-fold cross-validation to test these templates in three groups, respectively. The F-value is introduced to evaluate the accuracy of sequence labeling, and shown in formula (2).

$$F = \frac{2 \times P \times R}{R + P} \qquad (2)$$

Here, P is precision, and R is recall. The comparing results of precision, recall and F-values are shown in Fig. 5. The first group is template 1–3 to test the effect of text span on sequence labeling, the second one is template 4–6 for adding part of speech into the templates, and the third is 7–9 to verify multidimensional feature templates.

In the first group, since the span of template 2 is longer than template 1, the precision and recall rates are improved. Although the span of template 3 is larger than template 2, and the recall rate has increased, the precision rate is a little drop. The more information to learn will increase the number of information labeled, but it also causes confusion of learning rules resulting in reduced accuracy. In the second group, the average effect is better than the first one due to combining the part of speech with context span. However, as too much information in templates

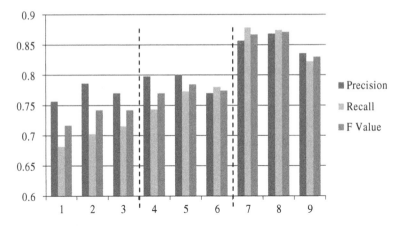

Fig. 5 Comparing three group of tests

Table 7 Evaluations of the optimal templates

Precision	Recall	F-values
0.905	0.943	0.923

lead to over fitting, the precision rate of template 6 is not higher. In the third group, the effect is improved more because of multidimensional features added in. Templates 7 and 8 presented good results, and the precision and recall rates are about the same, which indicate that the two-dimensional and three-dimensional templates can appropriately reflect the sentences characteristics of academic activities. The results of template 9 is not so satisfactory because of excessive jump training causing the confusion of learning rule in the multidimensional composite template, so that the precision and recall rates are declined significantly.

According to the three group tests, the optimal combination of templates is formed as the training templates. In the optimal templates, the context span is set 3, also the span of part of speech label is 3, and it includes two-dimensional and three-dimensional templates. Using the optimal feature templates and 382 projects applications to train CRF++, the evaluation of sequence labeling results are shown in Table 7. The precision and recall are greatly improved, and F-value reaches the maximum.

6 Conclusion

We adopted the forecast model based on CRF to extract academic activity transaction information from Chinese resumes of project applications. For the text sequence with Chinese word segmentation and part of speech labeling, the regular matching is used to process special academic words. By the conditional random prediction principle, we employed CRF++ toolkit to carry into execution, and designed the feature templates with characteristics of academic activity sentences. Furthermore, our analysis focuses on the labeling effect to three kinds of templates, which are with the context span, part of speech label and multidimensional composite template. With three groups of comparative tests, we evaluated the impact on the labeling results through comparing the precision, recall and F-value. The results show that adding more template features will produce a positive effect on the accuracy of labeling, but too much information can also cause confusion of learning rules. By experiments, we obtained the optimal combination of templates, and achieved the sequence label of academic activity trunk words. It is easy to form the academic activity transaction records. Since the training dataset are based on data collection in manual, it is time-consuming efforts, and it is very difficult to collect all sentence characteristics of academic activities. So, in our future work, we will focus on semi-supervised learning to improve the effectiveness of training data.

Acknowledgment This work was supported by Project 61073105 of National Natural Science Foundation of China.

References

1. Lafferty J, McCallum A, Pereira F (2001) Conditional random fields: Probabilistic models for segmenting and labeling sequence data. In: Proceeding 18th international conference on machine learning, pp 282–289
2. Hong M, Zhang K et al (2006) Chinese part of speech tagging method based on CRFs. Comput Sci 133(110):148–151,155 (In Chinese)
3. Zhou J, Dai X, Yin C et al (2006) Based on cascaded conditional random field model of Chinese organization names automatically identify. Electron J 34(5):804–809 (In Chinese)
4. Zhou X, Li H, Duan H et al (2010) Chinese medical text automatically extracts temporal relations based on CRFs. Chin J Biomed Eng 10:10–716 (In Chinese)
5. Zhu JN, Wen ZJ-R, Zhang B (2005) Conditional random fields for web information extraction. In: Proceedings of the 22nd international conference on machine learning, pp 1044–1051
6. Kudo T (2009) CRF++: yet another CRF toolkit [EB/OL]. http://crfpp.sourceforge.net/.2009

Predicting the Helpfulness of Online Book Reviews

Shaocun Tian and Xingwei Hao

Abstract With the rapid development of Internet and Web2.0, product review as a kind of user generated content plays an important role in people's lives. However, an overwhelming amount of reviews exist and they vary significantly in quality. The purpose of this paper is to alleviate this problem of information overload. We adopt a regression approach to predict the helpfulness of book reviews. A high quality subset of reviews is then selected accordingly. Since book review is typically longer and more diverse in content and style, we extract some complicate text features and propose a new way to weight word features. Experiment shows that our method performs better than the baseline TFIDF model.

Keywords Review helpfulness · Quality prediction · Data mining

1 Introduction

Recent years have seen a large growth of user generated content. One main form of UGC is user review on commercial (e.g., amazon.com) or noncommercial (e.g., imdb.com) websites. Reviews on these sites are of great importance for modern life. In fact, they play a very similar role as traditional effect of mouth, which has a major influence on people's decision. However, the difference is also obvious. For the offline situation, a few close friends are usually referred to for recommendations, while on Internet there are much more opinions we can find easily. In a typical scene, hundreds or even thousands of online reviews are available for a popular product. These reviews vary significantly in their quality, from good user experience to fraudulent spam. To address this problem, many sites provide a

S. Tian (✉) · X. Hao
School of Computer Science and Technology, Shandong University, Jinan, China
e-mail: tianshaocun@mail.sdu.edu.cn

Z. Wen and T. Li (eds.), *Knowledge Engineering and Management*,
Advances in Intelligent Systems and Computing 278,
DOI: 10.1007/978-3-642-54930-4_14, © Springer-Verlag Berlin Heidelberg 2014

voting system to users. Users can vote a review helpful or not, then reviews that receive more percent of helpful votes will be ranked high. One problem of this approach is that recent reviews attract little attention no matter whether it shows a good quality. Therefore, an automatic method of selecting high quality reviews will be useful in this situation.

In this paper, we focus on the task of finding best book reviews on user-centered Web community. Book review can be thought of as simple descriptions about the reviewed product, just as other kinds of reviews. But it usually aims to express writer's reading experience or understanding of the book content. It's more about communication and sharing. And writers have more freedom in choosing their content and style. Hence, we hold the assumption that the quality of book review relies heavily on text features such as choice of words and writing style. Based on this assumption, we propose a new scheme of weighting word features and use several other meta-text features such as topic relevance to predict the review quality. The result shows that our method performs better than the pure TFIDF model.

The next section describes related works on user review. The model and features used are described in Sect. 3. We report the experiment result in Sect. 4. And conclusions are presented in last section.

2 Related Works

There has been a lot of research works on user generated content. Most of them are about sentiment analysis [11, 15]. A small amount of works are on automatically assessing the quality of user review, most regarding it as a classification or regression problem. Kim et al. [7] used SVM regression to predict the helpfulness of a review. They utilized several kinds of features such as structural, lexical, syntax, and semantic features. Their experiment on an Amazon dataset including reviews for MP3 players and digital cameras proved that most useful features were text length, unigrams and product rating. Liu et al. [8] predicted the helpfulness of reviews with a nonlinear model based on radial basis functions. Three important factors used in their model were the reviewer's expertise, the writing style of the review, and the timeliness of the review.

All the works mentioned above used the helpfulness votes provided by Amazon or other sites as labeled data. Liu et al. [9] indicated that the user votes suffered from three types of bias: imbalance vote bias, winner circle bias, and early bird bias. They then employed human labors for annotation and proposed a classification-based method to recognize low quality reviews. The result was further used for opinion mining system, which typically summarized all the reviews into pros and cons without considering the quality of specific review [5]. To avoid using the biased votes, Tsur and Rappoport [14] presented an unsupervised algorithm called REVRANK to rank user reviews. They first constituted a virtual optimal review, and then reviews were ranked according to their distance to this virtual review.

Danescu-Niculescu-Mizil et al. [2] studied the factors in evaluation of product reviews from the perspective of social psychology. They found that a review's helpfulness was influenced by the relation of its rating and other ratings. Some other works also utilized social features to improve the quality prediction [1, 10]. Lu et al. [10] exploited social context by adding regularization terms and demonstrated that social context was helpful when a large training dataset was not available.

Another related area is computational stylistics. Most works done in this area are about author attribution, in which task function words often play an important role [12]. Stylistics is also used to recognize deceptions [4]. Kao and Jurafsky [6] recently did some interesting works on classifying poems. They analyzed factors that differentiated professional poems from amateur ones. The most significant one lied on the usage of abstract nouns and concrete nouns.

3 Data and Evaluation

3.1 Data

The dataset consists of reviews for the top 250 books crawled from Douban Reading.[1] For every book, top 200 reviews ordered by helpfulness were collected. If there were less than 200, all the reviews were collected. The user information of correspondent review writers was also crawled. In the preprocessing step, we removed the duplicates caused by different editions of same book and eliminated those reviews for which we couldn't get the user information. Finally, there were 218 books, 19,786 reviews and 16,178 reviewers left in our dataset.

After some statistical analysis of the dataset, we find that it suffers from the same biases as Liu et al. [8] indicates in their paper. At first, a few reviews receive lots of user votes, while most reviews get zero votes. For our dataset, since there are only six reviews displayed on the correspondent book page, it turns out that there are few received votes besides these top six reviews (see Fig. 1a). Second, most people tend to vote helpful rather than unhelpful, which results in a high overall average score of helpfulness. (Actually, for those reviews with at least one vote, the average score is 0.81.) At last, review length is considered as a very strong hint of review helpfulness. For our dataset, it shows a slight correlation between length and helpfulness. More details can be found in Fig. 1.

---

[1] These books are the most popular ones on Douban Reading. See http://book.douban.com/top250.

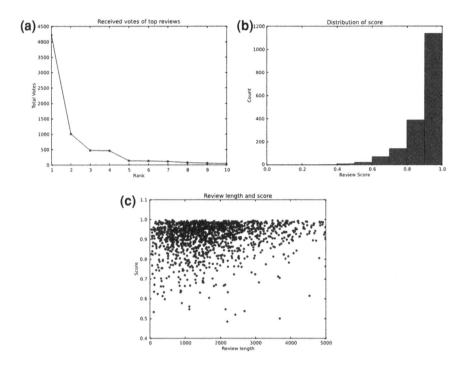

Fig. 1 **a** Top ten reviews for book *Into the White Night*. First few reviews receive a large number of votes. **b** Distribution of review score. Clearly people tend to vote helpful than unhelpful. **c** This graph shows a very slight correlation between review length and helpfulness

3.2 Model and Evaluation

It is clear that user vote is not an unbiased estimation of the real helpfulness of reviews, especially for those with few votes. However, we assume that for those reviews (typically top six) with enough votes user vote can be used as a reasonable reference. Given the number of helpful votes and unhelpful votes, we follow the definition of Kim et al. [7] and Liu et al. [9], then the helpfulness of a review is:

$$h(r) = \frac{N_+(r) + \lambda}{N_+(r) + N_-(r) + 2\lambda} \tag{1}$$

where $N_+(r)$ is the count of helpful votes and $N_-(r)$ is the count of unhelpful votes. λ is a smoothing parameter used to avoid the variance caused by too few votes. In this paper, we set $\lambda = 0.5$.

The main purpose of this paper is then proposing a method to automatically predict the helpfulness of a review, given the review text and reviewer's information. We adopt a regression approach to tackle this quality prediction problem. In fact, our final goal is selecting a high quality subset of reviews with size k. So we rank the reviews by predicted helpfulness score and select the top k reviews for every book. Prediction accuracy is then defined to be the ratio of correct predicted reviews. Assuming there are N_b books in test set, top k reviews set for book i is denoted as R^i_{tk} and in the same manner predicted top k reviews set as R^i_{pk}. Then we have:

$$p = \frac{\sum_{i=1}^{N_b} |R^i_{tk} \cap R^i_{pk}|}{k \times N_b} \qquad (2)$$

The absolute helpfulness scores of reviews from different books are of little value for this task. But for other tasks such as recommending recent reviews of high quality for general users, the helpfulness score can also be regarded as a useful indicator.

3.3 Features

Features used in this paper include word features, shallow syntax features, meta-text features, and social features. We introduce them below.

3.3.1 Word Features or Unigrams (UG)

ICTCLAS[2] is used to segment and tag the text. Since our dataset is not a very large one. Bigrams and Trigrams will suffer from the sparsity problem. So we only consider the unigrams. And words appear less than five times in our dataset are filtered out. We also filter out the most frequent 200 words, which typically are function words.

According to the intuition, a useful review should be closely related to the reviewed book. That is, if some words appear frequently in the reviews for one specific book, they are likely to be words speaking about the content of the book, thus should be given a higher weight. We take this factor into consideration and propose the following weight scheme:

$$w_{ij} = tf_i \cdot idf_i \cdot \log(tf_{ij} + 1) \qquad (3)$$

where tf_{ij} is the frequency of word i in all reviews for book j. Without this term, it becomes the normal TFIDF weighting scheme.

---

[2] ICTCLAS is a widely used tool for Chinese word segmentation and tagging. See http://ictclas.nlpir.org/.

3.3.2 Shallow Syntax Features (SS)

These features are thought to be about writing style. Many textbooks of writing are telling us to avoid using too many adjectives or adverbs. As it is explicitly said in Strunk's book: "Write with nouns and verbs, not with adjectives and adverbs" [13]. Hence the following features are considered: (1) Percentage of Nouns, Verbs, Adjectives, and Adverbs; (2) Percentage of Function Words; (3) percentage of punctuations. The reason that we add the percentage of punctuations is that good and correct punctuations usually mean good readability.

3.3.3 Meta-Text Features (MT)

Features included in this category are: (1) Review length; (2) Average Paragraph Length; (3) Type token ratio; (4) Mean frequency of words; (5) Topic relevance. Review length has been proved to be a very useful feature for predicting helpfulness [7]. Type token ration and mean frequency of words are adapted from Kao and Jurafsky [6]. This feature has positive impacts on filtering those low quality reviews which repeat the same sentence many times.

Topic relevance is a novel feature we use to measure the degree of how the review's content are related to the reviewed book. We defined all the reviews for one specific book as a complete review covering all related topics about this book. We calculate the similarity between every review and the correspondent book's virtual complete review in a transformed latent semantic space [3]. The similarity is then regarded as the topic relevance. This feature can be used to distinguish the good quality reviews for the reviewed book from those with too general content. (Although this feature seems to be an overlap with the weighting factor, it is proved to be useful in our experiment.)

3.3.4 Social Features (SC)

Under this category, all the information about the reviewer is considered: (1) Number of books user have read; (2) Number of reviews user have written; (3) Number of notes user have written; (4) Number of followers; (5) Number of followings; (6) Number of statues user have posted.

We expect that the helpfulness scores of all the reviews by one user are important indicators of user's expertise. Due to the inconveniences of Douban API our present model doesn't include this information. A detailed analysis of user network is also beyond the scope of this paper. We leave it for future research.

Table 1 Prediction accuracy and correlation coefficients

Features	p (k = 6)	p (k = 1)	Pearson	Spearman
UG	0.63	0.18	0.34	0.37
UG + SS	0.64	0.14	0.34	0.37
UG + SS + MT	**0.67**	0.23	0.36	0.38
UG + SS + MT + SC	0.59	0.23	0.37	0.39
SS + MT	0.63	0.14	0.22	0.28
SS + MT + SC	0.63	0.23	0.26	0.30
UG*	0.64	0.21	0.36	0.41
UG* + SS	0.65	0.18	0.37	0.42
UG* + SS + MT	**0.67**	**0.27**	0.36	0.42
UG* + SS + MT + SC	0.63	**0.27**	**0.40**	**0.43**

UG is the TFIDF weighted word feature
*UG** is the word feature with our new weight scheme

4 Experiment Result and Discussion

In this section, we describe the experiment result and its implications. Since Douban Reading only shows top six reviews on home page of every book, we also set the parameter $k = 6$. To test the performance on extreme cases, we also report the result on $k = 1$, i.e., predicting the most useful review. The regression method we adopt is Ridge regression. Although nonlinear approaches such as SVM regression with an rbf kernel are expected to have better performances, it is not the case in our experiment. Ridge method is proved to be more efficient. The absolute score of predicted helpfulness, as said in Sect. 2, might be useful in some situations. Thus, the overall Pearson and Spearman correlation coefficients are also listed. Table 1 summarizes the result.

The best result is achieved by combining word features, shallow syntax features, and meta-text features. Actually, shallow syntax features has little influence on the performance. This might result from the fact that we didn't distinguish different categories of books, while book reviews for fictions and technical books must be presented in very different styles. Another surprising fact is that social features decrease the accuracy, although the correlation coefficients are higher. This is a strong evidence of our assumption that the helpfulness of reviews mostly depends on review text. Besides, there is a noticeable improvement using our new weighting scheme.

When $k = 1$, combination of word features, shallow syntax features, and meta-text features also gives the best result. But in this case, the accuracy is much lower (only 0.27). Considering that $k = 1$ means selecting the most helpful review and the average number of reviews is larger than 10, this result is rather better than a random guess. On the other hand, helpfulness is really a subjective judgment; people usually have no agreement on which review is most helpful. Therefore, we suggest that $k = 6$ is a more reasonable choice.

At last, the Spearman correlation coefficient is higher than Pearson coefficient in all cases. This means that we can get a better rank performance without accurate prediction of the helpfulness score.

5 Conclusions

Book review is typically longer than other product reviews and more diverse in content and style, which motivates us to extract more text related features. In this paper, we introduced a novel way weighting word features and several new meta-text features. These features were then used to train a regression function to predict the helpfulness of book reviews. Experiment showed our method performed better than the baseline TFIDF model.

In the future, we plan to incorporate more semantic features such as sentiment and distinguish different categories of books. And a large dataset will also be helpful. Furthermore, we think it is interesting to apply this technique to other types of user generated content, where social context may play a more important role.

References

1. Agichtein E, Castillo C, Donato D, Gionis A, Mishne G (2008) Finding high-quality content in social media. In: Proceedings of the international conference on Web search and web data mining. ACM, pp 183–194
2. Danescu-Niculescu-Mizil C, Kossinets G, Kleinberg J, Lee L (2009) How opinions are received by online communities: a case study on amazon. com helpfulness votes. In: Proceedings of the 18th international conference on world wide web. ACM, pp 141–150
3. Deerwester SC, Dumais ST, Landauer TK, Furnas GW, Harshman RA (1990) Indexing by latent semantic analysis. JASIS 41(6):391–407
4. Feng S, Banerjee R, Choi Y (2012) Syntactic stylometry for deception detection. In: Proceedings of the 50th annual meeting of the association for computational linguistics: short papers (vol 2). Association for Computational Linguistics, pp 171–175
5. Hu M, Liu B (2004) Mining and summarizing customer reviews. In: Proceedings of the tenth ACM SIGKDD international conference on knowledge discovery and data mining. ACM, pp 168–177
6. Kao J, Jurafsky D (2012) A computational analysis of style, affect, and imagery in contemporary poetry. In: NAACL-HLT 2012, p 8
7. Kim S, Pantel P, Chklovski T, Pennacchiotti M, Disp ARTG, Politecnico V, Rey M (2006) Automatically assessing review helpfulness. In: Proceedings of EMNLP 2006, pp 423–430
8. Liu J, Cao Y, Lin CY, Huang Y, Zhou M (2007) Low-quality product review detection in opinion summarization. In: EMNLP-CoNLL, pp 334–342
9. Liu Y, Huang X, An A, Yu X (2008) Modeling and predicting the helpfulness of online reviews. In: Eighth IEEE international conference on data mining, ICDM'08. IEEE, pp 443–452

10. Lu Y, Tsaparas P, Ntoulas A, Polanyi L (2010) Exploiting social context for review quality prediction. In: Proceedings of the 19th international conference on world wide web. ACM, pp 691–700
11. Pang B, Lee L, Vaithyanathan S (2002) Thumbs up?: sentiment classification using machine learning techniques. In: Proceedings of the ACL-02 conference on Empirical methods in natural language processing (vol 10). Association for Computational Linguistics, pp 79–86
12. Stamatatos E (2009) A survey of modern authorship attribution methods. J Am Soc Inform Sci Technol 60(3):538–556
13. Strunk W (2007) The elements of style. Penguin, New York
14. Tsur O, Rappoport A (2009) RevRank: a fully unsupervised algorithm for selecting the most helpful book reviews. In: ICWSM
15. Turney PD (2002) Thumbs up or thumbs down?: semantic orientation applied to unsupervised classification of reviews. In: Proceedings of the 40th annual meeting on association for computational linguistics. Association for Computational Linguistics, pp 417–424

Word Sense Disambiguation Using WordNet Semantic Knowledge

Ningning Gao, Wanli Zuo, Yaokang Dai and Wei Lv

Abstract Word Sense Disambiguation (WSD) has been an important and difficult problem in Natural Language Processing (NLP) for years. This paper proposes a novel WSD method which expands the knowledge for senses of ambiguous word through semantic knowledge in WordNet. First, selecting feature words through syntactic parsing. Second, expanding the knowledge for the ambiguous word senses through glosses and structured semantic relations in WordNet. Third, computing the semantic relevancy between ambiguous word and context and achieving the purpose of WSD by semantic network in WordNet. Lastly, adopting the Senseval-3 all words data sets as the test set to evaluate our approach. Through a detailed experimental evaluation, the result shows that our approach achieves improvements over some classical methods.

Keywords Word sense disambiguation · Syntactic parsing · Semantic relevancy

1 Introduction

In Natural Language Processing (NLP), it is common that a word has multiple meanings. Word Sense Disambiguation (WSD) is to exploit an ambiguous word and determine which sense of the word should be assigned in the given context.

N. Gao (✉) · W. Zuo (✉) · Y. Dai · W. Lv
College of Computer Science and Technology, Jilin University, Changchun, China
e-mail: gaonn11@mails.jlu.edu.cn

W. Zuo
e-mail: wanli@jlu.edu.cn

Y. Dai
e-mail: mrdyk@126.com

W. Lv
e-mail: wlv@jlu.edu.cn

Z. Wen and T. Li (eds.), *Knowledge Engineering and Management,*
Advances in Intelligent Systems and Computing 278,
DOI: 10.1007/978-3-642-54930-4_15, © Springer-Verlag Berlin Heidelberg 2014

WSD is a long-standing problem in NLP, which has broad impact on many important NLP applications, such as machine translation, information retrieval, and question answering.

The human beings can distinguish ambiguous word through other words in context. To simulate the process of human thinking, we can extract useful information from the given context and then use it to achieve the purpose of WSD. Therefore, in order to determine which sense of ambiguous word should be adopted in the given context, we need two resources: ① the context in which the ambiguous word has been used, in general, can be represented by feature words set of it. There are two commonly used methods to select feature words: window-based methods and dependency-based methods; ② some kind of knowledge, related to senses of ambiguous word, constitutes the main basis for comparison with the context. A variety of machine-readable dictionaries and ontologies (e.g., WordNet [1]) can be used as knowledge sources of word sense's knowledge. Due to the advantages of WordNet (we will describe it later), we choose it as the knowledge source of our method. Also, we need a good method to compute the semantic relevancy [2] between ambiguous word and context.

In recent years knowledge-based WSD approaches have become quite popular. This paper proposes the SKW-WSD (SKW means semantic knowledge in the WordNet) method. This method utilizes nouns in the gloss and semantic relationships to expend the knowledge for each sense of ambiguous word. For different level semantic relevancy, we define different calculation methods. We select the feature words set with syntactic parsing. Then, we integrate the semantic relations, sense glosses, and word's frequency information in WordNet. Based on the semantic network in WordNet, we compute the semantic relevancy between ambiguous word and context, and select the meaning of ambiguous word which has the greatest semantic relevancy value as the right sense. We use the Senseval-3 all words data sets as test set to evaluate performance of the proposed method, the result shows that: this approach simplifies the traditional WSD algorithms and achieved good results.

2 Related Work

2.1 Word Sense Disambiguation

There are many works on WSD. Methods of WSD can be divided into corpus-based Methods and knowledge-based Methods.

Corpus-based Methods can be further divided into supervised methods and unsupervised methods. In a supervised WSD Method, WSD problem is formalized as a typical classification problem. Using a large sense-annotated corpus to extract the characteristic attributes of the specific meaning of ambiguous word, this method generates a classifier or classification rule by machine learning methods, which is used to judge the sense of new instances [3, 4]. The accuracy of this method is the highest until now, such as Memory-Based WSD [5] and WSD with support vector machines [6]. However, it is subject to a new knowledge acquisition bottleneck since they rely on substantial amounts of manually sense-tagged corpora, which is laborious and expensive to create. Therefore, its coverage is limited. In an unsupervised WSD method, WSD problem is formalized as the clustering problem. The underlying assumption is that similar senses occur in similar contexts. Thus, senses can be induced from text by clustering word occurrences and using some measure of similarity of context. Then, new occurrences of the word can be classified into the closest-induced clusters/senses. It is hoped that unsupervised learning will overcome the knowledge acquisition bottleneck, because they can work from raw unannotated corpora directly and do not depend on sense-annotated corpora. Such as graph-based WSD methods [7, 8], Unsupervised WSD methods which based on the Noisy Channel Model [9], the automatic disambiguation method which utilizes a dictionary and knowledge extracted from unannotated text [10], and Translation-based unsupervised methods which use a word-aligned parallel corpus in order to extract cross-lingual evidence for WSD [11] etc.

With the rapid development of the computer technology, a variety of machine-readable dictionaries and ontologies (such as WordNet) have appeared, and then Knowledge-based WSD methods began to develop rapidly. It is based on the hypothesis that words used together in text are related to each other. It is mainly using the dictionary knowledge (Such as annotations or semantic relatedness) to select the sense, which has the greatest semantic relevancy with context to achieve the purpose of disambiguation. On the basis of WordNet, Hwang Myunggwon et al. built a KB E-WordNet by using glossaries and relations in WordNet, and applied it to WSD [12]. Lee et al. present a method of using domain knowledge to disambiguate text. This method computed the domain relevance between ambiguous word and context's feature words. And then select the sense of ambiguous word which has the greatest domain relevance value as the right sense [13]. Huang and Lu propose a knowledge-based WSD method which selects feature words based on dependency relation and syntax tree [14]. Sinha and Mihalcea propose a graph-based WSD method [15]. When we talk about Knowledge-based WSD algorithms, we have to mention the classical Lesk algorithm [16], this algorithm is dependent on finding common words between definitions. The Adapted Lesk [17] is an adaptation of Lesk's dictionary-based WSD algorithm using the lexical database WordNet. The algorithm relies upon finding overlaps between the glosses of target words and context words in sentence, also their semantic related word' gloss, respectively. AAlesk [18] adapted the Adapted Lesk by giving all relationship a base score for the related words from WordNet relationship according to their importance.

2.2 *WordNet*

WordNet is a comprehensive dictionary and lexical database for the English language, developed by the Princeton University. It can be interpreted and used as a lexical ontology in the Computer science. It is an online lexical reference system, in that it groups words together based on their meanings. Nouns, verbs, adjectives, and adverbs are grouped into synsets, each of which expresses a distinct concept. Each synsets has an associated definition or gloss. This consists of a short entry explaining the meaning of the concept represented by the synsets. Each synsets can also be referred to by a unique identifier, commonly known as a sense-tag. Synsets are connected to each other through a variety of semantic relations. In our experiment, all relations are taken into consideration.

After years of development, WordNet has covered the whole English language in a detailed manner virtually. It has been applied in various works actively. There are rich semantic knowledge in WordNet: sense glosses, sense frequency, structured semantic relations, and so on. Those informations can be used for WSD. In recent years, WSD method which chooses WordNet as knowledge source has been developed rapidly and achieved unprecedented results.

In our methods, we integrate the advantages of the glosses and the semantic relations in WordNet. We use nouns in the glosses and semantic relations to expending the knowledge for the meaning of the ambiguous word. The calculation methods for different semantic relevancy can guarantee 100 % recall.

3 The SKW-WSD Method

This method employs sentence by sentence disambiguation. The framework of this method is shown in Fig. 1. It can be divided into three steps. The first step is to select context feature words set. According to the hierarchy relation and path distance between context words and the ambiguous word in the phrase structure tree, feature words are selected from the context and assigned different WSD weights. The second step can be divided into two parts: ① extract glosses of the ambiguous word from WordNet and then extract nouns set (the first part of expend knowledge for sense) for each sense from the gloss. So we make use of a tool to work out part of speech before extract nouns; ② extract synsets (the second part of expend knowledge for sense) for each sense corresponding to semantic relations. The third step is to select the right sense of the ambiguous word. We calculate the semantic relevancy between each sense of the ambiguous word and the feature words set, the semantic relevancy between expand knowledge for each sense and the feature words set. Then we compute weighted summation, and choose the sense with highest values as the right sense of the ambiguous word.

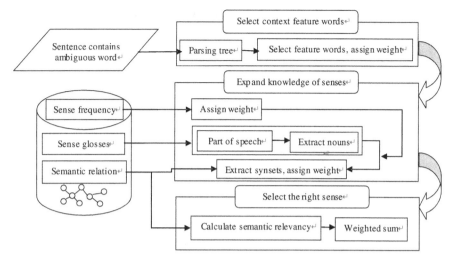

Fig. 1 The framework of the SKW-WSD method

3.1 Selecting the Context Feature Words

The window-based method may induce irrelevant noise words and miss some long-distance relevant words. The dependency-based method may be confused with the paucity of feature words. In this paper, our method adopts the method based on syntactic parsing [19] in feature words selection, which has overcome the shortcomings of the existing methods.

This method employs sentence by sentence disambiguation. First, every sentence is expressed as a phrase structure tree. Then, according to the hierarchy structure of the tree, from the leaf node of ambiguous word to the root node, the adjacent words on the tree are collected layer by layer as feature words. According to relative position between the feature word and ambiguous word, which includes hierarchical relation and path distance on the tree, WSD weight of the feature word is assigned according to (1).

$$\text{weight}(f_j) = \frac{1}{l^{\alpha}} \cdot \frac{1}{1 + \beta \log 10d} \tag{1}$$

where, f_j represents j-th feature word, l and d represent layer number and path distance on the tree, α and β are tuning factors, which adjust the effects of l and d.

Then, feature words are sorted on descending order by weight. Top-N feature words are selected to disambiguate the target word. The main idea of this method: the words that have the common ancestor and adjacent on the tree have stronger syntactic and semantic relationships. In our paper, the values of α, β and N are 0.4, 0.2 and 8, respectively, for a detailed description of parameter selection, please refer to the literature [19].

3.2 Expanding the Knowledge of the Word Senses

For annotation of one word, nouns in it play a major role. For example, the first sense of "car" in WordNet is "wheeled motor vehicle; usually propelled by an internal combustion engine; "he needs a car to get to work"". In this annotation, nouns "vehicle", "engine", and "work" play a major role in term of explanation of "car." Moreover, in WordNet, synsets are connected to other synsets via a number of semantic relations, such as synonymy, hyperonymy, hyponymy, meronymy, and so on. These relations express the meaning of the word comprehensively. For example, the gloss of the first sense of "car" is a kind of "auto, automobile, machine, motorcar". "Ambulance" is a kind of "car". "Car" has part: "accelerator, accelerator pedal, gas pedal, gas, throttle, gun." Therefore, we can expand the knowledge of the meaning of word by extracting the noun of gloss and synsets corresponding to semantic relations.

This expansion process is to expand the knowledge for senses of ambiguous word. We extract gloss and sense frequency from WordNet for each sense of ambiguous word, and assign different weight for each sense in accordance with sense frequency. After that, for each sense, we extract nouns from sense gloss as the first part of expand knowledge for the sense. Because noun' performance is better in terms of interpretation relatively. For each sense' synsets, according to corresponding relation (e.g., hyperonymy, hyponymy, Synonyms, Meronymy, etc.) we extract all synsets of the sense' synsets as the second part knowledge for the sense, we assign different weights for those synsets in accordance with the important degree of them.

The algorithm for expanding the knowledge of the meaning of word is as follows:

Input: W--ambiguous word, Type--set of all semantic relations
Output: GL--expand information from Gloss, SL--expand information from
*   semantic relations*
ExtendInformation (W, Type)
{   pos ← GetPOS(W)
*   gL ← Get(pos,W) /*get the glosses of W*/*
*   for i ← 0 to gL.size*
*   {   GL ← ExtractionNouns(Gloss)*
*   for j ← 0 to Type.size*
*   {   type ← Type[j]*
*   sL ← GetRelation(gL[i],type)*
*   SL.add(sL)*
*   } /* end for */*
*   } /* end for */*
*   return (GL, SL)*
}

3.3 Selecting the Right Sense

Definition 1 Sense–Sense semantic relevancy.
$r(S_1, S_2)$ is the semantic relevancy between sense S_1 and sense S_2 (S_1, S_2 are senses of different words). It can be boiled down to the semantic relevancy between the synsets Sy_1 of sense S_1 and the synsets Sy_2 of sense S_2. In WordNet, all synsets build a tree-like hierarchy structure depending on the hyperonymy and hyponymy relationship of synsets. Therefore, we can calculate the similarity by computing the shortest semantic distance, the depth and out-degree of the minimum ancestor between Sy_1 and Sy_2. Then the semantic relevancy of the two senses can be defined as (2).

$$r(S_1, S_2) = \frac{\text{dep}(S)}{\text{dist}(S_1, S_2) + \text{out}(S) + \varepsilon} \qquad (2)$$

where, S_1 and S_2 represent two senses, respectively. Sy_1 is the synsets of sense S_1 and Sy_2 is the synsets of sense S_2. dist(S_1, S_2) is the shortest path length of Sy_1 and Sy_2 in the synsets hierarchy structure. dep(S) and out(S) are the depth and out-degree of the minimum ancestor between Sy_1 and Sy_2. ε is an adjustable parameter, the value of which is 1 in our method.

Definition 2 Sense-Word semantic relevancy.
$R(S, W)$ is the semantic relevancy between sense S and word W. It can be computed by the semantic relevancy between S and every sense of W. Here, we can assign different weights for each sense of W in accordance with its sense frequency. This is because, to a certain degree, sense frequency reflects the usage rate of itself. The semantic relevancy of $R(S, W)$ is defined as (3).

$$R(S, W) = \sum_{i=1}^{|W_i|} v_i * r(S, W_i) \qquad (3)$$

where, S represents a word sense. W_i is the i-th sense of W. i is the number of W' sense. v_i is the weight of W_i.

Definition 3 Sense-WordSet semantic relevancy (WordSet means words set).
$R(S, C)$ is the semantic relevancy between sense S and word set C. Here, we can assign different weights for each word of C in accordance with its physical location. $R(S, C)$ is defined as (4).

$$R(S, C) = \sum_{i=1}^{|C|} V_i * R(S, C_i) \qquad (4)$$

where, S represents a word sense. C_i is the i-th word in C. V_i is the weight of C_i.

Definition 4 Word–Word semantic relevancy.

$R(W_1, W_2)$ is the semantic relevancy between word W_1 and word W_2. It can be computed by the semantic relevancy between every sense of W_1 and W_2. In this chapter, $R(W_1, W_2)$ is the maximum value of the semantic relevancy between every sense of W_1 and W_2.

Definition 5 WordSet–WordSet semantic relevancy.

$R(C_1, C_2)$ is the semantic relevancy between word C_1 and word C_2. For two word sets C_1 and C_2, if C_1 has n words: $C_{11}, C_{12}, \ldots, C_{1n}$; C_2 has m words: $C_{21}, C_{22}, \ldots, C_{2m}$. We require that the semantic relevancy of the C_1 and C_2 is the maximum value of the semantic relevancy of each word pairs.

We can assign different weights for the gloss and the expand knowledge of the sense of ambiguous word in accordance with its effect. Respectively, we compute the semantic relevancy $R(S, C)$ between the sense of ambiguous word and the feature word set, the semantic relevancy $R(GL, C)$ between the nouns set of the sense' gloss and the feature word set, the semantic relevancy $R(SL, C)$ between synsets corresponding to semantic relations of the sense and the feature word set, and then calculate weighted sum. The semantic relevancy of RS(S, C) is defined as (5), and $R(SL, C)$ is defined as (6).

$$RS(S, C) = V(R(S, C) + \gamma * R(GL, C) + \lambda * R(SL, C)) \tag{5}$$

$$R(SL, C) = \sum_{i=1}^{|SL|} R(SL_i, C) \tag{6}$$

where, S represents a sense. C represents the feature words set. SL_i is the i-th sysnets in SL. V is the weight of S. GL is the nouns set of the sense' gloss. SL is synsets set of the sense. γ and λ are tuning factors. Respectively, $R(S, C)$, $R(GL, C)$, and $R(SL, C)$ can be computed by the Sense-Word semantic relevancy, the WordSet–WordSet semantic relevancy and the Sense–Sense semantic relevancy. We choose the sense with highest values as the right sense of the ambiguous word.

4 Experiment

To verify the effectiveness of the proposed method, we adopt the English all words disambiguation task in Senseval-3 as test set. These data sets were manually annotated with the correct senses by human and used for competitions and evaluations of different systems. The test set contains three documents, 349 sentences, 1,969 words needs to be disambiguated. The Senseval-3 English-all words data sets provide the parse tree of text that needs to be disambiguated. Our method uses the part of speech tagging software developed by the Stanford University.

Table 1 The result which compared SKW-WSD method with AALesk

Algorithms	Precision (%)			
	Nouns	Verbs	Adjectives	Average
AALesk	33.3	26.2	47.5	33.4
SKW-WSD	42.5	27.9	51.3	43.5

In the proposed disambiguation model, there are parameters γ and λ. Therefore, we must find suitable value for them, and then evaluate its performance. In our paper, the values of γ, λ are 0.3, 0.4 respectively, We will further study and discuss the parameter optimization in the further work.

We also compared our method with the AALesk WSD methods which is the typical WSD method only based on gloss and semantic relations in WordNet. The result is shown in Table 1, which shows that our approach achieved a good result in precision.

5 Conclusion and Future Work

The SKW-WSD simplifies the traditional corpus-based WSD algorithms, which need to generate a classifier or classification rule by machine learning methods. In addition, the result of experiment also shows that extending the knowledge related to the ambiguous word by semantic knowledge of the word senses in the WordNet can help to identify the correct meaning. We only use WordNet as our data sources. In the future, we are interested in ① expressing word and context in a more effective way, and disambiguating a word by not only current knowledge but also past knowledge; ② improving the method of feature selection and parameter optimization.

Acknowledgment This work is supported by the National Natural Science Foundation of China under Grant (60973040), the National Natural Science Foundation of China under Grant (60903098), the basic scientific research foundation for the interdisciplinary research and innovation project of Jilin University under Grant (201103129) and the Science Foundation for China Post doctor under Grant (2012M510879).

References

1. Information on http://wordnet.princeton.edu
2. Budanitsky A, Hirst G (2006) Evaluating wordnet-based measures of lexical semantic relatedness. Comput Linguist 32(1):13–47
3. Azzini A et al (2009) Evolving neural word sense disambiguation classifiers with a letter-count distributed encoding. In: Artificial life and evolutionary computation, pp 111–120
4. Nguyen KH, Ock CY (2010) Margin perception for word sense disambiguation. In: SoICT, ACM, pp 64–70

5. Veenstra J, Van den Bosch A, Buchholz S et al (2000) Memory-based word sense disambiguation. Comput Humanit 34(1–2):171–177
6. Lee YK, Ng HT, Chia TK (2004) Supervised word sense disambiguation with support vector machines and multiple knowledge sources. In: Senseval-3: third international workshop on the evaluation of systems for the semantic analysis of text, pp 137–140
7. Navigli R, Lapata M (2010) An experimental study of graph connectivity for unsupervised word sense disambiguation. IEEE Trans Pattern Anal Mach Intell 32(4):678–692
8. Hessami E, Mahmoudi F, Jadidinejad AH (2011) Unsupervised weighted graph or word sense disambiguation. In: World congress on information and communication technologies (WICI), IEEE, 2011, pp 733–737
9. Yuret D, Yatbaz MA (2010) The noisy channel model for unsupervised word sense disambiguation. Comput Linguist 36(1):111–127
10. Chen P, Ding W, Choly M, Bowes C (2012) Word sense disambiguation with automatically acquired knowledge. IEEE Intell Syst 27(4):46–55
11. Lefever E, Hoste V, De Cock M (2013) Five languages are better than one: an attempt to bypass the data acquisition bottleneck for wsd. In: Computational linguistics and intelligent text processing, Springer, Berlin, pp 343–354
12. Hwang MG, Choi C, Kim PK (2011) Automatic enrichment of semantic relation network and its application to word sense disambiguation. IEEE Trans Knowl Data Eng 23(6):845–858
13. Lee WJ, Mit E (2011) Word sense disambiguation by using domain knowledge. In: IEEE 2011 international conference on (STAIR), pp 237–242
14. Huang H, Lu W (2011) Knowledge-based word sense disambiguation with feature words based on dependency relation and syntax tree. Int J Adv Comput Technol 3(8):73–81
15. Sinha R, Mihalcea R (2007) Unsupervised graph-based word sense disambiguation using measures of word semantic similarity. In: Proceedings of the IEEE international conference on semantic computing, pp 363–369
16. Lesk M (1986) Automatic sense disambiguation using machine readable dictionaries: how to tell a pine cone from an ice cream cone. In: Proceedings of the 5th annual international conference on systems documentation. ACM, pp 24–26
17. Banerjee S (2002) Adapting the Lesk algorithm for word sense disambiguation to WordNet. University of Minnesota
18. Chen YQ, Yin J (2005) Sense rank AALesk: a semantic solution for word sense disambiguation. In: Fuzzy systems and knowledge discovery. LNAI 3614. Springer, Heidelberg, pp 710–717
19. Wenpeng L, Huang H, Zhu C (2012) Feature words selection for knowledge-based word sense disambiguation with syntactic parsing. Przeglad Elektrotechniczny 88(1b):82–87

A Knowledge and Employee Evaluation Method for Knowledge Management System

Shikai Jing, Xiangqian Li, Haicheng Yang and Jingtao Zhou

Abstract Aiming at the problem of the low utilization rate of knowledge, low rate of staff dependence on knowledge management system, and the problem of lack of effective evaluation system and method, a comprehensive evaluation method of enterprise knowledge and employees was proposed. The knowledge evaluation system of multilevel and index was established based on the characteristics of knowledge resources and employee behaviors in knowledge management system. And then a knowledge evaluation model was constructed. The knowledge utility was used to represent the total index of knowledge. The method divided knowledge utility into knowledge quality and knowledge influence. The individual total score was divided into participation and contribution used to represent the total index of staff. According to this model, evaluation algorithm of each index was put forward. Finally, an example test was used to demonstrate the effectiveness and feasibility of the proposed method.

Keywords Knowledge management · Knowledge evaluation · Evaluation model · Evaluative index

1 Introduction

In the knowledge economy times, knowledge is becoming more and more important resources of enterprise. Knowledge has become the competitive advantage, and intangible assets. Although the rapid development of science and

S. Jing (✉) · X. Li · H. Yang
School of Mechanical Engineering, Beijing Institute of Technology, Beijing 100081, China
e-mail: jingshikai@bit.edu.cn

J. Zhou
School of Mechanical Engineering, Northwestern Polytechnic University, Xian 710068, China

Z. Wen and T. Li (eds.), *Knowledge Engineering and Management,*
Advances in Intelligent Systems and Computing 278,
DOI: 10.1007/978-3-642-54930-4_16, © Springer-Verlag Berlin Heidelberg 2014

technology information makes the knowledge acquisition more convenient, it also brings a series of problems: knowledge explosion, passing off fish eyes for pearls, low rate of knowledge sharing, and so on [1]. Therefore, many enterprises and software companies have developed knowledge management system.

However, the current knowledge management system is lack of effective evaluation system and evaluation method, making it difficult to build knowledge management standardization system. In addition, employees are reluctant to knowledge sharing because it is lack of corresponding incentive mechanism, and depends on the knowledge management system rarely. Therefore, scholars at home and abroad have carried out the research of knowledge contribution and evaluation field. Yang and Lai [2] analyzed the user's knowledge behavior in the virtual communists, pointed out that if users can get more benefits through sharing knowledge, then it's able to improve the user's motivation for knowledge sharing. Yu et al. [3] analyzed the user's motivation for participating in the contribution in Problem Solving Virtual Communities, and then put forward a motivation model. Wu [4] proposed staff knowledge structure and knowledge contribution index system based on balanced scorecard. Zhang and Liu [5] proposed a kind of assessment method and rewards and punishment scheme, which registers employees' knowledge contribution. Le et al. [6] build the appraisal model of the staff's knowledge contribution based on Web2.0 technology, which divided the knowledge contribution into knowledge participation, utility and innovation contribution three indicators.

In summary, the studies of knowledge evaluation are mainly focused on the virtual community user contributed knowledge analysis of influencing factors. However, for the knowledge management system, the current lack of appropriate knowledge evaluation method. Therefore, according to the characteristics of knowledge resources and staff behavior, combined with knowledge management system features, the evaluation index for knowledge and staff was proposed based on the existing research results. Evaluation system and index evaluation method was presented to improve the application effect of knowledge management system.

2 The Evaluation Model of Knowledge and Employee

In the knowledge management system, the total score of knowledge resources reflects employees' personal capacity, an employee's personal ability also affect the accuracy of its knowledge evaluation. In the professional field, the stronger an employee's ability is, the more his accuracy of knowledge will be. Therefore, different employees with different weights evaluation, and the knowledge resources' total score were got by knowledge evaluation. Knowledge resources total score will influence employee's score. Employees' score will affect their weights.

The knowledge and employee evaluation system model was established. There are three levels of evaluation index, as shown in Fig. 1. Take the utility of knowledge as the total index of knowledge, the employee's total score as the total

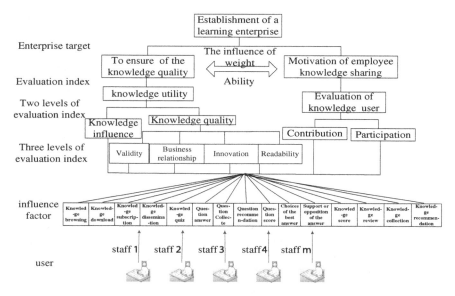

Fig. 1 The model of evaluation system

index of the staff. Then the utility of knowledge is divided into two indicators, they are knowledge quality and knowledge influence. Contain effectiveness, relevance, innovation, and readability is the tertiary indicators of knowledge quality. The employee's total score includes two indexes, they are contribution and participation.

3 The Evaluation Index and Influencing Factors

3.1 Knowledge Evaluation Indexes

As a total index of knowledge evaluation indexes, knowledge utility including knowledge quality and knowledge influence two secondary indexes. Knowledge quality is used to measure the value of knowledge itself. Knowledge influence is used to statistics the degree that knowledge influence on the knowledge users of the knowledge, and it is used to reflect the instructiveness of knowledge.

(1) knowledge quality

Knowledge quality is a general comprehensive evaluation index. In order to promote the fusion of knowledge evaluation and system, knowledge quality indicator can be further subdivided into validity, correlation, innovative, and readability. Knowledge quality is evaluated from the initiative scores that users give to the indexes.

Due to the comprehensive and complexity of engineering knowledge, some knowledge have relatively low commonality, but show a great influence in that situation, therefore, one of evaluation index is the effectiveness of engineering knowledge. Enterprise correlation is used to reflect the relevance of the knowledge resources and the application enterprise. In order to improve the quality of knowledge, it regards innovation as the third measure of knowledge quality, which is used to evaluate the innovation of knowledge resources and encourage employees for knowledge innovation. Readability will be regarded as the fourth index to reflect the knowledge quality.

(2) Knowledge influence

Influence will serve as other secondary indicators, which reflects the enlightening of knowledge to others and it is a utility for knowledge of the interpretation of a different angle. So knowledge influence index mainly reflect in the degree that knowledge effect on the management system. Intellectual influence factors mainly include employees' behaviors to the practice of knowledge in the system, mainly includes: knowledge collection, knowledge recommendation, knowledge download, knowledge score, knowledge reviews, knowledge comments, pros and cons.

3.2 Staff Evaluation Indexes

Staff evaluation indicators include user's total evaluation, participation, and contribution. User's total evaluation consists of the participation and contribution.

(1) Participation

Employee's knowledge behaviors (upload, download, comment, etc.) reflect the employee's participation. The frequency of knowledge behavior embodies the discretion of the employee participation in knowledge management system. At the same time, some system daily operation behavior has nothing to do with contribution, but it also embodies the user's participation, such as login system. The frequency of login system reflects the user's participation.

So this system set the influence factors of employee participation includes login system, subscription, release of knowledge collection, knowledge question and answer, question, recommend, support of the answer, answer opposition, selecting the optimal answer knowledge, knowledge scores and comments, knowledge collection, knowledge recommend, etc.

(2) Contribution

Contribution is used to measure the degree of an employee's contribution to the system, and it mainly includes three parts. The first part is the employee's knowledge score, the score's high or low reflects the contribution of employees. The second part is the employee's knowledge score of the knowledge question. The third part is the employees to participate in other users posted the contribution to knowledge or knowledge questions and answers. For example, in the behaviors such as knowledge reviews and knowledge interlocution embody the employees' contribution.

4 Evaluation Methods for Knowledge and Employee

For different evaluation index, both employees' active evaluation and system's automatically evaluation were proposed in this paper. Active evaluation is used to evaluate knowledge validity, business relevance, knowledge innovation, and knowledge readability since they are more subjectivity, while the indexes of knowledge influence, participation, and contribution are evaluated by automatically evaluation.

First, employees browsed the knowledge items and scored for the indexes. During the process, users' operations were recorded automatically. Then statistic employees' knowledge behavior to compute scores for indexes (include the indexes of knowledge influence, participation and contribution) by evaluate algorithm and evaluate rules. Finally, the final score of knowledge and employee was obtained through integrated evaluation method.

4.1 Active Evaluation

(1) Knowledge quality

Detailed grade rules could be given to cripple the subjective of active evaluation and improve the objectivity of knowledge evaluation. Five grade criteria are given for the four indexes (knowledge validity, business relevance, knowledge innovation, and knowledge readability), and knowledge quality is gained by

$$X_{1i} = \lambda_1 * X_{i1} + \lambda_2 * X_{i2} + \lambda_3 * X_{i3} + \lambda_4 * X_{i4} \tag{1}$$

where X_{1i} refers to the knowledge quality score of the knowledge item, $\lambda_1 + \lambda_2 + \lambda_3 + \lambda_4 = 1$, λ_1, λ_2, λ_3, and λ_4 refer to the weight of the index of knowledge validity, business relevance, knowledge innovation, and readability, respectively. The weight is specified and mirrors the enterprise's different attention for different indexes. X_{i1}, X_{i2}, X_{i3}, X_{i4}, respectively refer to the score of knowledge validity, business relevance, innovation, and readability graded. X_{i1}, X_{i2}, X_{i3}, $X_{i4} \in (1, 2, 3, 4, 5)$.

To distinct employees' capacity, different grade weight were given to different job levels. Generally speaking, users with higher knowledge level tend to higher evaluation weight, which means higher influence for the score of the knowledge. Both the Score and weight can be set according to the enterprise situation.

Knowledge quality score can be computed as follows through the comprehensive consideration of knowledge user scoring and scorer weight.

$$Y_{KQi} = \frac{\sum_{i=1}^{n} p_i \times X_{1i}}{\sum_{i-1}^{n} p_i} \tag{2}$$

where Y_{KQi} is the ith knowledge quality score of knowledge item. X_{1i} is the knowledge quality score of calculated, $X_{1i} \in (0, 5]$. p_i is the weight corresponding the evaluator' level.

4.2 Automatic Evaluation

(1) Knowledge influence

Among the affecting factors in the indexes of knowledge items influence, the factors like knowledge recommendation, knowledge comment, knowledge download, support, and opposition of comments reflect the influence of knowledge items directly. However, most of these behaviors are independent and the case that all or most of these behaviors appear is not necessary. Therefore, the scoring rules which are made for these affecting factors are independent and noninteraction.

Similar to actively grading, in the process of automatic grading, the weight of valuators should be taken into account for every affecting factor. Knowledge influence can be computed by

$$Y_{KEi} = \sum_{i=1}^{i=n1} a1 * P_i + \sum_{i=1}^{i=n2} a2 * P_i + \sum_{i=1}^{i=n3} a3 * P_i + \sum_{i=1}^{i=n4} b * N_i * P_i$$
$$+ \sum_{i=1}^{i=n5} c * P_i + \sum_{i=1}^{i=n6} d * P_i + \sum_{i=1}^{i=n7} j * P_i + \sum_{i=1}^{i=n8} k * P_i \tag{3}$$

where Y_{KEi} represents the score of ith knowledge item, $n1$ represents the number of users who mark it and grade this knowledge higher than three points. $n2$ represents the number of users who mark it but do not grade or grade this knowledge lower than three points. $n3$ represents the number of users who do not mark it but grade this knowledge higher than three points. $n4, n5, n6, n7, n8$ represent the numbers of users of knowledge recommendation, knowledge comment, knowledge downloading, pros and cons of comment. Ni represents the recommendation number.

(2) Participation

According to the importance of participators' system behaviors, every system behavior was given a specific score. The limit of a day should be set to control the increasing of users' degree of participation. Participation can be calculate by

$$Y_{PCi} = \begin{cases} \sum_{i=1}^{i=n} i * X_i, X_i \le 10 \\ \sum_{i=11}^{i=n} i * 10, X_i > 10 \end{cases} \quad (4)$$

where Y_{PCi} represents the ith participation degree. X_i is the number of knowledge behavior.

(3) Contribution

The contribution degree of knowledge users is made up of the statistics score of knowledge item score, knowledge Q&A score and knowledge behaviors of other users. Contribution can be calculated by

$$Y_{PGi} = \sum_{i=1}^{i=n_1} G_{1i} * P_i + \sum_{i=1}^{i=n_2} G_{2i} * P_i + \sum_{i=1}^{i=n_3} \left[\frac{G_{3i}}{10}\right] * P_i + P_i * G_{4i}$$
$$+ \sum_{i=1}^{i=n_5} \left[\frac{G_{5i}}{10}\right] * P_i \quad (5)$$

where G_{1i} is the score of knowledge i which is published by the knowledge user, $n1$ is the number of knowledge items published by the knowledge user. G_{2i} is the score of knowledge question, $n2$ is the total question number of the knowledge user. G_{3i} is the number of pros of knowledge Q&A, $n3$ is the number of comments. G_{4i} is the number of answers selected best among the answers published by the knowledge user, $n3$ is the number of comments. G_{5i} is the number of pros of comment i published by the knowledge user, $n5$ is the number of comments.

4.3 Comprehensive Evaluation of Knowledge and User

(1) Knowledge comprehensive score

Knowledge efficiency index includes two aspects of knowledge quality and knowledge influence, but they cannot obtain the knowledge benefit index through the simple stack processing. And the accuracy of knowledge quality evaluation will rise along with scoring number of the knowledge. In order to ensure that the influence of knowledge quality score will not be covered by the influence of knowledge influence index, a knowledge utility index was processed. Knowledge utility score can be computed as follows:

$$Y_{ki} = \begin{cases} (Y_{kqi} + 3) * Y_{KEi}, & R \ge 1000 \text{ and } P \ge 3 \\ (Y_{kqi} + 1) * Y_{KEi}, & R < 1000 \text{ and } P \ge 3 \\ Y_{kqi} * Y_{KEi}, & R < 1000 \text{ and } P < 3 \\ \frac{1}{2} * Y_{kqi} * Y_{KEi}, & R \ge 1000 \text{ and } P < 3 \end{cases} \quad (6)$$

where Y_{Ki} is the total score of ith knowledge item, Y_{KQi} is the knowledge quality score of ith knowledge item, Y_{KEi} is the knowledge influence scores of ith knowledge item. R is the scores number, P is scores.

(2) Employee comprehensive score

The total score of the knowledge users will be composed of the user participate and user contribution comprehensive score. According to the situation of enterprise, they can set the weight of two levels for two indexes to show their difference and attention. $\gamma 1$, $\gamma 2$ represent the weight of the enterprise for user participation and user contribution of knowledge was conferred. The individual total score can be calculated by

$$Y_2 = \gamma_1 * Y_{PCi} + \gamma_2 * Y_{PGi} \tag{7}$$

where

$$\gamma_1 + \gamma_2 = 1$$

5 Application Analysis

Based on knowledge evaluation system proposed in this paper, we have developed the evaluation module, aimed at the knowledge management system, in which users can rate each indicator according to the rules.

Take the knowledge Items as example. System monitored and recorded seven different users' behavior for three knowledge items within a day. The Table 1 displayed seven users' behavior for three knowledge items numbered from 001 to 007. The knowledge item A and C were published by 001, while the knowledge item B was published by 007.

The Table 2 showed knowledge utility score of knowledge A is the highest. Mainly due to weights of participate in knowledge A for evaluation workers is higher and give a higher score. In addition, employees participate in more knowledge behavior of knowledge A and it obtains to higher knowledge influence score. So the knowledge utility score of knowledge A is higher. However, participate in the knowledge B for evaluation with staff for higher weight with lower scores, and participation in behavior of the knowledge B is less. So the knowledge utility score of knowledge B is lower.

The Table 3 showed the record statistical results for knowledge behavior of 7 employees.

According to the statistics on the knowledge behavior of seven knowledge users and the scoring rules about knowledge behavior mentioned above, the indexes of knowledge users can be computed to realize the evaluation of knowledge users. The scoring results of staff 001 and staff 007 are listed in Table 4. As the

Table 1 The related knowledge behavior of knowledge A, B, and C

Knowledge behavior	User (weight)																				
	001(0.8)			002(0.4)			003(0.1)			004(0.4)			005(0.6)			006(0.1)			007(0.2)		
	K(A)	K(B)	K(C)	K(A)	K(B)	K(C)	K(A)	K(B)	K(C)	K(A)	K(B)	K(C)	K(A)	K(B)	K(C)	K(A)	K(B)	K(C)	K(A)	K(B)	K(C)
Effectiveness score (0.3)	4	2	4	5	1	5	1	4	4	4	2	2	4	3	0	5	5	5	0	0	0
Business relationship (0.3)	5	1	33	5	2	4	2	5	5	3	3	3	5	2	0	5	3	3	0	0	0
Innovative score (0.3)	5	2	33	3	3	3	2	2	3	3	3	3	4	3	0	5	2	2	0	0	0
Readability score (0.1)	4	2	44	3	2	4	2	2	4	5	5	2	3	1	0	5	5	5	0	0	0
Knowledge collection	1	0	00	1	0	1	0	0	0	1	1	1	1	0	1	0	1	1	0	0	1
Knowledge recommendation number	5	0	33	3	0	0	0	0	2	4	1	1	5	0	0	0	2	2	0	0	0
Knowledge review	1	1	1	0	0	0	0	1	1	1	1	1	0	0	0	0	0	0	0	1	0
Knowledge score	1	1	11	1	1	1	1	1	1	1	1	1	1	1	0	1	1	1	0	0	0
Knowledge download	1	0	00	0	1	0	0	0	0	0	1	0	0	0	0	0	1	1	0	0	0
Support of review	0	0	11	1	1	1	1	1	1	1	0	1	1	1	0	1	1	1	0	0	0
Opposition of review	1	1	00	0	0	0	0	0	0	0	0	0	0	0	0	0	0	0	1	0	0
The publisher of review	005	007	003	001	001	001	001	001	001	001	001	0	001	001	0	001	00	001	005	0	0

Table 2 Scores table of knowledge

Knowledge item	Knowledge quality	Knowledge influence	Knowledge utility
Knowledge A	4.15	24.4	124.44
Knowledge B	2.3	8.7	20.01
Knowledge C	3.39	15.9	69.801

Table 3 Behavior and participation scores of knowledge user

Knowledge behavior	User						
	001	002	003	004	005	006	007
Effectiveness score (1.0)	12	12	12	12	8	12	0
Knowledge collection (2.0)	1	2	0	3	2	2	2
Knowledge recommendation (1.00)	8	3	2	6	0	9	0
Knowledge review (4.00)	3	0	1	3	0	0	2
Knowledge download (3.00)	1	3	0	1	0	0	0
Support of review (1.00)	1	3	1	2	2	3	0
Opposition of review (1.00)	2	0	0	0	0	0	1
Participation scores	40	28	10	41	14	28	13

Table 4 Contribution and total score of employee

Staff	Source				
	Knowledge	Knowledge question	Other intellectual behavior	Contribution	User score
001	194.24	0	10	204.24	156.97
007	20.01	0	0	20.01	14.01

knowledge score, knowledge quiz score, and intellectual behavior score of other staffs are 0, the contribution score of other staves is 0.

The contribution score of knowledge user 001 is higher than 007. As knowledge user 001 published two knowledge items which gain higher knowledge quality evaluation. But knowledge user 007 releases lower indexes of knowledge items. So according to the knowledge contribution, user 001 releases more amount of knowledge, better quality of knowledge and higher degree of contribution. Besides, as user 001 participates actively in knowledge items released by other knowledge users, he gains 14 review supports as knowledge evaluation promulgator and gains 10 scores of contribution in knowledge items released by other knowledge users. Although user 007 releases knowledge evaluation too, he doesn't gain supports. So user 007 scores 0 in knowledge contribution items released by other knowledge users. This paper sets the weights of contribution degree and participation degree as 0.7 and 0.3, respectively. Above all, staff 001 gains higher scores than staff 007 as a whole which is the same as the evaluation result.

6 Conclusions

Reasonable evaluation methods and incentive mechanisms need to be constructed to encourage employees to make a positive contribution to the knowledge management activities of enterprises. A multilevel and multi-index knowledge evaluation system and a knowledge evaluation model were established based on the characteristics of knowledge resources and the behavior of employees in knowledge management system. Then the evaluation methods were proposed to include the active evaluation and the automatic evaluation aimed at different indexes. An example test was used to demonstrate the effectiveness and feasibility of the proposed method. Enterprises can set different evaluation index and weight according to different situations. This paper is still in the phase of theoretical research and system development. It need to improve evaluation index, algorithm efficiency in future.

Acknowledgment This paper is supported by International Cooperation Ministry of Science and Major Project (2011DFB10090), National Natural Science Foundation of China through approval (61104169) and Specialized Research Fund for the Doctoral Program of Higher Education through approval (2010610212002).

References

1. Li X, Yang H, Jing S et al (2012) Knowledge service modeling approach for group enterprise cloud manufacturing. Comput Integr Manuf Syst 18(8):1869–1880
2. Yang H, Lai C (2010) Knowledge-sharing dilemmas in virtual communities: the impact of anonymity [E B/ OL] (2008-07-11). http://academic-papers.org/ocs2/session/Papers/E5/484-2120-1-DR.doc. Accessed on 03 Mar 2010
3. Yu J, Jiang Z, Chan HC (2007) Knowledge contribution in problem solving virtual communities: the mediating role of individual motivations. In: Proceedings of the 2007 ACM SIGMIS CPR conference on computer personnel research. ACM, New York, pp 144–152
4. Wu J (2006) The research on employee knowledge contribution measurement. Tongji University, Shanghai
5. Zhang J, Liu Z (2004) Knowledge contribution inspiriting mechanism for knowledge management. J Tongji Univ Nat Sci 32(7):966–970
6. Le C, Xu F, Gu X, Pan K et al (2011) Evaluation model and algorithm for knowledge contribution of enterprise staff. Comput Integr Manuf Syst 17(3):662–671

Research on Knowledge Service for Product Lifecycle

Xiangqian Li, Shikai Jing and Jingtao Zhou

Abstract Aiming at the problem of the low utilization rate of knowledge in product lifecycle, and the problem of the knowledge that cannot be take full use to serve the business process, a construction scheme of the knowledge service oriented to product lifecycle was proposed. Firstly, the demands of knowledge service oriented to product lifecycle were analyzed and the basic concepts and characteristics of knowledge service were put forward. Secondly, the model of knowledge service and its structure was built; the integration of the system, the acquisition of the multi-source heterogeneous knowledge, and the technologies of knowledge push have been analyzed. Finally, part of the system function was given to illustrate the feasibility of the proposed method.

Keywords Knowledge service · Product lifecycle · Knowledge service platform

1 Introduction

With the coming of knowledge economy, knowledge is regarded as a kind of intelligence resource. How to make use of knowledge to provide services to human production has become a research hotspot. Knowledge service is a service based on knowledge. In recent years, knowledge service has been widely applied in libraries, consultation, tourism, finance, and many other areas. However, the research of knowledge service in manufacturing is just at the beginning. There is not a uniform definition about the concepts of knowledge services [1].

X. Li · S. Jing (✉)
School of Mechanical Engineering, Beijing Institute of Technology, Beijing, China
e-mail: jingshikai@bit.edu.cn

J. Zhou
School of Mechanical Engineering, Northwestern Polytechnic University, Xi'an, China

Z. Wen and T. Li (eds.), *Knowledge Engineering and Management*,
Advances in Intelligent Systems and Computing 278,
DOI: 10.1007/978-3-642-54930-4_17, © Springer-Verlag Berlin Heidelberg 2014

At present, it is generally believed that the concept of knowledge service was first put forward by Ren Junwei in his paper "Knowledge economy and knowledge service of library" published in 1999. This paper lead knowledge service into library and information science [2]. The reference [3] believes that knowledge service system for virtual product development is a kind of intelligent design assistant; it captures and analyzes the designer's behavior in the designing process, and provide proactive knowledge service in time. The reference [4] points out that knowledge service changes the knowledge assets into knowledge products and services, which utilizes the Internet to sale and promote the knowledge production and service. The reference [5] describes knowledge service as a package in knowledge management activities, which can achieve knowledge processing function. The reference [6] holds the point that knowledge service is a commercial method of mining knowledge assets, which can help enterprises to enhance competitive advantage and expand market coverage.

Although the researches on knowledge management and knowledge service have made great achievements, the research of knowledge service oriented to product lifecycle is relatively less. The problem of how to realize the combination of knowledge and business process has not been solved well. At present, people pay more attention to the management and application of the lifecycle knowledge. This trend proposed a higher requirement to knowledge management.

Therefore, a kind of knowledge service technology of product lifecycle was studied in this paper, and a knowledge service platform was constructed to support business activities in product lifecycle. It is very significant to improve the employee's working efficiency and enhance the core competitiveness of enterprises.

2 Demand Analysis of Lifecycle Knowledge Service

In the constant depth fusion process of industrialization and information, many enterprises have accumulated more and more knowledge. However, how to obtain and reuse the whole lifecycle knowledge resources is an urgent problem that need to be solved. At present, knowledge management and knowledge application exit the following problems [7, 8].

(1) Cannot Store the Innovation Knowledge

Innovation knowledge of enterprises are mostly contained in the engineering drawings, 3D models, and the design schemes, but the function of current CAD systems usually can only support drawing or model of detailed design, and record the results knowledge. It cannot record and store the knowledge of design basis and design thinking. A large amount of design intents and design process information are ignored or lost, which bring great difficulty to design knowledge reuse.

(2) Cannot Find Knowledge in Time

With the constant development and expansion of enterprises, enterprises will accumulate product information data, which are generally in a structured, semi-structured, and unstructured stored in different systems. Research illustrates that approximately 70 % of time will be spent on the literature review for new product design, which seriously affects the design efficiency. Therefore, in order to improve work efficiency, it is urgent need to provide knowledge resources in the form of service.

(3) Low Satisfaction of Knowledge Retrieval

At present, knowledge retrieval is a common method of knowledge acquisition in the knowledge management system. Along with expansion of the knowledge sources and individual demand change, the single way of knowledge retrieval will be difficult to meet the needs of users. So personalized knowledge services and the other service models needed to be put forward to improve user satisfaction.

(4) Low Rate of Knowledge Reuse

Although the PLM system has been widely implemented in many enterprises, the management software has not risen to the level of knowledge service, still confined to the traditional classification and retrieval of data storage, resulting in lack of interaction with the business process activities. Therefore, knowledge and product business process need to be combined to increase the knowledge reuse rate.

(5) Knowledge is Independent of Each Other

Generally speaking, designers not only want to obtain design knowledge in the research stage, but also need to obtain the other process knowledge, such as process knowledge, and maintenance process. Sales and maintenance personnel also need to master the knowledge of product's design stage. So, employees want to get the knowledge resources of product life cycle at any time to raise their working efficiency. At present, however, knowledge stored separately in different systems in the product lifecycle of business activities, independent of each other, lacks system organization, management, and application.

In conclusion, it is significant and urgent to establish the knowledge service model of the whole lifecycle through the knowledge in the entire business process, improve knowledge sharing and reuse, and create value for the enterprise.

3 The Concept and Characteristic of the Whole Lifecycle for Knowledge Service

The knowledge service of product lifecycle based on knowledge discovery, acquisition and integration, take the product development process as the service object. An active, real-time, and personalized knowledge service needs to be provided according to the users' environment and requirements. Knowledge will be used for all business process to meet the needs of different users. The characteristics and meaning of product lifecycle knowledge service:

(1) Take the business process of product lifecycle as the service object. The ultimate goal of knowledge service is through knowledge in the entire business processes, support the activities of product lifecycle, and finally to realize knowledge innovation.
(2) Provide personalized service. In order to meet the users' different demands, it is needed to provide the real-time, active, and personalized knowledge resources for users by knowledge mining and knowledge fusion.
(3) The intelligent service. In the knowledge service process, intelligent agent tools need to be developed by the modern information technology, such as manufacturability analysis and thinking navigation to make the whole course more intelligent.
(4) The diversification of knowledge. Knowledge source of the lifecycle in the knowledge service process is more comprehensive, including the structure knowledge, semi-structured, and unstructured knowledge.

4 The Architecture of the Lifecycle Knowledge Service Platform

4.1 Service Model

The knowledge service model of whole lifecycle is shown in Fig. 1. Knowledge source is the basis of knowledge service, and the product lifecycle is the object of knowledge service, effective organization and management of knowledge is guaranteed of the knowledge service.

In the process of knowledge service, firstly, knowledge of the data centers and experts was extracted through the integration service and cooperative service. Then, knowledge was scientific, and reasonably managed by effective organization system and the knowledge engine. Finally, knowledge was timely and intelligently provided to users through the push and innovation service, which combine knowledge and business process.

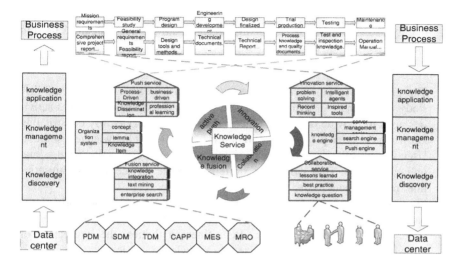

Fig. 1 The knowledge service model of product lifecycle

(1) Knowledge Source

The knowledge comes mainly from two aspects in the product lifecycle. One source is the information systems of enterprise. There are many information systems in the process of information construction, such as PDM, SDM (simulation data management system), and TDM (test data management system). Large amount of structured and unstructured data is stored in these systems. This data is one of the important knowledge sources. Another source is product engineer. They have rich engineering experience, and this hidden knowledge is also an important knowledge source.

(2) Knowledge Discovery

Knowledge service is based on knowledge discovery and acquisition. Aiming at multisource heterogeneous knowledge resources, such as documents knowledge, 3D models knowledge, and design thinking process knowledge, the unified logical data layer was established through the whole enterprise search technology, text mining, 3D annotation, thinking record, and knowledge integration to discovery and acquisition of the multi-source heterogeneous data.

(3) Knowledge Organization

Knowledge organization aims to save and extract the knowledge, which can provide data support for knowledge collection system and knowledge application system. The knowledge organization system was established through the concepts, lemmas, and knowledge items. Knowledge engine consists of management engine, search engine, and push engine. Management engine is responsible for the item lifecycle management. Search engine is mainly used in knowledge retrieval and sort. And the push engine is used in pushing knowledge resources to users.

(4) Knowledge Application

The products development process includes task needs analysis, feasibility study, preliminary design, and detailed design. The different business activities need to provide different content of knowledge service. By constructing knowledge service platform, different users can get their knowledge resources. Users can benefit with real-time recording, storage experience knowledge, and online inquiries. They can also acquire knowledge service through subscription, retrieval, and push.

From what has been discussed earlier, knowledge service network of the entire lifecycle is based on the integration of data center, discovery knowledge, and exquisite knowledge. In business activities, the integration of knowledge and process was realized by knowledge service.

4.2 Service Architecture

Based on the knowledge service model, the knowledge service system architecture of product whole lifecycle was put forward and shown in Fig. 2. The whole structure was divided into data resource layer, basic services layer, tool layer, system application layer, and knowledge presentation layer.

(1) Data Resource Layer

Data resource layer is located at the bottom of the whole system, and mainly includes ontology database, dictionary database, knowledge base, and external database. The knowledge base includes specification, experience, model, and method. External database includes database of external integration system.

(2) Basic Service Layer

Basic services layer on top of the data resource layer provides various support services for knowledge service system. In the layer, it not only meets the various network protocols of the communication needs between system components, but also provides support technology for tool layer, such as knowledge base management, knowledge retrieval engine, tool engine and integration engine.

(3) Tool Layer

This layer provides various application tools, including fusion tools, collaborative tools, push tools, and innovative tools. The fusion tools include knowledge acquisition, modeling, classification, and clustering. Collaborative

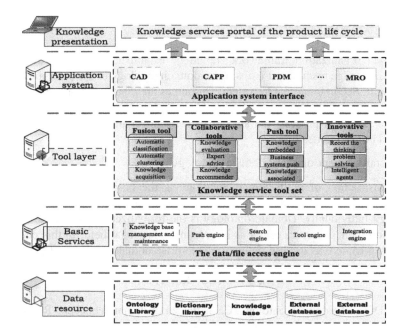

Fig. 2 Knowledge service architecture of product lifecycle

tools mainly include knowledge exchange, knowledge evaluation, and expert consultation. The push tools are used to push the relevant knowledge to personal working table by analyzing task. Knowledge innovation is implemented by applying the intelligent agent of innovation tools. These tools provide lifecycle management from knowledge acquisition, organization and management, and application.

(4) Application System Layer

Application system layer includes software system of the full lifecycle for business activities, such as product design system, manufacturing system, and management system. Knowledge will be through various business systems by system integration.

(5) Knowledge Presentation Layer

Knowledge presentation layer is the system portal, including application client and enterprise entrance. The user interface should be highly personalized and intelligent, and users can get knowledge service based on personal needs.

5 Knowledge Service Methods

A unified organization framework should be established to organized internal static knowledge resources. First, a multi-domain ontology of the enterprises needs to be built. Second, internal knowledge resources are distributed indexed, and then add ontology information to the indexed knowledge. On this basis, by combining with the business process, the system extracts information about users' task to obtain knowledge demand.

(1) Establishment of Multi-domain Ontology

The construction of multi-domain ontology is the basis of knowledge service methods, which make internal knowledge systematic and establish the domain ontology individually. Domain ontology is the formalized expression of the concepts and their semantics relations in certain professional field. In this paper, the following triple was used to define the enterprise domain ontology [9].

$$DO = \ <CS, \ RS, \ H>$$

where DO is the domain ontology, CS is the set of domain concepts, RS is the set of semantics relations between concepts, H is the structure of concepts.

(2) Distributed Index

After the construction of multi-domain ontology of enterprises, the index library should be established by distribution indexing the knowledge base of the subordinate enterprise. By adding ontology information, index knowledge is semantically tagged and they are managed in the knowledge classification system. When the system needs to provide knowledge service, the index management center confirms the URL of index server containing index words, and sends the related requests to the server. The system calculated feedback results from every server, and returned the final search results. The distributed architecture of Map/Reduce and the Lucene can be used to construct the distributed index [10].

For each business system data source, every knowledge resource is segmented semantically in the domain ontology and the system extracted the keywords as the basis of knowledge retrieval. And then, other related attributes and knowledge source address are extracted to form index, which will be put into index library. The index library in the system only manage knowledge resources characteristic attributes and knowledge source address, but not knowledge itself and structure. The index knowledge can be defined using the following method:

$$IK = \ <KA, \ URL>$$

where IK is the index knowledge, KA is the knowledge attributes, URL is the knowledge source address. Especially, KA and URL include:

Fig. 3 Knowledge retrieval service for enterprises

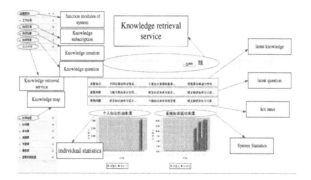

<KA>: {Document number, Name, Date of creation, creator, Profession, Set of keywords}, <URL>: {Affiliated unit, Database}.

(3) Knowledge Service Demands Perception

The knowledge service demand is mainly used for users' demands perception. Based on workflow, when the task is dispatched to the user, the system extracts real-time task attributes such as description, profession, abstraction, and numbers. The structural expression of knowledge service is formed by semantic extension analysis based on ontology, which will be used in the following retrieval process.

6 The Development of Platform Function

In order to verify the effectiveness of the proposed method in this paper, a knowledge service platform based on B/S structure was developed for an enterprise application. Because of the fact that the research content of this paper is in the theory research and system development stage, only a part of knowledge service function has been completed.

(1) Knowledge Retrieval Service

   The knowledge retrieval service for enterprises can be realized based on the integrated system. At present, four business systems (PDM, TDM, SDM, and quality management system) have been integrated in this paper. The process of knowledge retrieval and browse was completed and shown in Fig. 3.

(2) Knowledge Subscription Service

   The knowledge subscription service is used to achieve on time and active service by submitting the subscription application to system. Users need to fill in the subscription subject and subscription time, as shown in Fig. 4, the system will automatically send the result on time to users, and the results are shown in Fig. 5.

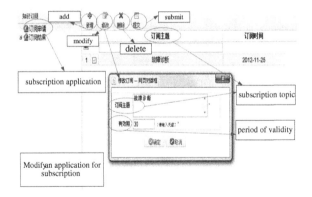

Fig. 4 Knowledge subscription service

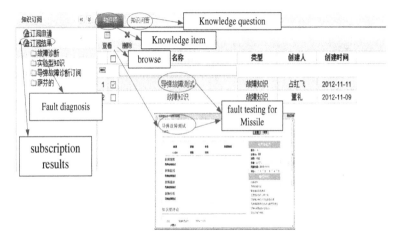

Fig. 5 The results of the subscription

(3) Knowledge Retrieval Service

As shown in Figs. 6 and 7, the knowledge question service is used to the cooperative communication between users. The user can ask and answer questions at any time. The knowledge question service helps to achieve the knowledge conversion from the implicit to the explicit.

Fig. 6 Knowledge question service

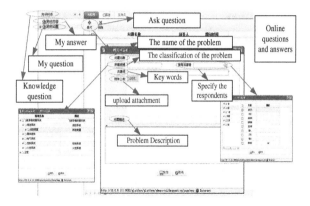

Fig. 7 Browse the problem

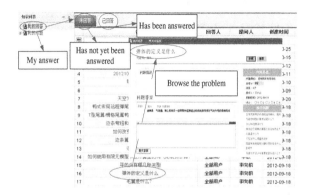

7 Conclusions

Knowledge service concept of the product whole lifecycle was presented in this paper. The idea of knowledge service was introduced into product lifecycle. Knowledge service demands of the product lifecycle were analyzed. The basic concept of knowledge service of the lifecycle was presented, and the knowledge service model construction of the lifecycle was put forward. Some sections of the system function have been given, and the other research contents of this paper are in the theory research and system development stage.

Acknowledgments This paper is supported by International Cooperation Ministry of Science and Major Project (2011DFB10090), National Natural Science Foundation of China through approval (61104169), and Specialized Research Fund for the Doctoral Program of Higher Education through approval (2010610212002).

References

1. Sheng W (2009) Research on key technologies of knowledge service for product design process. Zhejiang University, Zhejiang
2. Yang T, Wang Y, Xiao T et al (2002) Research on personalized active design knowledge service systems. Comput Integr Manuf Syst 8(2):950–959
3. Dong Y (2003) Research on the mechanism for knowledge services. Chinese Academy of Sciences, Beijing
4. Wang J, Pan X, Liu L (2006) Research on agile knowledge reuse based on multi-Agent and knowledge service. Comput Integr Manuf Syst 12(6):840–846
5. Mentzas G, Kafentzis K et al (2007) Knowledge services on the semantic web. Commun ACM 50(10):53–58
6. Li X (2008) Research on several problems for platform construction of knowledge service. Northeastern University, Shenyang
7. Liu Y (2011) Research on key technologies of knowledge providing service for product design process. Zhejiang University, Zhejiang
8. Li X, Yang H, Jing S et al (2012) Knowledge service modeling approach for group enterprise cloud manufacturing. Comput Integr Manuf Syst 18(8):1869–1880
9. Yu X, Liu J, He M (2011) Design knowledge retrieval technology based on domain ontology for complex products. Comput Integr Manuf Syst 17(2):225–231
10. Lu Q (2010) Research and application of distributed index in donghua search engine. Donghua University, Shanghai

Technological Diversification Effect on Business Performance: A Probing into Intermediaries Role of Product Innovation Strategy

Dajun Li, Ling Li and Yong Huang

Abstract There is an intrinsic correlation between technological diversification and enterprise product strategy. In this study, on the basis of product strategy, the impact on innovation performance of technology diversification is investigated and practical data of product innovation is also analyzed by structural equation modeling (SEM) methods. The research found that the intermediary role of product strategy is remarkable in technology diversification impacts on both business performance and innovation performance, especially significant on business performance.

Keywords Technology diversification · Product innovation strategy · Innovation performance

1 Introduction

Technology diversification refers to the technological variety of one enterprise. At present, more and more scholars begin to pay attention to the differences caused by diversified technology of various organizations and business performance because of the increasing importance of technology and innovation for gaining competitive

D. Li (✉)
School of Economics and Management,
Xi'an University of Technology, Xi'an 710054, China
e-mail: yk-ldj@163.com

L. Li
School of Economics and Management, Xi'an University of Science and Technology,
Xi'an 710054, China

Y. Huang
Faculty of Humanity and Foreign Language, Xi'an University of Technology,
Xi'an 710054, China

Z. Wen and T. Li (eds.), *Knowledge Engineering and Management*,
Advances in Intelligent Systems and Computing 278,
DOI: 10.1007/978-3-642-54930-4_18, © Springer-Verlag Berlin Heidelberg 2014

advantage for businesses [1]. The current research focuses on the differences of diversification technological strategy, specialization and variety of corporations, which prefer to the consequences of different alternatives.

Scholars have noted that with economy of scope and knowledge sharing knowledge, technology diversification has the positive effect while with and the additional conflicts and coordination of other aspects, technology diversification display a certain negative impact [2]. These two factors will influence the relationship between diversity of technology and innovation jointly and cause a certain degree of uncertainty, thus not resulting in uniform conclusion of current technological diversity on innovation performance. Moreover, because of the different focus of perspectives and research priorities and the lack of path research of exploring the impact of diversity on firm performance, there are no overall analyses on the relationship among diversification of enterprise technology, product innovation strategy and firm performance. Therefore, this study will combine diversify enterprise technology, product innovation strategy and corporate performance into an unified analytical framework to explore the relationships between diversified technology and innovation and the role of innovative in this process.

2 The Literature Review

Technology diversification means the extent of fundamental technological variety of an enterprise. Diversification of technology may prevent the negative effects of lock-in effect and continuous evolution and enterprise transformation of the company with a particular technology. Breschi pointed out the existence of two alternative technologies which can be applied by an enterprise [1]. One is to take the advantage of market opportunities by the application of diversified technology, and another is to develop new products in particular market minority level through strategic expertise. In the long run, Huang and Chen [2] pointed out that an inverted U-shaped relationship exists between diversity and innovation performance of technology.

In the view of coordination costs, coordination cost perspective, previous studies have found that excessive caused adverse effects in business performance [3]. A high level of diversified technology will increase the coordination costs, and technological innovation costs have increased for more coordination and integration of technical knowledge across a variety of technical disciplines, falling into the trap of over-diversification [4].

Therefore, when the diversity of technology is beyond a certain level, it is likely that the inner treatment cost is high enough, and the economic scope and knowledge spillover benefit from it, resulting in a decline in innovation performance. In the view of innovation activities, technological diversification can provide enterprises with a new technology tracks, which is not only conducive to business exploration and identification of new technological opportunities, but also conducive for enterprises to multi-combine knowledge of different technologies in various fields, lowering the negative effects causing by technology lock-in effect and improving innovation success rate by the means of diversification sharing of research and development combination and reducing the potential risks of each R&D projects. For above reasons, hypothesis 1a and 1b are proposed:

Hypothesis 1a: Enterprise technology diversification has a positive impact on its business performance.

Hypothesis 1b: Enterprise technology diversification has a positive impact on its innovation performance.

Enterprises with a variety of technological capabilities, often consider how to develop new product markets or achieve greater competitiveness in the original product areas by applying their diversity of technical knowledge, namely, product innovation strategy. The development of new products is an integration process combining all kinds of knowledge and technology within the enterprise resource, in essence, a process of integrating knowledge in different fields into a relatively stable product. The technical exploration among several technology areas is the precondition of new products development and production, therefore, technology-based diversified product strategy is considered to be a higher success rate [5]. Gambardella [6] studies have shown that in the industries of electronics, chemical and other technology-intensive industries, many companies were in the accelerated expansion of technological capability system, whereas, at the same time, reducing or stabilizing their operating products ranges. For these reasons, hypothesis 2a and 2b are proposed:

Hypothesis 2a: Technology diversification within enterprises can be helpful for them to adopt relevant various product innovation strategy.

Hypothesis 2b: Technology diversification within enterprises can be helpful for them to adopt platform serialization product innovation strategy.

Technology diversification product strategy enable enterprises to gain more synergistic effect than developing a single product, which can dispersed product development risks, and the more diversified technologies are, the more obvious related economies of scope and economies of scale are, which can improve financial performance and profitability [7]. In addition, with the accelerated shortening of product life cycles and the diversification of customer needs, technology diversified product platform can enhance existing products in the market competitiveness. Companies can greatly shorten new product development cycle, reduce production costs and extend the product range of vitality and market competitiveness by product strategy platform serialization, and improving technical efficiency of resource allocation as well. For reasons above, hypothesis 3a, 3b, 4a and 4b are proposed:

Hypothesis 3a: Diversification product strategy concerned has a positive impact on business performance.

Hypothesis 3b: Diversification product strategy concerned has a positive impact on innovative performance.

Hypothesis 4a: Platform serialized product strategy has a positive impact on business performance.

Hypothesis 4b: Platform serialized product strategy has a positive impact on innovative performance.

Based on above analysis, product innovation strategy seems to play an intermediary role in the influence of technology diversification on business performance: the accumulation of technology diversification can promote enterprises to adopt certain product innovation strategy, and consequently, the implementation of product innovation will lead to changes of business performance and innovation performance. For reasons above, hypothesis 5a, 5b, 6a and 6b are proposed:

Hypothesis 5a: Related diversified product strategy is the mediating variables, by which technology diversification affect business performance, that is, technology diversification affect the business performance by affecting related diversified products strategy.

Hypothesis 5b: Related diversified product strategy is the mediating variables, by which technology diversification affect innovative performance, that is, technology diversification affect the innovative performance by affecting related diversified products strategy.

Hypothesis 6a: Platform serialized product diversification strategy is the mediating variables, by which technology diversification affect business performance, that is, technology diversification affect the business performance by affecting related diversified products strategy.

Hypothesis 6b: Platform serialized product diversification strategy is the mediating variables, by which technology diversification affect innovative performance, that is, technology diversification affect the innovative performance by affecting related diversified products strategy.

3 Research Approaches, Questionnaire Design and Variable Measurement

In this paper, research is conducted based on sampling technology-intensive manufacturing, which included high-tech industries except software industry (aerospace, pharmaceuticals, computer and office equipment, electronics and communications equipment, medical equipment and instrumentation), traditional manufacturing industries requiring large technical investment in product development and production (automotive, electrical machinery, chemical industry, etc.), and by using interviews with senior management in the enterprise, questionnaire design, research methods to expand the assumptions empirical research.

Questions of scale diversification refer to Grover and Goslar [8] and Nieto and Quevedo [9] mainly; questions of scale product innovation strategy refer Meyer [10] and Li and Kwaku [11] etc.; and questions of enterprise performance include questions as follows: 'comparing with domestic competitors, the business innovation performance situations in five years; new product sales revenue of total sales revenue, the average annual number of new products, the average annual number of patent applications, the rate of new product development, with domestic competitors, your company's operating performance during the past five situations: sales growth, ROE, market share growth, the product's overall market competitiveness' and so on.

According to research above, the questionnaire is accomplished combined with interviews with some enterprises, which conduct a small-scale pre-test in up to 40 high-tech zones enterprises selected from Shaanxi, Zhejiang, Jiangsu and other places. Then more than 20 senior managers of companies and some experts are interviewed to consult on the terms of measurement accuracy, reasonableness and terms. According to the feedbacks, the questionnaire is revised to form a formal questionnaire. The official release time for the questionnaires is from early September, 2012 to March, 2013. The author distributed 650 pieces of questionnaires through direct delivery (mail, e-mail, site release) and indirect delivery (commissioned release) and recovered 331 questionnaires finally, in which 58 unqualified pieces were excluded and 191 valid questionnaires were obtained with an effective rate of 42 %.

4 Data Analysis

To analyze overall after merging samples from different sources, analysis of variance on the enterprise samples performance measurement indicators is conducted to test whether there is a significant difference between the means of data collected from two groups of samples, in order to test the effectiveness of the combined sample. It was found that the significance probability values of all indicators Levene statistic are greater than 0.05, which means that the enterprise performance evaluation measure index of samples collected in two approaches have homogeneity of variance. The significance probability values of all indicators F statistic are greater than 0.05, which indicating that there are no significant difference between the enterprise performance evaluation measure index of samples collected in two approaches and can be analyzed after data merger. In addition, the article also apply normality test on collected data, resulting in normal distribution sample data with high quality.

Five factors were extracted to conduct sample factor analysis, in which the corresponding factor load factor were greater than 0.5, and a total of 78.5 % variation were explained. The Cronbach α of each variable is greater than 0.70, the acceptable level in a study, which indicates a high internal consistency. These five factors were named sequentially as technology diversification, serialization product platform strategy, related diversified product strategy, business performance and innovation performance.

According to assumption, structural equation model is proposed. In this structural model, $\chi2/df = 1.695$, RMSEA $= 0.060$, GFI $= 0.880$, AGFI $= 0.844$, NFI $= 0.874$, TLI $= 0.934$, CFI $= 0.943$, IFI $= 0.944$, PGFI $= 0.680$. Table 1 shows the exact path coefficients. In this model, the proposed hypothesis 1a, 1b, 2a, 2b, 3a, 3b, 4a, 4b are supported.

Furthermore, we use structural equation modeling (SEM) to test the hypothesis about several mediating effect. Firstly, the initial inspection on overall effect of intermediaries structural equation model is conducted and the model fitting

Table 1 The path coefficients of setting model

Active route	Standardized path coefficients	S.E.	C.R.	Hypothesis	Test results
Business Performance ← technology diversification	0.547***	0.133	7.256	1a	Support
Innovation performance ← technology diversification	0.667***	0.151	5.935	1b	Support
Related diversified product strategy ← technology diversification	0.657***	0.139	6.325	2a	Support
Platform serialization product strategy ← technology diversification	0.595***	0.108	5.923	2b	Support
Business performance ← related diversified product strategy	0.403**	0.110	4.824	3a	Support
Business performance ← platform serialization product strategy	0.348**	0.124	4.242	4a	Support
Innovation performance ← related diversified product strategy	0.438**	0.092	5.306	3b	Support
Innovation performance ← platform serialization product strategy	0.367**	0.103	4.572	4b	Support

indicator of the initial model are $\chi2/df = 2.601$, RMSEA $= 0.099$, GFI $= 0.885$, AGFI $= 0.849$, NFI $= 0.890$, TLI $= 0.849$, CFI $= 0.857$, IFI $= 0.858$, PGFI $= 0.606$ separately. The key indicators of CFI and IFI did not reach the reference standard (>0.90), indicating that the initial model and sample data is also not fit well and need amending. Through residual analysis, it showed abnormal residuals concerning technological diversification and business performance, which means the emergence of negative residuals -0.62. Therefore, this path should be deleted to make this model more fitting with sample data. After re-examined with the fitting degree test of amended model, the fitting index are $\chi2/df = 1.259$, RMSEA $= 0.037$, AGFI $= 0.896$, NFI $= 0.901$, TLI $= 0.978$, CFI $= 0.973$, IFI $= 0.977$, PGFI $= 0.679$, GFI $= 0.906$ separately. Above fitting indicators meet the requirements. Table 2 shows the exact path coefficients. According to test results, assuming 5a, 5b, 6a and 6b are correct and there are slight difference in the intermediary role among the product innovation strategy and technological diversification impacting on innovation performance and operating performance.

5 Results and Discussion

This empirical results show a significant positive impact ($\beta = 0.547$, $\beta = 0.667$) of technological diversity on business performance and innovation performance, indicating that if the enterprise expand its range of technical knowledge and

Table 2 Revised overall structural equation models of path coefficients

Active route	Standardized path coefficients	S.E.	C.R.	Significance level
Platform serialization product strategy ← technology diversification	0.677	0.130	6.768	***
Related diversified product strategy ← technology diversification	0.586	0.112	5.749	***
Innovation performance ← related diversified product strategy	0.230	0.107	2.550	*
Business performance ← platform serialization product strategy	0.339	0.120	3.548	***
Business performance ← related diversified product strategy	0.344	0.142	3.621	***
Innovation performance ← platform serialization product strategy	0.263	0.099	2.638	**
Innovation performance ← technology diversification	0.319	0.129	3.195	**

ability, it can promote an intersection fusion between technical knowledge and product management effectively, reduce the decrease in profits caused by technical lock-in and enable enterprises to obtain a stable operating return; The diversity of technology infrastructure can improve the absorptive capacity of the enterprise, enable enterprise benefit from knowledge spillovers from other firms and also enable enterprises to integrate knowledge from different technical areas to create a brand-new technology, products or services.

Technology diversification has a positive impact on enterprise adoption of technology platform series products innovative strategies and related product diversification technologies ($\beta = 0.595$, $\beta = 0.657$), indicating that technology diversified technology integration can enable enterprises to improve integration capabilities, promote enterprises to build a solid technical platform, improve platform innovation capability, and thus release a series of competitive products; Technology diversification is an important prerequisite for enterprises to enter new product field, since successful companies tend to follow a sequential strategy, which is to conduct technical diversification, and then carry out technology-related product diversification strategy, and all these 'offensive diversified enterprise' generally have higher sales growth, so technology-based diversified product diversification strategy is more likely to succeed.

Platform strategy series products help enterprises improve business performance ($\beta = 0.348$) and innovation performance ($\beta = 0.367$), indicating that the enterprise share resources through the platform establishment of design ideas and elements in product development process, reduce the product innovation cycle greatly, shrink the cost of new product development and manufacturing costs, which can launch new products quickly and consistently and dominate the market differentially. Technology-related product diversification strategies help enterprises improve

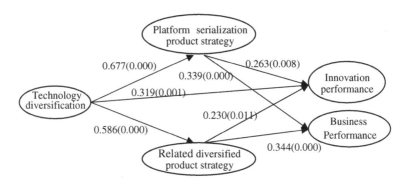

Fig. 1 Revised structural equation model and the overall relationship between the variables

business performance ($\beta = 0.403$) and innovation performance ($\beta = 0.438$), indicating that the development of a new product or business is built on enterprise technological accumulated capabilities based on the diversification strategy, which make it easier for companies to enter new markets successfully and gain a foothold, and they can get better access to a range of economic, financial performance and higher profit.

From the view of intermediary role of product innovation strategy, on the one hand, SEM empirical results show partly that technology diversification have a significant positive effect on product innovation and business operating performance and product innovation strategy has a significant positive impact on the business operating performance. However, in the initial overall SEM model, when operating performance conducted the overall regression analysis on technology diversification and product innovation strategy, we found that product innovation strategies still has a significant positive impact on business performance and discovered that although product innovation strategies has a positive impact on business performance, the regression coefficient decreased into a insignificant levels ($\beta = 0.212$, $P = 0.066 > 0.05$), that is, the emergence of mediating variables (Product Innovation Strategy) 'digest' the impact of technology diversification on business performance, which show that technology diversification achieve influence on business performance through the fully meditation of innovative strategies for both products.

On the other hand, SEM empirical modeling results show partly that technology diversification has a significant positive effect on innovation performance and product innovation strategy has a significant positive effect on innovation performance, but in the final overall SEM model (Fig. 1).

When innovation performance conducted the overall regression analysis on technology diversification and product innovation strategy, we found that technology diversification still has a significant positive impact on innovation performance and reach a significant levels ($\beta = 0.319$, $P = 0.000 > 0.001$), that is, the emergence of mediating variables (Product Innovation Strategy) 'weaken' the impact of technology diversification on innovation performance, which show that technology diversification achieve influence on innovation performance through the meditation in part of innovative strategies for both products.

6 Conclusions and Prospects

This research formed technology diversification, product innovation strategy, business performance into a research framework to confirm the intermediary role of product innovation strategy between diversification and innovation, which, to a certain extent, solve the question on technical diversity impact on competitive advantage. Currently, technology-intensive enterprises are facing important issues such as the rising labor costs, the oppression of Southeast Asia cost advantage, skilled and professional labor shortages and cost advantages loss. This research indicates that in the global technology innovation practice, with the reference of developed innovative experiences, technology diversification has important implications for our high-tech enterprises in technological innovation to improve innovation performance. As energy and focus are limited, sample in this study excluded regional differences and industrial clusters technical capabilities, thus making the conclusion covering a certain narrow applicability. In subsequent studies, more objective measurement methods need considering, the scope of the sample need expanding, and different industry characteristics and business properties need comparing in order to obtain more valid conclusions.

Acknowledgments This study is funded by Xi'an Science and Technology Project (Program NO. SF1225-3): Science and Technology Development Service Based on Overall Scientific and Technological Resources Needs in Xi'an.

References

1. Breschi S, Lissoni F, Malerba F (2003) Knowledge-relatedness in firm technological diversification. Res Policy 32(1):69–87
2. Huang YF, Chen CJ (2010) The impact of technological diversity and organizational slack on innovation. Technovation 30(7–8):420–428
3. Vila LE, Perez PJ, Morillas FG (2012) Higher education and the development of competencies for innovation in the workplace. Manag Decis 50(9):1634–1648
4. Lin BW, Chen CJ, Wu HL (2006) Patent portfolio diversity, technology strategy, and firm value. IEEE Trans Eng Manage 53(1):17–26
5. Miller DJ (2006) Technological diversity, related diversification and firm performance. Strateg Manag J 27(7):601–619
6. Gambardella A, Torrisi S (1998) Does technological convergence imply convergence in markets? Res Policy 27(5):445–463
7. Robins J, Wiersema M (1995) A resource-based approach to the multibusiness firm: empirical analysis of portfolio interrelationships and corporate financial performance. Strateg Manag J 16(4):277–299

8. Grover V, Goslar MD (1993) The initiation, adoption, and implementation of telecommunications technologies in US organizations. J Manag Inf Syst 1(10):141–163
9. Nieto M, Quevedo P (2005) Absorptive capacity, technological opportunity, knowledge spillovers, and innovative effort. Technovation 25(10):1141–1157
10. Meyer M, Utterback JM (1993) The product family and the dynamics of core capability. Sloan Manag Rev 34(3):29–47
11. Li HY, Kwaku A-G (2001) Product innovation strategy and the performance of new technology ventures in China. Acad Manag J 44(6):1123–1134

Paraphrase Collocations Extraction Based on Concept Expansion

Maoyuan Zhang, Wang Li and Hong Zhang

Abstract This paper proposes a method based on concept expansion to extract paraphrase collocation. Collocations of forms of $\langle V, OBJ, N \rangle$ (verb-object collocations) and $\langle V, SUB, N \rangle$ (subject-predicate collocations) are extracted after syntactic analysis is done to the sentences. Then the words used in the collocations are expanded based on related words getting from concept semantic to get the candidate of paraphrase collocations. In order to filter these paraphrase collocations, following four features are chosen: part of speech feature, mutual information feature, HowNet-based semantic similarity feature, and context-based semantic similarity feature. Compared to existed method, this method does not restrict the word in paraphrase collocation to synonym. The experiment shows that every feature exploited is useful for improving the performance.

Keywords Paraphrase collocations · Information extraction · Concept expansion · Binary classification

M. Zhang · W. Li (✉) · H. Zhang
Academy of Computer Science, Central China Normal University, Wuhan, China
e-mail: wangli0147@gmail.com

M. Zhang
e-mail: zhangmyccnu@126.com

H. Zhang
e-mail: QJzhanghong@gmail.com

Z. Wen and T. Li (eds.), *Knowledge Engineering and Management*,
Advances in Intelligent Systems and Computing 278,
DOI: 10.1007/978-3-642-54930-4_19, © Springer-Verlag Berlin Heidelberg 2014

1 Introduction

Paraphrases are alternative ways to convey the same information [1]. It has already been proved that paraphrase findings could be directly used in various domains of NLP such as question answering (QA) [2], information extraction (IE) [3], machine translation (MT) [4], etc.

This paper addresses the problem of automatically extracting Chinese paraphrase collocation pairs. A paraphrase collocation pair includes two collocations which convey the same meaning, but are not identical in words utilized. In this paper, the term collocation follows the definition of Wu and Zhou and refers to a lexically restricted word pair with a certain syntactic relation [5].

Previous work on paraphrase collocations has few relations to Chinese. This paper proposes a method for Chinese paraphrase collocations extraction which is a beneficial attempt. Besides, compared to existed method, this method does not restrict the word in paraphrase collocation pair to synonym which performs more flexible.

Our method is based on concept expansion. Collocations of forms of $\langle V, OBJ, N \rangle$ (verb-object collocations) and $\langle V, SUB, N \rangle$ (subject-predicate collocations) are extracted after syntactic analysis is done to the sentences. Then the words used in the collocations are expanded based on related words getting from concept sematic to get the candidate of paraphrase collocations. In order to filter these paraphrase collocations, following four features are chosen: part-of-speech feature, mutual information feature, HowNet-based semantic similarity feature, and context-based semantic similarity feature.

The remainder of this paper is organized as follows. Section 2 reviews related work on paraphrase collocations extraction. Section 3 presents our method specifically. Section 4 describes the experiment. And finally this paper is concluded in Sect. 5.

2 Related Work

Wu and Zhou first propose the concept of synonymous collocation, which is equivalent to the concept of paraphrase collocation in this paper. Their method for paraphrase collocation extraction comprises of three steps: (1) extract collocations from large monolingual corpora; (2) generate candidates of synonymous collocation pairs with a word thesaurus WordNet; (3) select synonymous collocation candidates using translations.

After Wu, Zhao formally come up with the concept of paraphrase collocations and view paraphrase collocations extraction as binary classification problem [6]. In this method, Zhao utilized a English-Chinese parallel corpus and used seven features to distinguish paraphrase collocations. In this method, they restrict the object word utilized in verb-object collocations to the same which degenerate their research to extract paraphrase words in particular context.

3 Proposed Method

Our method for paraphrase collocations extraction comprises of three steps: (1) extract collocations from corpora; (2) generate paraphrase collocation candidates based on concept expansion; (3) filter candidates through chosen features.

3.1 Extracting Collocations

In this step, collocations of forms of $\langle V, OBJ, N \rangle$ (verb-object collocations) and $\langle V, SUB, N \rangle$ (subject-predicate collocations) are extracted after syntactic analysis is done to the sentences and then filter out collocations which occur less than 3 times.

3.2 Getting Paraphrase Collocation Candidates

For a collocation $\langle$Head, Relation Type, Tail$\rangle$, we get its paraphrase collocation candidates through expanding Head and Tail. The expansion of Head and Tail are obtained from Concept Expansion. Concept refers to a semantic knowledge unit referred by semantic tags, defined by semantic fingerprint. Zhang [7] computes words similarity based on structured information extracted from Chinese Wikipedia. Zhou arranges and get some concepts'-related word group [8]. Zhang and Liu [9] get some concepts' semantic fingerprint through HowNet. Simply speaking, candidates getting from concept expansion contain not only synonymous but also the other associated nonsynonymous. Table 1 gives some example of concepts. Table 2 describes the algorithm of getting paraphrase collocation candidates based on Concept Expansion in detail.

Through the above method, we obtained candidates of paraphrase collocation pairs. The next step is to filter out inappropriate candidates.

3.3 Filtering Out Inappropriate Candidates

Following four features are chosen as paraphrase collocation feature to filter out inappropriate candidates: (1) part of speech (POS) feature; (2) mutual information feature; (3) HowNet-based semantic similarity feature; (4) context-based semantic similarity feature.

(1) *POS feature*: In paraphrase, collocation pair word and its related word group should have the consistent POS. For collocation Col_i and its candidate S_{ij}, the computational formula is shown in formula:

Table 1 Some examples of concepts

Semantic tags	Semantic fingerprint
人类	(人类, 1.000), (男性, 0.842), (女性, 0.840), (自然人, 0.837), (女人, 0.819), (男人, 0.813), (智人, 0.808),…
基因	(基因, 1.000), (遗传因子, 1.000), (基因工程, 0.978), (DNA, 0.975), (脱氧核糖核酸, 0.975),…

Table 2 The algorithm of getting paraphrase collocation candidates

For each collocation ($Col_i = \langle \mathrm{Head}, R, \mathrm{Tail}\rangle \in U$), do the following:
Use concept expansion to expand Head and Tail and get sets C_{Head} and C_{Tail}
Generate the candidate set of paraphrase collocation
$\quad S_i = \{\langle w_1, R, w_2\rangle \backslash w1 \in \{\mathrm{Head}\} \cup C_{\mathrm{Head}} \,\&\, w_2 \in \mathrm{Tail} \,\&\, \langle w_1, R, w_2\rangle > \neq Col_i\}$

$$Eq(w_1, w_2) = \begin{cases} 1 & \text{if}(\mathrm{POS}_{w1} = \mathrm{POS}_{w2}) \\ 0 & \text{otherwise} \end{cases}$$

$$S_{\mathrm{pos}} = Eq(Col_i, S_{ij}) = Eq\left(\langle \mathrm{Head}_i, R_i, \mathrm{Tail}_i\rangle, \left\langle C_{\mathrm{Head}_{ij}}, R_i, C_{\mathrm{Tail}_{ij}}\right\rangle\right)$$
$$= Eq\left(\mathrm{Head}_i, C_{\mathrm{Head}_{ij}}\right) \times Eq\left(\mathrm{Tail}_i, C_{\mathrm{Tail}_{ij}}\right)$$

Here, POS(w) stands for the POS of word w.

(2) *Mutual Information Feature*: Mutual information feature bases on such a fact: if two words could constitute a collocation, these two words must have certain dependency relationship. The bigger the score of mutual information get, the more possible of two words to constitute a collocation. For two words w_1, w_2 in a collocation Col, we compute the mutual information score as shown in formula:

$$S_{\mathrm{MI}}(\mathrm{Col}) = I(w_1, w_2) = \log_2 \frac{p(w_1, w_2)}{p(w_1) \times p(w_2)} \tag{1}$$

$$p(w_1, w_2) = \frac{\#(w_1, w_2)}{N}, \ p(w_1) = \frac{\#(w_1)}{N}, \ \mathrm{p}(w_2) = \frac{\#(w_2)}{N} \tag{2}$$

Here, $I(w_1, w_2)$ stands for word w_1 and w_2's mutual information, $\#(w_1,w_2)$ stands for the number of concurrence of w_1 and w_2, N stands for the total number of sentences in corpus, $S_{\mathrm{MI}}(\mathrm{Col})$ stands for the mutual information of collocation Col.

To set the score we get into $(0 \sim 1)$, we amend the formula as follows:

$$S_{\mathrm{MI}}(\mathrm{Col}) = \frac{I(w_1, w_2) - \min(I(X, Y))}{\max(I(X, Y)) - \min(I(X, Y))} \tag{3}$$

Here, max($I(X, Y)$) stands for the maximum of gotten mutual information, min($I(X,Y)$) stands for the minimum of gotten mutual information.

(3) *HowNet-based semantic similarity feature:* Lots of work indicated that we could get word similarity through HowNet [10]. Here, we assume that word Head and Tail contribute independently to the semantic of a collocation. We compute the HowNet-based semantic similarity score as shown in formula:

$$
\begin{aligned}
S_{\text{Hownet}} &= \text{Sim}\left(\text{Col}_i, S_{ij}\right) = \text{Sim}\left(\langle \text{Head}_i, R_i, \text{Tail}_i \rangle, \langle C_{\text{Head}_{ij}}, R_i, C_{\text{Tail}_{ij}} \rangle\right) \\
&= \frac{\text{Sim}\left(\text{Head}_i, C_{\text{Head}_{ij}}\right) + \text{Sim}\left(\text{Tail}_i, C_{\text{Tail}_{ij}}\right)}{2}
\end{aligned}
\tag{4}
$$

Here, S_{Hownet} stands for the score of HowNet-based semantic similarity feature, Sim stands for two words' similarity.

(4) *Context-based semantic similarity feature:* Context-based semantic similarity feature's basic idea is that words tend to appear in similar context have similar meaning [11]. In this paper, we extend this idea to believe collocations which have similar meaning tend to appear in similar context.

For a collocation Col_i, its feature vector is $V_{CT}(\text{Col}) = \langle (w_1, p_1) (w_2, p_2), \ldots, (w_n, p_n) \rangle$. Here, w_i stands for the context word of a collocation, p_i stands for the weight of w_i. The formula of computing p_i is shown as follows:

$$
p_i = 1 + \log TF
\tag{5}
$$

Here, TF stands for the number of occurence of w_i.
For two feature vectors, we compute the similarity as follows:

$$
S_{\text{Context}}(V_{CT_1}, V_{CT_2}) = \cos(V_{CT_1}, V_{CT_2}) = \frac{V_{CT_1} \cdot V_{CT_2}}{|V_{CT_1}| \times |V_{CT_2}|}
\tag{6}
$$

Here, S_{Context} stands for the context-based semantic similarity.
With the above four features, we define the similarity of two collocations as follows:

$$
S = S_{\text{pos}} \times (\lambda_1 \times S_{\text{Context}} + \lambda_2 \times S_{\text{hownet}} + \lambda_3 \times S_{\text{MI}})
\tag{7}
$$

Here, $\lambda_1 + \lambda_2 + \lambda_3 = 1 (\lambda_1 > = 0, \lambda_2 > = 0, \lambda_3 > = 0)$. When the score of S over threshold T, we believe two collocations as paraphrase collocation.
Table 3 shows the algorithm of filtering out inappropriate collocations.

Table 3 The algorithm of filtering out inappropriate collocations	Input: Collocation Col$_i$ and paraphrase collocation candidate S$_i$
	Output: Paraphrase collocation pairs
	For collocation Col$_i$ and collocation S$_{ij}$:
	Compute the two collocations's similarity S
	If S >= T: Reserve Col$_i$ and S$_{ij}$ as paraphrase collocation

4 Experiments and Results

4.1 Getting Paraphrase Collocation Candidates

We choose und2000 part of simplified Chinese in NTCIR-5 as our corpora. In collocation extraction stage, ctbparser is chosen as syntax analysis tool and we extract 57,675 collocations. Table 4 shows the detail of extracted collocations. For these got collocations, Algorithm 1 is adopted to get paraphrase candidates.

4.2 Data Annotation and Determine Threshold

We randomly pick up 1,050 pairs of candidates and deliver these to two markers. A consistent ratio of the two reaches 87.3 %. For those which are inconsistent, we deliver them to the third marker to mark the data. The result shows that we get 316 pairs of positive candidates and 734 pairs of negative candidates.

To determine threshold T, we compute precision P, recall R, and F-value F:

$$P = \frac{A \cap B}{A} \quad R = \frac{A \cap B}{B} \quad F = \frac{2PR}{P+R}$$

Here, A stands for set of paraphrase collocation pairs getting from above method, B stands for set of paraphrase collocation pairs marked by human markers.

Under the circumstances of $\lambda_1 = 0.2, \lambda_2 = 0.5$ and $\lambda_3 = 0.3$, Table 5 shows different P, R, and F under different T. According to Table 5 and Fig. 1, we get the biggest F-value under the circumstances of $T = 0.54$.

With the above method, we extract 145,793 pairs of paraphrase collocation. Table 6 shows some examples of extracted paraphrase collocations.

4.3 Comparison with Binary Classifier Method

Table 7 shows the comparison between our method and binary classification method. According to the table, we can tell that our concept expansion method's precision slightly lower than binary classification method, but both recall and F-

Table 4 The detail of extracted collocations

Type	Number	Ratio (%)
⟨V, OBJ, N⟩	30397	52.7
⟨V, SUB, N⟩	27278	47.1

Table 5 Different P, R, and F-value under different T

T	P (%)	R (%)	F (%)
0.4	35.77	89.87	51.17
0.5	44.60	78.48	56.88
0.6	69.64	49.36	57.78
0.7	85.29	18.35	30.21

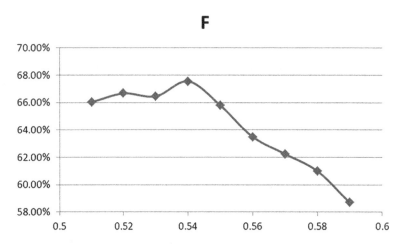

Fig. 1 Different F-value under different T

Table 6 Examples of extracted paraphrase collocations

外界 SUB 期盼 versus 外头 SUB 希望
战线 OBJ 延长 versus 阵地 OBJ 拓展
走势 OBJ 呈现 versus 动向 OBJ 显露
规定 SUB 实施 versus 规程 SUB 贯彻

Table 7 Comparison result

Method	P (%)	R (%)	F (%)
Binary classification method	72.60	61.30	66.46
Concept expansion method	70.06	65.19	67.54

Table 8 Different feature combination experiment result

Feature combination	P (%)	R (%)	F (%)
Remove context feature	67.97	65.82	66.88
Remove HowNet feature	58.28	64.56	61.26
Remove mutual feature	69.29	61.39	65.10
Reserve all features	70.06	65.19	67.54

value exceed binary classification method. Besides, binary classification method restricts the object word utilized in verb-object collocations to the same which degenerate their research to extract paraphrase words in particular context.

4.4 Feature Analysis

In order to actually validate our selected features are useful, we also design three experiments, each feature is removed every time and computes the corresponding precision, recall, and F-value. Table 8 shows the result. From the table, we can tell that every feature we choose is useful.

5 Conclusion

This paper proposes a method based on concept expansion to extract paraphrase collocation. Collocations of forms of $\langle V, OBJ, N \rangle$ (verb-object collocations) and $\langle V, SUB, N \rangle$ (subject-predicate collocations) are extracted after syntactic analysis is done to the sentences. Then the words used in the collocations are expanded based on related words getting from concept semantic to get the candidate of paraphrase collocations. In order to filter these paraphrase collocations, following four features are chosen: part of speech feature, mutual information feature, HowNet-based semantic similarity feature, and context-based semantic similarity feature. Compared to existed method, this method does not restrict the word in paraphrase collocation to synonym. The experiment shows that every feature exploited is useful for improving the performance. For future work, we plan to take more features into consideration to improve the precision.

Acknowledgment This work was supported by the National Natural Science Foundation of China (No. 61003192), the self-determined research funds of CCNU from the colleges' basic research and operation of MOE(No. CCNU13A05014, No. CCNU13C01001), the Major Project of State Language Commission in the Twelfth Five-year Plan Period (No. ZDI125-1), the Project in the National Science & Technology Pillar Program in the Twelfth Five-year Plan Period (No. 2012BAK24B01), the Program of Introducing Talents of Discipline to Universities (No. B07042) and the NSF of Hubei Province (No. 2011CDA034).

References

1. Barzilay R, McKeown KR (2001) Extracting paraphrases from a parallel corpus. In: Proceedings of the 39th annual meeting and the 10th conference of the European chapter of association for computational linguistics (EACL) 2001, pp 50–57
2. Lin D, Pantel P (2001) Discovery of inference rules for question-answering. Nat Lang Eng 7 (4):343–360
3. Sekine S (2005) Automatic paraphrase discovery based on context and keywords between NE Pairs. In: Proceedings of IWP 2005
4. Zhang YJ, Yamamoto K (2002) Paraphrasing of Chinese utterances. In: Proceedings of COLING. Morristown: association for computational linguistics, 2002, pp 1163–1169
5. Wu H, Zhou M (2003) Synonymous collocation extraction using translation information. In: Proceedings of ACL 2003, pp 120–127
6. Zhao S, Lin Z, Ting L, Sheng L (2010) Paraphrase collocation extraction based on binary classification. J Softw 21(6):1267–1276
7. Zhang H (2011) Extracting structured information from the Chinese Wikepedia and measuring relatedness between words. Central China Normal University, China
8. Zhou K (2012) Related words of concept study based on Chinese Wikipedia. Central China Normal University, China
9. Zhang M, Liu M (2009) The retrieval system based on concept extending. In: 2009 Second Asia-Pacific Conference on computational intelligence and industrial applications, pp 365–368
10. Liu Q, Li S (2002) Lexical semantic similarity computation based on Hownet. Comput Linguist Chin Lang Process 7(2):59–76
11. Bannard C, Callison-Burch C (2003) Paraphrasing with bilingual parallel corpora. In: Proceedings of ACL 2003, pp 120–127

A Probabilistic Method for Tag Ranking in Tagging System

Peng Zhang, Liang Yao, Yin Zhang, Baogang Wei and Cheng Gao

Abstract Since WEB2.0, more and more online communities began to use tag—words selected by users or generated by computer algorithms—to help people find or organize data resources. Unfortunately, the tags are generally in a random order without any importance or relevance in information, which seriously limit the effectiveness of these tags in tag-based applications. In this paper we present a tag ranking method which first computes the probability of each tag associated with a given book, and then adjust the probability as well as the tags' order based on users' tag-click behaviors. Then an initial strategy which provides a better initial probability is described to improve our method. Experimental results show that users' tag-click behaviors can reflect the relevance between books and tags to some extent and our approach is both efficient and effective.

Keywords Tag ranking · User behaviour · Probabilistic model · Search

P. Zhang · L. Yao · Y. Zhang (✉) · B. Wei · C. Gao
School of Computer Science and Technology, Zhejiang University, Hangzhou, China
e-mail: zhangyin98@zju.edu.cn

P. Zhang
e-mail: pengzhang1991@zju.edu.cn

L. Yao
e-mail: yaoliang@zju.edu.cn

B. Wei
e-mail: wbg@zju.edu.cn

C. Gao
e-mail: hoopdog@zju.edu.cn

Z. Wen and T. Li (eds.), *Knowledge Engineering and Management*,
Advances in Intelligent Systems and Computing 278,
DOI: 10.1007/978-3-642-54930-4_20, © Springer-Verlag Berlin Heidelberg 2014

1 Introduction

Recent years have witnessed an explosion of online collaborative tagging appli-cations in the World Wide Web. These applications encourage users to assign tags to describe data resources (e.g. *Flickr* [13], *YouTube*). At the same time, a number of machine learning algorithms have been proposed to automatically generate or recommend high-quality tags. With plenty of tags as metadata, users can conve-niently organize and search data resources.

Ames and Naaman [8] point out that the principal purpose of tagging is to make data resources better accessible to the public. However, tags are often imprecise and incomplete [9–11], which should be ranked. Higher ranked tags are better than lower ranked tags in quality and relevance. To address the problem, we proposed a probabilistic tag ranking method based on users' tag-click behaviors. We call it Click-Based Tag Ranking (*CBTR*).

We observed a phenomenon on *CADAL* [14]—one of the most important digital libraries in china—that, one user searches "data" and gets interested in book "data network" written by Dimitri Bertsekas. He hopes to find more relative books written by Bertsekas, and then clicks the tag "Bertsekas" for a second search. Similarly, such user behavior also appears in Flicker. For example, Tom likes cat, so he searches "cat" on Flickr and finds a picture of a pretty black cat. He wants to see more pictures by the uploader, then clicks the tag "black". The reason Tom clicks the tag "black" is that he saw a black cat in the original picture. Obviously the tag "black" is relevant to the picture.

Based on above phenomenon, we propose a hypothesis that users' tag-click behaviors can reflect the relevance between tags and data resources to some extent. We conduct a preliminary experiment to validate our hypothesis. Then we propose a probabilistic tag ranking method based on users' tag-click behaviors. Another experiment is conducted to show the performance of our method.

The rest of this paper is organized as follows. First, we introduce some related work in Sect. 2. Second, we describe the process of our method in Sect. 3. After that, a series of experiments are conducted to prove our hypothesis and measure the performance of our method. Detailed experimental procedure is described in Sect. 4. Finally, we conclude this paper in Sect. 5.

2 Related Work

Extensive research efforts have been dedicated to address tag related problems. Roughly, they can be summarized to the following four aspects.

- To automatically annotate images with tags, Li and Wang [5] developed a real time system (ALIPR) which is fully automated and high-speed to annotate tags for online pictures. The key technologies in their system are D2-clustering method and HLM.

Table 1 Symbols and their meanings

Symbol	Definition
B	A book
T	A tag
BT	The tags associated with book B
BT_i	The i-th tag of book B
BK	Number of times that book B's tags are clicked
|B|	Tag's number of book B

- Although, a lot of encouraging results have been reported, the automatically annotated tags are still far away from satisfactory. Then manual tagging or folksonomy is proposed to let people freely assign tags to data resources. Sigurbjornsson et al. [4] investigated how to assist users to mark tags to photos on Flickr. Song et al. [1] investigated user behavior in collaborative tagging system. They tried to find users' interests who can help the system to recommend better tags.
- The user provided tags are orderless which significantly limits their applications. Liu et al. [12] proposed a tag ranking scheme aiming to automatically rank the tags associated with images. They first estimate initial relevance scores for the tags based on probability, density, estimation, and then perform a random walk over a tag similarity graph to refine the relevance scores.
- As introduced in introduction, tags are often imprecise and incomplete. Sen et al. [6] proposed implicit and explicit mechanisms for determining tag quality. They hold an opinion that the more a tag has been selected by users, the more relevant the tag is to data resources. Awawdeh and Anderson [7] described how users' tags can be enhanced with metadata in the form of additional tags automatically extracted from the original document.

3 Our Approach

Our approach mainly focuses on mining information from users' tag-click behaviors to rank existing tags. In this section, we will introduce our tag ranking method. Table 1 lists the symbols and their meanings used throughout this paper.

3.1 Overview

According to our hypothesis, we can use BK to measure tag's relevance to the book. If tag "A" has been clicked more times than tag "B", we hold the opinion that tag "A" is more relevant to the book than tag "B". We adopt this opinion in a probabilistic model and introduce an initial strategy to optimize this model.

3.2 Probabilistic Model

We regard every tag-click behavior as an independent event. The probability that the tag BT_i will be clicked is $P(BT_i)$. Initially, we set $P(BT_i) = 1/|B|$, which means at the beginning we don't know which tag is more relevant to the book. Then when a tag-click behavior happens, if the tag BT_i is clicked, we should increase $P(BT_i)$ and at the same time decrease $P(BT_j)$ where $j = i$. This update process can be performed by the following formulae:

$$P(BT_i) = P(BT_i)\frac{BK}{BK + 1} + \frac{1}{BK + 1} \tag{1}$$

$$P(BT_j) = P(BT_j)\frac{BK}{BK + 1} \tag{2}$$

After the update process, we rank tags BT according to $P(BT_i)$ and $P(BT_j)$. Formula (1) has an intuitive meaning. Suppose book B's tags have been clicked BK times, now the $(BK + 1)$-th tag-click event happens and tag BT_i is clicked. This means that the current $P(BT_i)$ is less than the real value it should be, so we need to increase $P(BT_i)$ and the $(BK + 1)$-th tag-click event should take a weight of $1/(BK + 1)$. At the same time, the other tags of book B should be decreased. To illustrate the meaning of the formula (2), we can change it to an equal form:

$$P(BT_j) = P(BT_j)\frac{BK}{BK + 1} + \frac{0}{BK + 1} \tag{3}$$

Formula (3) is similar to the formula (1) both in format and meaning.

Figure 1 gives an example of the evolution of the probabilistic model. Figure 2 shows a real case of how the users' tag-click behaviors influence the rank of tags.

3.3 Initial Strategy

In our method, we just give every tag an equal probability at the beginning. This is not a good strategy because there are a lot of excellent algorithms proposed by other researchers to measure the relevance of tags and data resources. Here we adopted PTR algorithm proposed by Liu et al. [12] as an initial strategy. We use $S(BT_i, B)$ to denote the relevance score between tag BT_i and Book B:

$$S(BT_i, B) = \frac{1}{|B_i|}\sum_{b_k \in B_i} K\delta(B - b_k) \tag{4}$$

$$K_\delta(B - b_k) = \exp(-\frac{||B - b_k||^2}{\delta^2}) \tag{5}$$

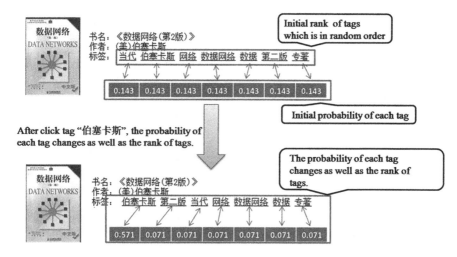

Fig. 1 An example of our probabilistic model

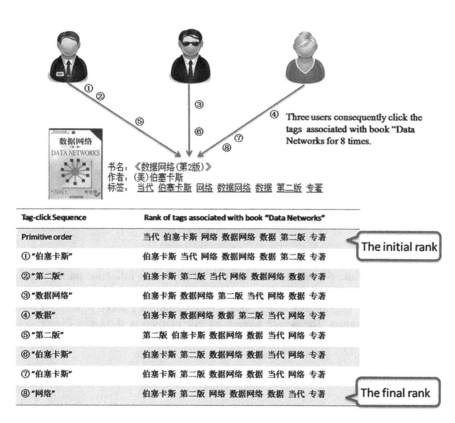

Fig. 2 A real case of how the users' tag-click behaviors influence the rank of tags

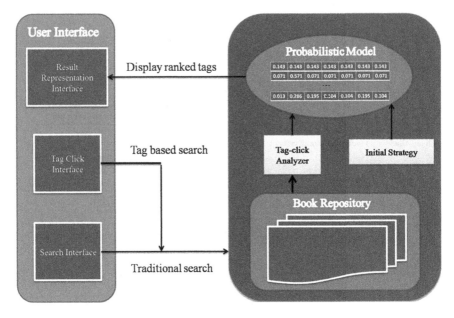

Fig. 3 The whole process of our tag-ranking method

B_i is the set of books which contains tag BT_i, $K\delta$ is the Gaussian kernel function with the radius parameter δ(0.5 in our system). For each book B which contains tag BT_i, B_i can be viewed as its neighbors. The Gaussian kernel function can be regard as the soft voting from B_i and the relevance score is the sum of the soft voting. So, the result is based on "collective intelligence" from neighbor books. However, the sum of $S(BT_i, B)$ is not 1. So we add a step to normalize it. The initial value of $P(BT_i)$ can be calculated by:

$$P(BT_i) = \frac{S(BT_i, B)}{\sum_{i=0}^{|B|} S(BT_i, B)} \tag{6}$$

According to law of large numbers, with the increasing of BK, $P(BT_i)$ will reach a stable value. So the initial value of $P(BT_i)$ has no influence on the stable value in theory. On the other hand, a good initial value which is closer to the stable value will reduce the number that a book's tags should be clicked to make it own statistical significance. The whole process of our method is shown in Fig. 3.

4 Experiment

To make our hypothesis convincing, we build a tag-based search system to do a series of experiments. The structure of the system is shown in Fig. 3.

4.1 Dataset and Experimental Settings

We get about 400,000 books' metadata form *CADAL*. The metadata contains book's name, author, publisher and tags. All the tags are in random order. Then we build a book search system. Users can find books in three manners: (1) Type in query keywords to do a traditional search. (2) Click a tag of a book to find all the books which contain the clicked tag. (3) Mix use of the above two manners.

Because our approach is based on a probabilistic method, the result will have no statistical significance until the tags of a book have been clicked many times. In our system, the average number of tags per book is approximately five. Suppose the tags of a book should at least be clicked 10 times, it's easy to figure out that for all the 400,000 books in our system, we need to collect more than 4 million tag-click behaviors. This is impossible for an experimental system. So we focus on a small part of books (2,215 books) which belong to topic DATA and POEM. The two topics are the most and second most searched topics in our system.

4.2 Analysis of User's Tag-Click Behavior

To prove our hypothesis that users click tags because they subconsciously think the tag and the book is related, we collect 562 tag-click behaviors and then analyze them. We invited 17 volunteers to participate in our experiments. We asked them to set a level for every clicked tag, which is similar to *NDCG* [2]. For each book, the clicked tag is labeled as one of the five levels: Most Relevant (score 5), Relevant (score 4), Partial Relevant (score 3), Weakly Relevant (score 2), and Irrelevant (score 1). Then we calculate the average level of each tag. The result can be shown in Fig. 4. Figure 4 shows that 72 % (score 5, score 4, score 3) of the tags are relevant to the book and only 4 % (score 1) is not relevant to the book. The result shows that our hypothesis is reasonable.

4.3 Evaluate Tag Ranking Method

We use two methods to evaluate our method: *NDCG* and *Best-N Score*. We record BK and pick out the top 40 books to calculate. We also compared our method with other popular tag ranking methods such as PTR, RWTR, PTR + RWTR [12].

First, we use NDCG as the performance evaluation measure. For a given book, each of its tags will be given a relevant score by volunteers. For a book with a ranked tag list t1, t2, t3,…,tn, the NDCG is computed as:

$$\mathrm{N}n = \mathrm{Z}n \sum_{i=1}^{n} (2^{r(i)} - 1) / \log(1 + i) \tag{7}$$

Fig. 4 The average
percentage of the five levels

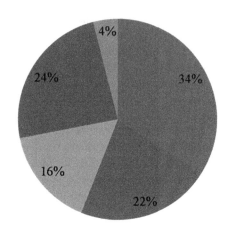

Average Level Percentage
■ Level 5 ■ Level 4 ■ Level 3 ■ Level 2 ■ Level 1

Fig. 5 NDCG of different
tag rank algorithm

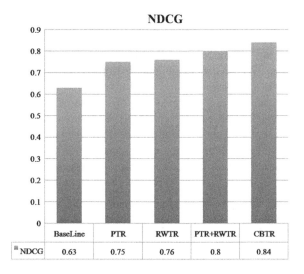

where r(i) is the relevance level of the i-th tag and Zn is a normalization constant
that is chosen so that the optimal ranking's NDCG score is 1. The result is shown
in Fig. 5. We take the tags' original order as baseline. The result shows our method
achieves a NDCG of 0.84, which is the biggest among all the five methods.

Then, we use *Best-N score* [3] to evaluate our method. Best-N score is often
used to measure the relevance of the top-N results returned by a ranking method.

Fig. 6 Best-N score of baseline, PTR, RWTR, PTR + RWTR, CBTR

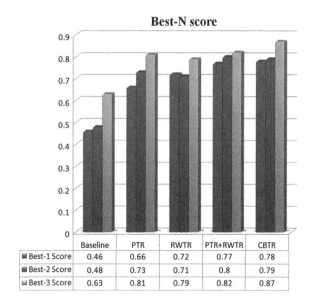

Best-N score

	Baseline	PTR	RWTR	PTR+RWTR	CBTR
Best-1 Score	0.46	0.66	0.72	0.77	0.78
Best-2 Score	0.48	0.73	0.71	0.8	0.79
Best-3 Score	0.63	0.81	0.79	0.82	0.87

$$Best - Nscore = \frac{\sum_{i=1}^{N} r(i)}{\sum_{i=1}^{N} rIDCG(i)} \tag{8}$$

where $r(i)$ is relevance level of the i-th tag ranked by our method, $r_{IDCG}(i)$ is relevance level of the i-th tag ranked by IDCG [2]. The experiment result is shown in Fig. 6. According to Fig. 6, our method performs better than Baseline in all cases. When N = 1 and 3, our method is the best one. However, when N = 2, our method performs a little worse than PTR + RWTR, but the gap is very small.

5 Conclusion

In this paper, we proposed a hypothesis that users' tag-click behaviors can reflect the relevance between tags and data resources. We conducted an experiment to validate it. Then we introduced a probability tag ranking method based on users' tag-click behaviors (*CBTR*). We built a probabilistic model and introduced an initial strategy to provide a better initial probability of each tag for this model. The experimental results show that our approach really boosts the performances of tag ranking.

Acknowledgment This work was supported by the China Academic Digital Associative Library Project, the Special Funds for Key Program of National Science and Technology (Grant No. 2010ZX01042-002-003), the Program for Key Innovative Research Team of Zhejiang Province

(Grant No. 2009R50009), the Program for Key Cultural Innovative Research Team of Zhejiang Province, the Fundamental Research Funds for the Central Universities, Zhejiang Provincial Natural Science Foundation of China (Grant No. LQ13F020001) and the Opening Project of State Key Laboratory of Digital Publishing Technology.

References

1. Song S, Yu L, Yang X (2010) LDA-based user interests discovery in collaborative tagging system. In: International conference on intelligent systems and knowledge engineering(ISKE)
2. Jarvelin K, Kekalainen J (2002) Cumulated gain-based evaluation of IR techniques. ACM Trans Inf Syst (TOIS) 20(4):422–446
3. Lau JH, Newman D, Karimi S, Baldwin T (2010) Best topic word selection for topic labelling. In: Proceedings of the 23rd international conference on computational linguistics: posters (COLING)
4. Sigurbjornsson B, van Zwol R (2008) Flickr tag recommendation based on collective knowledge. In: Proceedings of the 17th international conference on World wide web (WWW)
5. Li J, Wang JZ (2008) Real-time computerized annotation of pictures. IEEE Trans Pattern Anal Mach Intell 30(6):985–1002
6. Sen S, Harper FM, LaPitz A, Riedl J (2007) The quest for quality tags. In: Proceedings of the international ACM conference on Supporting group work(GROUP)
7. Awawdeh R, Anderson T (2009) Improved search in tag-based systems. In: International conference on intelligent systems design and applications (ISDA)
8. Ames M, Naaman M (2007) Why we tag: motivations for annotation in mobile and online media. In: Proceeding of the SIGCHI conference on human factors in computing system
9. Golder S, Huberman BA (2006) The structure of collaborative tagging systems. J Inf Sci 32(2):198–208
10. Sen S, Lam SK, Rashid AM, Cosley D, Frankowski D, Osterhouse J, Harper FM, Riedl J (2006) Tagging, communities, vocabulary, evolution. In: Proceedings of the ACM 2006 conference on CSCW
11. Bischoff K, Firan CS, Nejdl W, Paiu R (2008) Can all tags be used for search? In: Proceedings of the 17th ACM conference on information and knowledge management (CIKM)
12. Liu D, Hua X-S, Yang L, Wang M, Zhang H-J (2009) Tag ranking. In: Proceedings of the 18th international conference on World wide web(WWW)
13. Flickr, http://www.flickr.com/
14. CADAL, http://www.cadal.zju.edu.cn/

Research on Construction Technology Innovation Platform Based on TRIZ

Zhikun Ding, Shuanglong Jiang and Jinchuang Wu

Abstract Considering the insufficient knowledge management and technology innovation in the domestic construction industry, this paper proposes the design framework of the technology innovation platform based on TRIZ by taking advantage of available patent knowledge in the industry. The function modules and their roles in technology innovation are illustrated through the graphic user interfaces and the underlying database. Based on extracted construction patent knowledge, the development of construction technology innovation platform enables a heuristic environment to help the industry improve the innovation capacity and efficiency by motivating knowledge worker's innovative thinking.

Keywords Technology innovation · Construction patent · TRIZ · Technology innovation platform

1 Introduction

Construction industry is knowledge-intensive, ill-structured, and one of the most challenging industries due to the complicated, uncertain, and dynamic nature of construction operations. Besides, as one of the key industries for national economy, it is of high demand of innovation. However, the construction innovation performance in our country is not satisfying compared to that in other countries as shown by the patent statistics published by the State Intellectual Property Office. Usually, a project team is setup to manage a construction project from the beginning to the end. Different designs, construction, and management will lead to the uniqueness of each construction project, which contributes to the generation of

Z. Ding (✉) · S. Jiang · J. Wu
College of Civil Engineering, Shenzhen University, Shenzhen, China
e-mail: ddzk@szu.edu.cn

Z. Wen and T. Li (eds.), *Knowledge Engineering and Management*,
Advances in Intelligent Systems and Computing 278,
DOI: 10.1007/978-3-642-54930-4_21, © Springer-Verlag Berlin Heidelberg 2014

technical innovation and new knowledge. But the new knowledge and innovations in the construction industry are not managed well in our country [1]. With the rapid development of computer and network applications, computer technology has become the important method of managing construction knowledge, information and technology innovation. For this reason, providing the knowledge sharing platform is an effective way to assist knowledge workers in technological innovation and to promote the construction technology innovation by applying computer technology [2].

TRIZ is an acronym for Russian meaning-the theory of inventive problem solving. At present, the invention principles of TRIZ have been widely used to generate suggestions for technology innovation problems in various fields. Related research on the application of TRIZ theory as well as computer-aided innovation platform system is available worldwide. However, most of them are constrained to the field of mechanical design, while in other areas, especially in the construction industry, technology innovation practice and research have not been taken seriously. Therefore, it is critical to improve the efficiency and capability of technology innovation in the construction industry by taking into account innovation theories such as TRIZ.

This paper will design and develop construction technology innovation platform by integration of knowledge management, technology innovation theory of TRIZ and computer technology. The platform takes TRIZ theory as the pinpoint of technology innovation and builds a database of the available patent knowledge of construction industry. The development of the platform can not only help knowledge worker make full use of existing patent knowledge but also achieve technology innovation more efficiently by application of TRIZ.

2 Literature Review

2.1 Knowledge Management

Knowledge management, in short, is to build a quantitative and qualitative knowledge system to enable a loop through the process of knowledge integration, knowledge communication, knowledge innovation, and knowledge accumulation [3, 4]. Its focus is to discover, obtain, and create knowledge according to the prior successful experience and lessons. Among the processes, the most important one is knowledge communication. The reason is that knowledge worker can communicate with others, summarize similar innovation and existing experience to conduct knowledge innovation. Under the condition of knowledge communication, the accumulation of knowledge by individuals or organizations will never stop in the dynamic cycle, as shown in Fig. 1.

Luo Chuan stated that besides a successful knowledge management method, the effective environment and platform provided by advanced information technology are required as well in knowledge sharing [5]. In this study, the construction

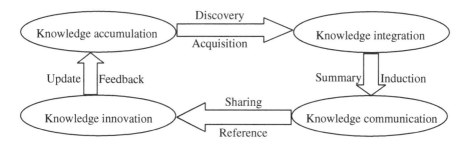

Fig. 1 The process of knowledge management

technology innovation platform serves as a media of knowledge sharing and communication, and provides integrated management of general knowledge and construction patent knowledge so that knowledge workers can utilize and share the existing knowledge to achieve technology innovation.

2.2 Triz

TRIZ is proposed by Altshuller. The theory is built on a foundation of extensive research covering hundreds of thousands of inventions across different fields to identify generalizable patterns from inventive solutions and the distinguishing characteristics of the problems that these inventions have overcome. An important part of the theory has been devoted to revealing patterns of evolution and one of the objectives pursued by leading practitioners of TRIZ has been the development of an algorithmic approach to the invention of new systems, or the refinement of existing ones [6]. In this case, the pattern and inherent law of technology innovation are extracted for knowledge workers' future inventions. In addition, TRIZ theory was extracted from technology innovation problems in different industries, therefore it can also be used in various industries. The solution mode of TRIZ to invention problems is shown in Fig. 2 [7]. Utilizing conventional solutions to explore innovations, the problems may be solved after many trials but no innovations may occur despite the hard endeavor. While using the basic principles of TRIZ, special problem can be converted to general questions in order to get the general solution, namely, the standard scheme of TRIZ. Then inventive solutions become available by combining with knowledge and experience in the specific issues. Consequently, the objectives and efficiency of getting the innovative solutions have been improved considerably through application of TRIZ.

Niu Zhanwen et al. as pioneers comprehensively introduced TRIZ to China which included 40 inventive principles, 39 general engineering parameters, its contradiction matrix table and other important contents. In addition, he pointed out that the application of TRIZ is of great potential in the future [8]. Since then, many universities and enterprises have been learning, applying, and promoting TRIZ theory. As a result, the number of journal articles about TRIZ theory is increasing

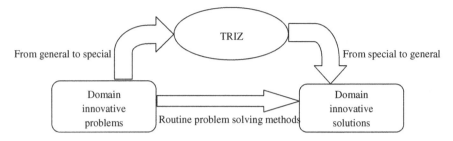

Fig. 2 The solution model of TRIZ to invention problems

every year, some of which are related to the basic content of TRIZ theory and the research progress. Beyond that, though, most of them are about research on application of TRIZ in the mechanical design fields [9]. Compared with domestic research, TRIZ theory had been widely applied in various fields abroad. In supply networks, the Multi-Agent Tool Management System was developed based on TRIZ, which clarified invention directions based on heuristics or principles to resolve the contradictions between product demand and cost minimization [10]. In the economic field, in order to choose the most suitable solution of TRIZ, the research classified and subdivided the TRIZ methods to analyze and compare with different schemes, in terms of time, costs, and resources. Besides, the proposed classification was tested and validated by means of a case study concerning a printing problem of small- and medium-sized enterprises [11]. In mathematics, researchers utilized the corresponding innovation principles of TRIZ to mark out standard ways of numerical development methods namely TRIZ-fractal map, which illustrated the basic ideas of the evolution of the numerical method. Taught procedure of systematization of mathematic knowledge by TRIZ-Fractal map, the students increased the efficiency of learning numerical methods and enhanced the understanding of nature [12]. In the field of technology innovation in Chinese construction industry, some research identified the frequently utilized TRIZ engineering parameters and innovation principles by analyzing construction template and scaffolding patents. The research findings could help the construction industry to improve the innovation capacity [13, 14]. However, other related issues in the construction industry still remained to be further researched from the perspective of TRIZ. Therefore, the research on construction technology innovation platform based on TRIZ has strong practical significance.

2.3 Innovation Platforms Based on TRIZ

There are related research about innovation platform based on TRIZ theory in literature. The existing innovation platform covered basic module of TRIZ, contradiction matrix, and browsing module of patent cases, which fulfilled the corresponding functions, as shown in Table 1. Based on QFD and TRIZ, Liao Zhiping

Table 1 Innovation platform based on TRIZ

Years	Authors	Titles	Modules	Functions	Application fields
2006	Liao Zhiping	Research on innovative product design system based on QFD and TRIZ	1. Conflict resolution 2. Case browse	1. Help user to solve conflicts 2. Inspire users	Mechanical design
2007	Tian Xin	Computer aided product innovative design and software implementation based on TRIZ	1. Knowledge base system 2. Case base system 3. Operating system	1. Related knowledge introduction 2. Related instances management and query 3. Interface system application	Mechanical design
2007	Liu Zhentao	Research on proprietary knowledge services and its platform for small and medium-sized enterprises	1. TRIZ theory 2. Patent knowledge 3. Software presentation 4. BBS	1. Introduce the source, process and cases of TRIZ 2. Introduce the patent knowledge and provide patent network link 3. Provide some information about some auxiliary software 4. Learn TRIZ and ask questions	Enterprise knowledge management

(continued)

Table 1 (continued)

Years	Authors	Titles	Modules	Functions	Application fields
2009	Zhang Yujing	Research on product innovation design and software development based on TRIZ	1. TRIZ introduction 2. TRIZ matrix 3. Technique evolution 4. Case query	1. Introduce TRIZ theory 2. Solve conflicts in the process of design 3. Help users forecast the development direction of products 4. Look for more related case	Mechanical design
2011	Tang Zhituo	Research on mechanical product creative design based on TRIZ	1. User demand investigation 2. Internal staff survey 3. TRIZ matrix 4. Product function optimization	1. Check survey questionnaire of user 2. Survey the product evaluation of employees 3. Contradiction Matrix query 4. Consider the product function optimization	Mechanical design
2012	Zhao Qian	The creation and software development of bags innovative design system based on TRIZ evolution theory	1. Query 2. Modification 3. Addition 4. Delete	1. Bag characters query 2. Modify the bags all kinds of characters 3. Add bag sample according to the characteristics 4. Delete the inconformity bag sample	Fashion design

developed the product innovation design system, which realized the structure of quality house, the call of quality house, query of conflict matrix, case browsing, maintenance of TRIZ case, and other functions. However, the structural elements, quantitative evaluation of the quality house as well as the software module function remained to be further improved [15]. According to the strength of the development tools, Tian Xin built a platform by adopting Delphi as the developing tool and Microsoft Access as the database manager. Focusing on bolted flange connection and the supply chain, the research discussed the application of the platform in the engineering management [16]. Liu Zhentao put forward the online service platform to help small- and medium-sized enterprise use patent knowledge in order to shorten the development cycle and improve development efficiency during innovation design [17]. Zhang Yujing developed a platform which provided knowledge workers with guidance and design directions according to problems. Besides, the research verified the effectiveness of the evolutionary theory and the software system through an analysis of the wrench instance [18]. From the viewpoint of mechanical product innovation design, Tang Zhituo developed a computer-aided innovation design platform of mechanical products that proposed a four-module system framework, such as the user demand investigation, internal staff survey, contradiction matrix, and the product function optimization [19]. Based on the viewpoint of luggage innovation design, Zhao Qian put forward a set of design systems for bag design by means of VB development and knowledge database. The research mainly presented the study on the realization of searching the product characteristics, evolution mode and invention principle, modifying, adding, and deleting the information of the system, which could afford good reference value for bag designer and improve the design efficiency [20]. But most studies are limited to the field of mechanical design, the research in other fields are not much. With the rapid development of construction industry in recent years, the amount of solid building patents in China grew from 18,878 in 2009 to 36,054 in 2011 according to the statistics published by State intellectual property office. As the growth during the past years is almost doubled, research on construction technology innovation platform based on TRIZ theory is particularly important to systematically manage construction patent knowledge and encourage further innovation.

3 The Development of the Construction Technology Innovation Platform

3.1 The Function Modules of the Construction Technology Innovation Platform

The development of construction technology innovation platform involves construction patent knowledge management, the knowledge of TRIZ theory, and computer programming technology. Construction patents have been published in

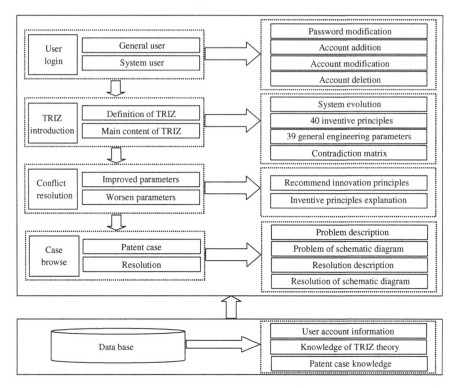

Fig. 3 The function modules of platform

the patent information service platform which contains all public available patents of the construction technology innovation [21]. The platform takes Visual C++ 6.0 as the development tool and adopts SQL Server 2000 database management system to save the available construction patent, knowledge of TRIZ theory and user account information. The function modules of the platform are shown in Fig. 3. In the user login module, users are required to input account information before entering, and they can modify their account information after login. The module of TRIZ introduction can help users to know more about TRIZ theory, including the explanation of system evolution, 40 inventive principles, 39 general engineering parameters and contradiction matrix. Then, in conflict resolution, users can find the corresponding recommended principles by choosing improved parameters and worsen parameters related to their own problem. Moreover, the available patent knowledge of recommended principles can be referred by users in the module of patent case browse. Therefore, user's innovative thinking may be motivated by the description and schematic diagram of the reference patent.

Fig. 4 The initial interface

3.2 Visualization

The following graphical user interfaces (GUI) will illustrate the operations of the platform. After inputting the user name and password, the platform enters the initial interface, as shown in Fig. 4. If a user logs in as the system administrator, he can modify, add, and delete account information while the ordinary users can only change his password, as shown in Fig. 5.

Users can choose general engineering parameters including improved and worsened parameters according to their own problems. In Fig. 6, 39 general engineering parameters in TRIZ theory are listed in the two list boxes. Users determine the contradiction between general parameters in TRIZ on the basis of the problem description in a specific field. After clicking on the "OK" button, the TRIZ recommended principles appear. There are about 4 recommended principles with the details listed on the right. If there is not recommended principle in the TRIZ contradiction matrix, the dialog "Sorry as there is no recommended innovation principle" will appear. Users can redefine the problem so as to acquire the TRIZ recommended principles. If these recommended principles are so abstract for the user that it is not operational to solve the problem, users can click on the case browsing button to read related application principles and patent cases, as shown in Fig. 7.

3.3 Database

The successful implementation of the above user interfaces depends critically on the database storing related information and knowledge. There are four tables (namely, user information table, innovation principle table, contradiction matrix

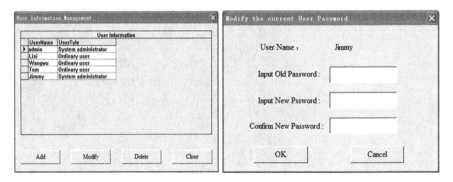

Fig. 5 Account information modification

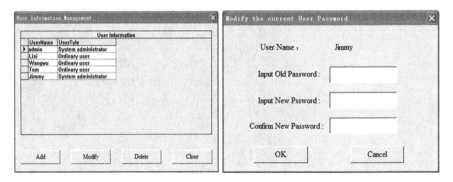

Fig. 6 Contradiction matrix

table, and patent information table) created in the SQL Server 2000 database
management system. User information table is shown in Fig. 8. It saves account
information and modifies information in Fig. 5. Figure 9 shows the index number
and the knowledge of TRIZ contradiction matrix. Based on this information, Fig. 6
can properly display the recommended principles and instructions, which is shown
in Fig. 10.

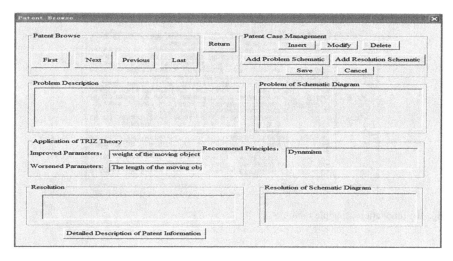

Fig. 7 Patent browse

Fig. 8 User information table

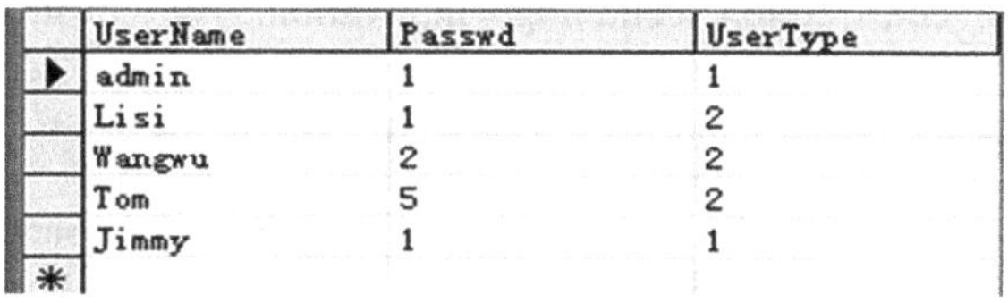

	UserName	Passwd	UserType
▶	admin	1	1
	Lisi	1	2
	Wangwu	2	2
	Tom	5	2
	Jimmy	1	1
*			

Fig. 9 Contradiction matrix table

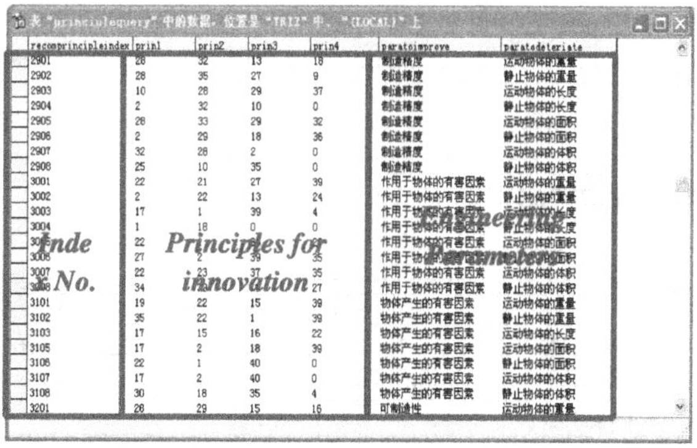

4 Conclusions

This paper discusses the application of TRIZ theory and proposes the importance of patent knowledge management in the construction industry. By integrating the function modules of some existing platforms with the current construction technology innovation practice, a design framework of construction technology innovation platform based on TRIZ is presented. The construction technology

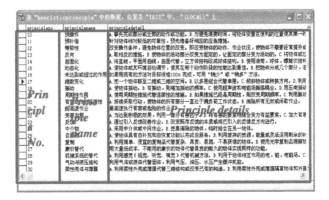

Fig. 10 Innovation principle table

innovation platform is developed by applying object-oriented programming and relational database design. The objective of the innovation platform is to strengthen construction patent knowledge management and improve the capability of construction technology innovation. The construction technology innovation platform is helpful to improve the efficiency and accomplishments in the process of technology innovation problem solving.

Acknowledgments This research is conducted with the support of the National Science Foundation for Young Scholars of China (Grant No. 71202101); Humanities and Social Science Research Funding (No. 10YJCZH025), Ministry of Education of P.R.C; Scientific Planning Research Grant (No. 2009-K4-17, No. 2011-K6-24), Ministry of Housing and Urban-Rural Development of P.R.C; Shenzhen University Education Research Grant. The authors are also thankful to those who participated in this study and supported the research work.

References

1. Hua L, Mei L, Mei S (2011) Shortcomings and innovative strategies of technology innovation in Chinese construction industry. J Eng Manage 25:359–363
2. Huiling L, Liang Z (2007) The key and countermeasures research on the construction of technology innovation in our country. Ind Technol Econ 4:35–36
3. Congying W, Qing L, Jianping W (2010) Analysis of impact factors of knowledge management for construction enterprises. J Eng Manage 1:96–101
4. Junxiang Y, Jinsheng H (2013) Study on relation between knowledge management internal driving force and knowledge management dynamic ability. Stud Sci 2:258–265
5. Chuan L (2005) Knowledge management and the development of information technology. J chongqing jiaotong univ 4:145–148
6. Wikipedia TRIZ. http://en.wikipedia.org/wiki/TRIZ
7. Mingqun Z, Cunli F, Rijun W, Shijun Z (2012) Introduction to TRIZ 100 q—TRIZ innovative tool guide. China Machine Press, Beijing, pp 9–11
8. Zhanwen N, Yanshen X, Yue L, Ianqiang GJ, Li LI (1999) Inventive scientific methodology-TRIZ. Chin J Mech Eng 1:92–97

9. Qian L, Fujun Z, Yinming L, Yabo W (2012) The existing problems and deficiencies in the domestic TRIZ theory research. Sci Technol Innov Herald 18:221–222
10. Teti R, Addona DD (2011) TRIZ based tool management in supply networks. Procedia Eng 9:680–687
11. Filippi S, Motyl B, Ciappina FM (2011) Classifying TRIZ methods to speed up their adoption and the ROI for SMEs. Procedia Eng 9:172–182
12. Berdonosov V, Redkolis E (2011) TRIZ-fractality of mathematics. Procedia Eng 9:461–472
13. Zhikun D, Jiefu W, Manbo H, Jiayuan W (2011) Innovation principles of construction template patents based on TRIZ. J Eng Manage 2:143–146
14. Zhikun D, Jiehui L, Jian G, Jiayuan W (2011) Research about innovation principles of construction scaffold patents. J Guangzhou Univ (Nat Sci Ed) 6:84–88
15. Zhiping L (2006) Research on innovative product design system based on QFD and TRIZ. Master thesis, Zhejiang University of Technology
16. Xin T (2007) Computer aided product innovative design and software implementation based on TRIZ, Master thesis, Shaanxi University of Science and Technology
17. Zhentao L (2007) Research on proprietary knowledge services and its platform for small and medium-sized enterprises. Master thesis, Shanghai Jiao Tong University
18. Yujing Z (2009) Research on product innovation design and software development based on TRIZ. Master thesis, Shandong University
19. Zhituo T (2011) Research on mechanical product creative design based on TRIZ. Master thesis, Tianjin University of Technology
20. Qian Z (2012) The creation and software development of bags innovative design system based on TRIZ. Master thesis, Shaanxi University of Science and Technology
21. Guangdong Intellectual Property Office (2013) Guangdong patent information service platform. http://www.gdzl.gov.cn/

The Improvement Research of Mutual Information Algorithm for Text Categorization

Lu Kai and Chen Li

Abstract Mutual information (MI) algorithm has many shortages in the feature selection of text categorization compared to other selection algorithms. For these shortages, this article introduces some important factors like term frequency or something that MI has not yet considered, and then puts forward the improved MI algorithm based on the composite ratio factor. And by the experiment the improved method can get a good improvement effect.

Keywords Feature selection · Text categorization · Mutual information · Composite ratio

1 Introduction

Text categorization is that according to the content of texts, assigns texts to the predefined categories with some kinds of text categorization algorithm [10] by computer, and it is an important issue in the IR and data mining, etc. Text categorization is generally divided into four stages: text pretreatment, text characters representation, feature dimension reduction, and training classification stage. In the text characters representation stage, we generally adopt vector space model (VSM) to represent texts. Among the four stages, the high dimensionality of vector space is one big difficulty. How to reduce the dimension effectively, remove the redundant features, and improve the efficiency and precision of categorization is the first problem to solve. Current feature selection algorithms are document

L. Kai (✉) · C. Li
School of Computer Science, Central China Normal University (CCNU), Wuhan, China
e-mail: 329408650@qq.com

C. Li
e-mail: 403609111@qq.com

Z. Wen and T. Li (eds.), *Knowledge Engineering and Management*,
Advances in Intelligent Systems and Computing 278,
DOI: 10.1007/978-3-642-54930-4_22, © Springer-Verlag Berlin Heidelberg 2014

frequency (DF), information gain (IG), card square, and mutual information (MI). Of all the methods, the effect of MI is the worst [1, 6].

Mutual information is one algorithm of information theoretic method of feature selection for text categorization [7]. In recent years, there are many improvements about MI, such as the literature [2] imports the dispersion among categories to MI algorithm; literature [3] puts forward to introduce the term frequency, cohesion factor among categories and coupling degree inside a category to MI algorithm; literature [4] mainly takes the term frequency in the training set into consideration to improve MI. According to the general strategy of the improvement, this article introduces the composite ratio (CR) factor to MI algorithm, and the improved MI algorithm makes some improvement on time performance and the categorization accuracy by the experiment.

2 The Research and Improvement

In this section, we first introduce the traditional MI algorithm and its shortages. Then we come up with the improved MI algorithm.

2.1 Introduction of Mutual Information

In the feature selection stage, MI shows the relevant extent between term t and category C_i [8]. The larger MI value is, the larger the relevant extent between class C_i and term t. The formula is shown below:

$$\text{MI}(t, C_i) = \log \frac{p(t, C_i)}{p(t)p(C_i)} = \log \frac{p(t/C_i)}{p(t)} \approx \log \frac{AN}{(A + C)(A + B)} \tag{1}$$

Among the formula (1), the parameter A is that the number of docs that include term t in category C_i, the parameter B is the number of docs that include term t in the other categories except category C_i, and the parameter C is in category C_i the number of docs that do not include term t.

For different categories, we need to calculate the MI value of term t in every category, respectively, and then calculate the MI value of term t in the entire training set by different algorithms. Generally there are two algorithms to calculate the total MI value, and they are weighted mean square algorithm and maximum value algorithm. They are showed as follows:

$$\text{MI}_{\text{avg}}(t) = \sum_{i=1}^{|c|} p(C_i)\text{MI}(t, C_i) \tag{2}$$

$$MI_{max}(t) = \max[p(C_i) * MI(t, \ C_i)] \tag{3}$$

This article uses the weighted mean square algorithm (the formula (2)).

2.2 The Shortages of Mutual Information

Of all the feature selection methods, the effect of MI is the worst. Empirical studies show that there are some factors that influence the result of feature selection and they are term frequency and categories information [9]. But MI does not take these important factors into consideration.

1. One of the shortages is that MI is affected by the critical characteristics significantly. According to the formula (1), when the $P(t/C_i)$ value is equal, rare terms have larger MI value than the common term. So it tends to select the low frequency term. For example, as follows in Table 1.

   According to the MI formula (1) and Table 1, it shows that the MI value of the term t_1 is equal to the MI value of term t_2 in category C_i. But in fact the ability of distinguishing category of term t_2 is higher than term t_1, because in category C_i the term t_2 has a higher term frequency. So if we use the traditional MI formula, the ability of distinguishing category of the term t_1 and term t_2 is equal. So it is inconsistent with the fact.

2. According to the MI formula (1), if $P(t)$ value is larger and $P(t/C_i)$ is smaller, then the MI value is much smaller, and the MI value may even be negative, and then the probability that selecting the term is smaller. The negative MI value explains that the term appears in the category rarely or not appears in the category, but may appear in other categories. So the term will dislodge by the traditional MI formula, and in fact these terms are also important.

2.3 The Improvement

The general improvement strategy of MI is to introduce some useful factors to the MI algorithm, and these factors may contain some important information that the traditional MI does not consider. The literature [5] puts forward CR algorithm to select characteristic terms, and fully considers some important information between term and category. So this text introduces CR information into the MI algorithm, and puts forward the MI improvement algorithm MI_CR.

CR factor takes some related elements between terms and categories into consideration. Some parameters are shown in Table 2.

In Table 2, the parameter A is in category C_i the number of document that include term t, the parameter C is the number of document that include term t in

Table 1 The times that one term appears in every doc in category C_i

Term	Category C_i		
	Doc_1	Doc_2	Doc_3
t_1	2	3	0
t_2	5	0	10

Here we suppose that category C_i only includes three documents (Doc_1, Doc_2, Doc3), and just list two terms. The number in the table is the times that the term exist in the doc

Table 2 Some parameters that are used in the MI_CR formula

Term	C_i	$\overline{C_i}$
t	A	C
$\bar{t}$	B	D

Parameter A, B, C, and D are all the number of documents. $\overline{C_i}$ are the categories that except category C_i

other categories except category C_i, the parameter B is the number of document that do not include term t in category C_i, and the parameter D is in other categories except category C_i the number of document that do not include term t.

About the CR factor, there are four important elements that can estimate the ability of distinguishing categories of terms:

1. If term t is appeared in the most documents of category C_i, then it suggests that in the category C_i the term t is widely distributed. So the term t has a strong relevance with category C_i, and can represent category C_i. The formula is $A/(A + B)$. The larger this value is, the more relevance with C_i the term is.

2. If term t has a higher DF in the category C_i, and has a lower DF in other categories, and then it suggests that there is a strong correlation between term t and category C_i. The formula is $A/(A + C)$. The larger this value is, the more relevance with C_i the term is.

3. If the most documents of category C_i include the term t and in other categories most documents also include the term t, then term t has a lower capacity to distinguish from category C_i. The formula is $A/(A + B) - C/(C + D)$. The larger the value is, the stronger the ability of distinguishing the categories of term t is.

4. From the aspect of term frequency, if term t has a higher average term frequency in category C_i and a lower average term frequency in other categories, then the term t is more representative for Category C_i.

Considering the four factors above, we can know that the third factor considers the disparity among categories and is almost same to the second factor. So this text only considers the 1, 2, and 4 factors.

So, based on the factors above, we can get the CR factor formula as follows:

$$CR(t, C_i) = \frac{A}{A+B} \times \frac{A}{A+C} \times \frac{TF(C_i)}{TF(\overline{C_i})} \qquad (4)$$

This article introduces the CR factor into MI formula, and then can get the improved algorithm based on CR factor. The formula is shown below:

$$MI\_CR(t, C_i) = \log \frac{AN}{(A+C)(A+B)} + \log CR(t, C_i) \qquad (5)$$

The specific steps about term selection:
Input: term set T and category set C
Output: characteristic term set T'

Step 1:   In the pretreatment stage, work out the document frequency DF_{ij} and term frequency TF_{ij} of term t_i ($t_i \in T$) in the Category C_j ($C_j \in C$).

Step 2:   According to the result of the 1st step, work out the parameter A, B, C, D; According to the TF_{ij}, work out the average of term frequency of every term in every category.

Step 3:   According to the improved MI formula (MI_CR), work out the MI_CR value of every term in every category, and work out the average of MI_CR value of every term.

Step 4:   According to the average value of MI_CR, arrange terms from large to small, and then select the TOP-K (K is given in advance) terms as the result characteristic word set T'.

3 The Experiment and Result Analysis

In this section, we first do experiments to investigate the effect of these feature selection algorithms including MI, TF_CA_CI, and MI_CR. Then we do the time complexity analysis and make experiments to validate the analysis.

3.1 The Experiment Result and Analysis

This article uses the news dataset of SouGou lab, and selects four categories: finance, education, travel, and military category. From every category, we select 400 documents as the training set, 100 documents as the test set randomly. This article uses the word segmentation tool of Chinese Academy of Sciences (ICT-CLAS) as the Tokenizer, and the categorization method is KNN categorization algorithm. Feature selection methods are respectively the traditional MI algorithm,

Table 3 Result of the experiment

Categories	Methods		
	MI	TF_CA_CI	MI_CR
Finance	0.459/0.62	0.489/0.64	0.492/0.84
Travel	0.776/0.59	0.79/0.64	0.785/0.62
Education	0.6/0.6	0.705/0.55	0.756/0.59
Military	0.685/0.61	0.62/0.67	0.797/0.55
Macro-average accuracy/recall	0.63/0.605	0.651/0.625	0.71/0.65/
Macro-average F1	0.617	0.638	0.68

The number before the slash is accuracy rate (p), and the other number is recall rate (r). The F1 value is a synthetically value and the formula is $2pr/(p + r)$. TF_CA_CI is also an improved mutual information algorithm and is latest, so we can compare MI_CR algorithm with the TF_CA_CI and MI algorithm to embody the effect of MI_CR algorithm

the TF-CA-CI-based improved MI algorithm of literature [3], and MI-CR algorithm of this article. In the feature selection step, we select the top 20 % as the final terms and use accuracy rate (p), recall rate (a), and F1 value as the performance evaluation method of text classifier. The result of the experiment is shown in Table 3.

From the result of the experiment, the MI_CR algorithm shows a good improvement effect on the whole except the travel category. So the improved algorithm for the traditional MI algorithm is effective and attains a better improvement effect compared with the traditional MI and the other improved MI algorithm of literature [3].

3.2 Time Complexity Analysis

The TF_CA_CI-based improved MI algorithm of the literature [3] is also an improved MI algorithm; it takes the term frequency, cohesion factor among categories and coupling degree inside a category into consideration.

The formula of TF_CA_CI factor is

$$\text{TF\_CA\_CI} = \sum_{i=1}^{j} \text{tf}_{ik} + \frac{(\text{df}_i - \overline{\text{tf}_i})^2}{\overline{\text{tf}_i}} + \text{df}_i \qquad (6)$$

$$\overline{\text{tf}_i} = \frac{\sum_{i}^{c} \text{df}_i}{|C|} \qquad (7)$$

According to the traditional MI formula (1), the parameters in Table 2 are worked out in the pretreatment stage, so we can use them directly. In the MI_CR algorithm, the first and second parts are the same order of magnitude to the third

Table 4 Result of experiment about the time complexity analysis

Algorithm	MI	TF_CA_CI	MI_CR
Time (s)	146	219	185

MI is the traditional MI algorithm, and TF_CA_CI is also an improved MI algorithm based on TF_CA_CI. MI_CR is the improved algorithm in this article and we introduce some important factors, so MI_CR needs more time to calculate compared to the traditional MI algorithm

part of the TF_CA_CI. And the calculation of term frequency is also same. So according to the formula, document and term frequency is known in advance, and the two algorithms have the same time complexity O (nm). But in the TF_CA_CI algorithm it needs to calculate the formula (7), so it needs to calculate the document number of every term in categories. For this reason, the MI_CR is a little better than the TF_CA_CI algorithm on time performance and the experiment result also can confirm the correctness of the argument.

So according to Table 4, we know that MI_CR algorithm uses less time than the TF_CA_CI algorithm. So the experiment proves that our time complexity analysis is right.

4 Conclusions

According to the general improvement strategy, this article introduces CR factor into the traditional MI to improve the MI algorithm, and takes some important factors that the traditional MI does not consider into consideration, and proves that the improved MI algorithm is superior to the traditional MI by the experiment. But in the experiment result the travel category does not show the improvement effect. So next job is to find which category is most suitable for the improved MI algorithm, and for these categories do the further improvement, and make the improved MI algorithm can show the best results than other feature selection algorithms for these categories.

References

1. JingShu S, BoFeng Z, Xin X (2006) Advances in machine learning based text categorization. J Softw 17(9):1848–1859
2. Jian L, WeiMing Z (2008). Study and improvement of mutual information based text feature selection method. Comput Eng Appl 44(10):135–137
3. JiaJia C, DeXian Z (2013) Improvement of TF-CA-CI algorithm-based mutual information selection. Comput Appl Softw 30(3):255–257
4. Zhilong F (2011) Improvement of mutual information of feature extraction. Comput Mod 7:49
5. YuFang Z, Yong W (2013) New feature selection approach for text categorization. Comput Eng Appl 49(5):132–135

6. LiLi S, BingQuan L (2011) Comparison and improvement of feature selection method for text categorization. J HARBIN Inst Technol 43:2011

7. Kumar P, Babber S (2013) Information theoretic method of feature selection for text categorization. Int J Math Arch (IJMA) 3(12):2229–5046

8. XiaoLi F, XiaoXia L (2010) Study on mutual information-based feature selection in text categorization. Comput Eng Appl 46(34):123–125

9. Xu Y, Li JT, Wang B, Sun CM (2008). A category resolves power-based feature selection method. J Softw 19(1):82–89

10. Li S, Xia R, Zong C et al (2009) A framework of feature selection methods for text categorization. In: Proceedings of the joint conference of the 47th annual meeting of the ACL and the 4th international joint conference on natural language processing of the AFNLP, vol 2. Association for Computational Linguistics, pp 692–700

Service-Oriented Knowledge Acquisition Paradigm and Knowledge Cloud Platform

Yuan Rao and Shumin Lu

Abstract Cloud computing is an emerging new computing paradigm for delivering some computing services to consumers, which provides scalable and inexpensive service-oriented computing infrastructures with good quality of service levels. In this paper, based on the studies about the characteristic of knowledge service, a feature model of knowledge service is proposed with three features, such as service requirement, knowledge service process, and the quality of knowledge service (QoKS). Furthermore, the architecture about service-oriented knowledge acquisition is built to provide a fundamental for knowledge service, and the lifecycle process about knowledge service is studied under the service-oriented architecture. Then, a best practice about Knowledge Cloud, named Eknoware, is developed based on the architecture above-mentioned, which can provide some knowledge service patterns and reorganize the knowledge clusters to suitable customers.

Keywords Service-oriented knowledge acquisition · Knowledge as a service · Cloud computing · Eknoware · Knowledge cloud platform

1 Introduction

Cloud Computing is a model for enabling convenient, on-demand access to a shared pool of configurable computing resources (e.g., networks, servers, services, applications, and storage) that can deliver an open platform to integrate knowledge as a kind of cloud service in the next four-to-eight years [1]. Murray [2] thought

Y. Rao (✉)
College of Software Engineering, Xi'an Jiaotong University, Xi'an 710049, Shannxi, China
e-mail: yuanrao@163.com

S. Lu
College of Social science, Xi'an Jiaotong University, Xi'an 710049, Shannxi, China

Z. Wen and T. Li (eds.), *Knowledge Engineering and Management*,
Advances in Intelligent Systems and Computing 278,
DOI: 10.1007/978-3-642-54930-4_23, © Springer-Verlag Berlin Heidelberg 2014

that the knowledge as a service (KaaS) is becoming "the future of the future". Knowledge or semantic-based techniques and technologies combined with conventional IT are promoting a new "wave" to provide the understanding, insight, and experience of knowledge-related services and knowledge innovation products. All these actionable knowledge and services are produced by staff, customers, or suppliers in one enterprise and by cloud computing systems to form a knowledge sharing platform with more opportunities to discover and purchase the service or knowledge, which can be acquired directly from producers by search engine with relevant overwhelmed data [3], by social network with crowd-sourced answers [4] and by online consulting with domain knowledge experts [5]. Therefore, Knowledge service needs new approach from a common knowledge delivery to a personalization acquisition for end users. According to the characteristics of service with intangibility, heterogeneity, and inseparability, a public knowledge service mechanism and application platform are investigated to integrate professional knowledge acquisition process and knowledge service strategies together under cloud computing environment in the paper.

The rest of this paper is organized as follows: Sect. 3 presented an overview of service-oriented knowledge acquisition and related works. Section 3.1 researched on the features and existed problems about knowledge service and proposed a new mechanism of service-oriented knowledge acquisition based on cloud computing environment. Section 3.2 illustrated the architecture about knowledge service cloud platform with multi-element and analyzed the knowledge services invocation operations to "push" the relevant services together into the composited knowledge. Section 4 provided a case study, EKNOWARE platform with a new knowledge service, to illustrate the merits of service-oriented knowledge acquisition. Finally, the conclusion and future works is presented in Sect. 5.

2 Related Research Works

Recently, more and more researchers [6, 7] are focusing on the KaaS in cloud computing paradigm to forge a new thinking and powerful methodologies for service innovation and operational improvement. From the definition of knowledge [2], the four viewpoints, such as (1) personal perception (2) output acquired from information (3) organizational resources, and (4) combination of personal perception and output of information, also can build a knowledge management process comprised of knowledge creating, organizing, sharing, and using tacit and explicit knowledge. Wong [8] proposed a knowledge engineering to utilize the repository information and knowledge to inform the future decision and activities with four key activities, i.e., knowledge creation, knowledge mapping, knowledge retrieval, and knowledge use. Therefore, how to bind the viewpoints of knowledge and activities of knowledge engineering together to find a suitable knowledge acquisition mechanism for different users is very important. Moreover, the Intellectual Capital Index (ICI) model [9] is described by the Intellectual Capital dimension tree and divided intellectual capital into human capital with the sub-elements of competence, attitude, and intellectual property, and structural capital with relationship capital, organizational capital, and the renewal and development value. Furthermore, Some companies, such as RightAnswers, Salesforce, CumulusIQ and NTT utilized these value model of knowledge and tools of knowledge service to manage and share enterprise knowledge capital and gain competitive advantage by extending human capital and enhancing the effectiveness of knowledge service strategies intra- or inter-enterprise. In general, the knowledge service strategies [10] can be divided into three aspects: knowledge management service, knowledge value-added flow and creative service, and socialized knowledge exchange and trade in cloud. In particulate, the service-oriented knowledge acquisition and knowledge cloud platform technology [11] will provide a challenge to resolve the semantic problems for automated knowledge service without much manual intervention, which can put knowledge into all the business process to win more value-added.

3 Service-Oriented Knowledge Acquisition Paradigms

3.1 Service and Knowledge Service Characteristics

A service, as an intangible experience performed for a customer to pay the producer for a performance or promise of a performance, is composed of the service providers and service consumers, the service process, and physical fundamental environment. The different services may be selected and organized by different customer to satisfy their certain requirements. The Service Value should be balanced by the quality of process and the costs of acquiring the service. Therefore, a service can be illustrated as follow:

$$\Psi(s) = (\Omega(\text{person}) + \Phi(\text{process}))^{p(t)} \tag{1}$$

where:

- $\Psi(s)$ denotes a provided service, which can be managed and classified in various domains for satisfying the user's service requirement.
- $\Omega(\text{person})$ denotes a limited set of persons, who is related to certain service in one organization, such as staff, customers, or suppliers and needs the service to consume or to create new service for consume. Therefore, all these persons can be divided into two kinds of roles: service provider and service consumer.
- "+" denotes a physical transfer channel. IT technique, today, including service management software system or knowledge service cloud, is becoming the more and more important mechanism for service acquisition and utilization, which can decrease the cost of service acquiring and knowledge clustering for different person's requirements over Internet.

- Φ(process) denotes the process of service, which defines an operational flow of activities with a serial of standard steps in the process.
- $p(t)$ denotes the degree of service's performance with time. It means that the service is time-perishable and the performance of service will change with time and utilization method.

By this definition, we can understand the kernel value about service is to provide the best process to right person with the best satisfaction by suitable channel. The second one is to share and propagate the right knowledge service to certain persons who need it. But, in the era of knowledge economic and cloud computing, especially, knowledge service is focusing on the personalized requirements by suitable mobile devices and more applications, which not only changes the interactive relationship between person and service process, but blurs the border between service provider and service consumer. All users have the same responsibility for coproducing knowledge service, depending exclusively on help desk analysts, developers as an alternative to traditional consulting and systems integration engagements. Everyone is an owner to obtain the more relevant knowledge service combined with his personalized requirement and to share something with others, including contents, valuation, and recommendation. Therefore, knowledge service not only changes the process to satisfy personalization knowledge acquirement and to yield the best results for specially question, but also changes the knowledge sharing mechanism from passive reading to active creating by cloud pattern. A feature model of knowledge service is proposed as follow:

As a new paradigm of resource sharing under cloud computing environment, knowledge service also can be built from three different features in Fig. 1, such as knowledge service requirement, the quality of knowledge service, and the process of knowledge service. From the aspect of service requirement, knowledge service is a process driven by user's knowledge requirement, which aims to solve user's personalized question from simple keywords to semantic understanding under the business scenario. From the aspect of the service lifecycle, the knowledge service lifecycle should be supported from creation and mapping to retrieval from simple knowledge to aggregation knowledge, and end to combination service with complex knowledge under user's context environment. From the aspect of the quality of knowledge service (QoKS), the related personalization solution is provided by the filter of person's profile model and to satisfy the user's questions and requirements.

3.2 Service-Oriented Knowledge Acquisition Architecture

Service-oriented knowledge acquisition is on demand, where approach needed to skill acquisition and knowledge reorganization, which decouples customization activities into helps, training, and solutions from monolithic project. The

Fig. 1 The feature model
about knowledge service

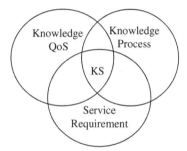

knowledge service can be acquired on a browser, self-service, and socially-net-worked basis in the cloud. In general, the strategies about service-oriented knowledge acquisition can be divided into three aspects: knowledge resources, knowledge service objects management, the knowledge service roles in cloud environment. The definition about Services-Oriented Knowledge Acquisition Architecture can be represented by a three tuple:

$$\xi = (\Omega, \Sigma, \Pi) \qquad (2)$$

where:

- Ω is a finite set of knowledge resource objects with $\Omega = \{r_i | 1 \le i \le n, r_i \notin \varphi\}$, which includes the vital information, knowledge and experts about program, and process resources of enterprise.
- Σ is a finite set of knowledge service objects (e.g., knowledge service) with $\Sigma = \{s_j | 1 \le j \le m, s_j \notin \varphi\}$, which encapsulated the knowledge resources into knowledge service index and stored them into the knowledge service base (KS Base) for user's query and acquisition;
- Π is a finite set of roles with $\Pi = \{p_f | 1 \le f \le k, p_f \notin \varphi\}$, which includes three kinds of role in traditional architecture, such as knowledge service provider (KSP), knowledge service broker center (KSBC), and knowledge service requester (KSR). Each role provides the different functions in whole architecture.

For instance, knowledge service provider, as a knowledge service creator, not only can provide standard service process and value-added trade methods for requester, but also publish the profile information about service and provider to KSBC over Internet. As a limited filtering condition, these profile information are very important for knowledge extracting and filtering. The knowledge service broker center (KSBC) as a catalog of knowledge services and knowledge reor-ganized center, which introduces a lightweight mechanism to discover the different categories of knowledge services in this architecture, such as content service, experts services and crowed-sources, service by search engine, instant message tools, and social network platform, respectively. All the knowledge services need to build a service index in KSI by Service Agent and can store all information into

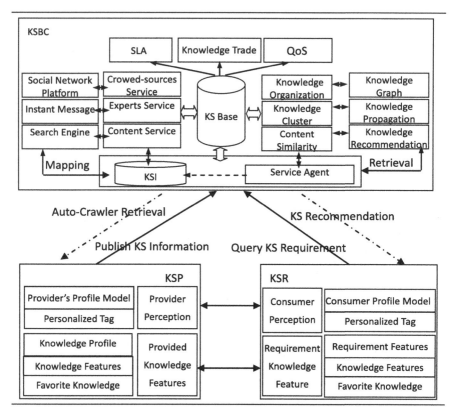

Fig. 2 The architecture of service-oriented knowledge acquisition

the Knowledge Service Base for query and acquisition. The knowledge service requester (KSR), as knowledge service customer, can acquire the knowledge services filter by the personalization requirements and design specifications.

Based on the definition above-mentioned, the architecture of service-oriented knowledge acquisition is illustrated as Fig. 2. On the one hand, the person in KSR client can query the knowledge service by catalog of service in KSBC to satisfy his personalization requirement in this mechanism. On the other hand, KSBC also can acquire the information of consumer perception and the knowledge requirement features, and then can analyze and organize these relative knowledge services together, and end to recommend some knowledge service package, such as a knowledge graph or knowledge clustering, back to user automatically. Meanwhile, the providers (KSP) have two mechanisms to publish new knowledge service information into KSBC, one is to publish the information about provider's profile model and service features information into KSBC and build a Knowledge service Index, and another is to retrieve the knowledge service by auto-crawler in KSBC to acquire suitable service information automatically.

4 The Case Study About Knowledge Cloud Platform

4.1 The Process of Service-Oriented Knowledge Cloud

KaaS, as a service-oriented knowledge cloud platform, can provide the best reorganized knowledge, in the best context, at the best time, with the best experts, in best service mechanism, as needed to satisfy different user's personalization requirement with a standard service process. Moreover, KaaS needs to leverage the cloud to access and make available all sorts of knowledge, provided by knowledge service provider and stored in KS Base, from static repositories to more interesting collective knowledge and to real-time customer contribution. Therefore, based on the architecture of service-oriented knowledge acquisition, the detailed lifecycle of knowledge service acquisition with some main steps should be analyzed from KSBC perspective in this section and shown in Fig. 3.

The first step is to publish the knowledge service and KSP's profile information. The knowledge service or knowledge content and some subsidiary information about KSP, such as profile model and service publish specification, should be published together into KS Resources Bases (KSRB) for increasing the accurate of knowledge filtering and clustering in the KSBC.

The second step is to reorganize and cluster the knowledge and services in KSBC. The service engine, composed by knowledge aggregation engine, service context engine, and profile analysis engine, can retrieve the knowledge service feature and store these service brief index information into Knowledge Directory, which is the global catalog of knowledge service information.

The third step is to map and utilize the knowledge service. When users utilize the KSR Client to access the KaaS Cloud, the requirement and some profile information of KSR should be published into KS Resources Bases firstly. Then, the Knowledge Directory is utilized by KSRs to search and query what knowledge services they want. Meantime, the service engine is used to analyze and aggregate the related knowledge services limited by service QoS and SLA, which can satisfy to the KSR's requirement and can push some valuable knowledge to KSR clients automatically.

4.2 The Case Study of Service-Oriented Knowledge Cloud Platform

Based on the architecture and the lifecycle of service-oriented knowledge acquisition, this paper provides a knowledge cloud platform, named Eknoware (http://eknoware.com, as shown in Fig. 4), to various knowledge consumers and providers, such as personal users and some enterprise or organization users, by social networks for promoting the knowledge propagation and consumption. EKNOWARE

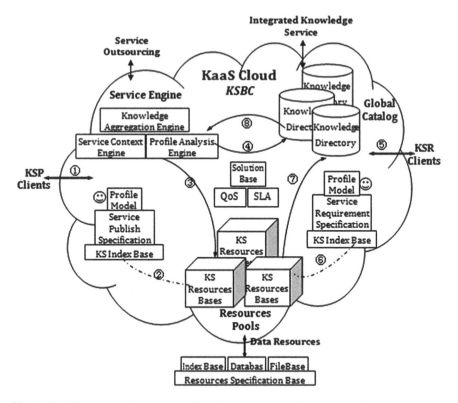

Fig. 3 The lifecycle and key process of service-oriented knowledge acquisition cloud

platform provides some basic functions for knowledge sharing and learning, such as expert portal, KS activity, KS project management, online questions and answers, discussions forums, and instant messages to communicate with multi-users to construct an online knowledge community. Knowledge Service Resources Base, as a fundamental service, can store and provide various relative knowledge resources for the knowledge engine to classify and aggregate together. A knowledge service dashboard is utilized to select the suitable service functions to provide the best knowledge content to KSR. In addition, in order to reuse these knowledge resources, the Eknoware platform also can provide an OPEN API by Oauth 2.0 to third parts application, which not only can exchange knowledge between different users in inner platform, but also share some knowledge services to different knowledge groups in third part's applications. Some core functions about Eknoware can be illustrated in Fig. 4 with more detailed information.

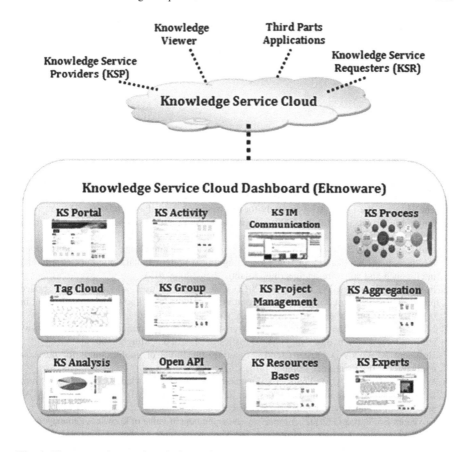

Fig. 4 Eknoware: the practice platform of service-oriented knowledge acquisition cloud

5 Conclusions and Future Work

In cloud environment, the purpose of knowledge service, however, is to reuse knowledge resources and create a new value-added service method with more efficiency. Firstly, the characteristic of knowledge service is studied and the three features, such as service requirement, knowledge service process, and the quality of knowledge service, is analyzed in this paper. Then, a feature model of knowledge service is proposed as an infrastructure for further research. Secondly, the architecture about service-oriented knowledge acquisition is built to provide a fundamental of knowledge service, and the lifecycle process about knowledge service is studied under the service-oriented architecture to promote the implementation of knowledge service. In the end, a best practice about Knowledge Acquisition Cloud, named Eknoware platform, is developed based on the architecture above-mentioned, which can provide some knowledge service patterns and push the best knowledge to suitable customers.

Acknowledgments This paper is jointly supported by "2012 National Torch Plan Project in China (2012GH571817)", "2012 the Fundamental Research Funds for the Central Universities in China(08143003)," "2012 Shanxi Province Key Scientific and Technological Project (2012K11-08, 2012K06-18)," and "2012 XI'AN Science and Technology Project (CX12178(3))".

References

1. Knowledge as a Service (KaaS) (2013) A proposal, Available http://www.stonecobra.com/knowledge-as-a-service-kaas-a-proposal/ (Cited 1, Step. 2013)
2. Murray A, Mohamed M (2010) The future of the future: are you ready for the coming knowledge cloud? Available http://www.kmworld.com/Articles/Column/Future-of-the-Future/The-Futureof-the-FutureAre-you-ready-for-the-coming-knowledge-cloud3f-66075.aspx
3. Singh AA, Srivatsa M, Liu L (2009) Search-as-a-service: outsourced search over outsourced storage. ACM Trans Web 3(4):13–33
4. Rao Y, Chen S (2013) Research on KAAS-based social knowledge collaboration service mechanism and application. In: The proceeding of 2013 international conference on artificial intelligence and software engineering. Atlantis Press, June 2013
5. Fu-Ren L, Cong-Ren W, Hui-Yi H (2013) Can a wiki be used as a knowledge service platform? In: The proceeding of 7th international conference on knowledge management in organizations: service and cloud computing, vol 172, pp 365–376
6. Malcolm D (2010) The five defining characteristics of cloud computing. ZDNet, http://www.zdnet.com/news/the-five-defining-characteristics-of-cloud-computing/287001
7. Hofmann P, Woods D (2010) Cloud computing: the limits of public clouds for business applications. IEEE Internet Comput 14(6):90–94
8. Wong SC, Crowder RM, Wills GB, Shadbolt NR (2006) Knowledge engineering—from front-line support to preliminary design. In: Brailsford DF (ed) Proceedings of ACM symposium on document engineering. Amsterdam
9. Govender SS, Pottas D (2007) A model to assess the benefit value of knowledge management in an IT service provider environment. In: The proceedings of the 2007 annual research conference of the South African institute of computer scientists and information technologists on IT research in developing countries, ACM, New York©, pp 36–45
10. Aurich JC, Mannweiler C, Schweitzer E (2010) How to design and offer services successfully. CIRP J Manufact Sci Technol 2:136–143
11. Chew EK (2011) Towards a methodology framework for designing a KAAS system. IEEE, pp 2823–2826

Visualization Analysis of Subject, Region, Author, and Citation on Crop Growth Model by CiteSpace II Software

Hailong Liu, Yeping Zhu, Yanzhi Guo, Shijuan Li and Jingyi Yang

Abstract A detailed visual analysis of publications related to crop growth model development and trend was carried out based on the latest version of information visualization and analysis software CiteSpace II. A total of 6,079 publication data between 1995 and 2011 were collected from Thomson ISI's SCI (Web of science in the Science Citation Index Expanded Edition). After analysis by CiteSpace, the most productive countries related to Crop Growth Model, as well as institutes, key scholars, co-citation patterns, etc., were visualized and identified from 1995 to 2011.

Keywords Mapping knowledge domains · Crop growth model · Information visualization · Co-citation analysis · CiteSpace II

1 Introduction

For a scientist or researcher, large-scale literature reviews from extremely large corpora of literature are big challenges. Therefore, a practical, an efficient, and a systematic method is needed to be built to assist in narrowing down the references [8]. With the increasing development of modern information technology and statistics, mapping knowledge domains are becoming a new study subject, which can visually show the structural relationship and development approach of

H. Liu · Y. Zhu (✉) · S. Li
Agricultural Information Institute, Chinese Academy of Agricultural Sciences/Key
Laboratory of Agri-information Service Technology, Ministry of Agriculture,
Beijing 100081, China
e-mail: zhuyeping@caas.cn

Y. Guo
Institute of Food and Nutrition Development, Ministry of Agriculture,
Beijing 100081, China

J. Yang
Greenhouse and Processing Crops Research Centre, Agriculture & Agri-Food Canada,
Harrow, ON N0R 1G0, Canada

Z. Wen and T. Li (eds.), *Knowledge Engineering and Management*,
Advances in Intelligent Systems and Computing 278,
DOI: 10.1007/978-3-642-54930-4_24, © Springer-Verlag Berlin Heidelberg 2014

scientific knowledge [13]. Accordingly, the software for knowledge mapping was developed to facilitate such studies as CiteSpace [5], NetDraw [2], VOSviewer [19], and Bibexcel [16]. By calculating the sharp growth rate of word or terms' frequencies, CiteSpace II can extract the burst terms from titles, abstracts, descriptors, and identifiers of bibliographic records. The current research front can then be displayed by labeling these terms in heterogeneous networks of terms and articles [5]. Using visualization software, the development trend, research fronts, and pivotal researcher and work in specific research area could be reflected, detected, and exhibited by visually showing the relationship between knowledge structure and evolution process from a vast accumulation of literature [5]. Based on scientometric analysis of 1,064 journal articles from 1990 to 2010, Niazi and Hussain [15] analyzed the research areas of agent-based computing using Network Workbench and CiteSpace and identified the largest cluster, found out the timeline of publication of index terms, and analyzed the core journals and key subject categories [8]. In China, Zhou [23] did visual analysis of the information visualization of its evolution, recent development and the research institutions in China using the CiteSpace software. Zhao and Wang [21] visualized the research on pervasive and ubiquitous computing to investigate the most productive countries and institutes, distribution of core journals and research foci between 1995 and 2009. Zhao and Zhang [22] identified the research models in Chinese digital libraries, and compared the results with international digital library development with the assistance of visual knowledge software, UCINET and Netdraw. Chen et al. [7] used CiteSpace to analyse the emerging trends in regenerative medicine.

Crop growth models have been employed extensively and efficiently in many fields, such as agro-ecosystems, crop management, and decision making. Therefore, they are useful tools for management decisions, such as cultural practices, fertilization, irrigation and pesticide application, and to make predictions, such as agricultural chemical leaching, climate change impacts, and yield estimation.

The main objectives of this study are to identify the most productive research countries and institutes, to find core researchers as well as their co-citation patterns and to detect the subject category, country and institution, author co-citation network, and the most cited articles in the research area of Crop Growth Model from 1995 to 2011. The research results can provide useful information to facilitate modeling research and application.

2 Materials and Methods

2.1 Research Tools

In 2004, CiteSpace was first developed by Chaomei Chen to facilitate the visual analysis of trends, patterns, and critical changes in a changeable information environment [4]. In CiteSpace, Timeline views and time-zone views display the publication time and peak time of articles and terms, Cluster views is node and link diagrams, where the nodes present author, institution, country, term, keyword, cited reference, cited journal, paper, and so on [5, 6]. The node size represents the overall citation frequency. Link represents co-citation or co-occurrence, the line's thickness represents the strength proportion of co-citation or co-occurrence. Each color corresponds to a time slice following the legend bar above the visualization area. However, if a node has a purple ring, which means that the node has a high betweenness centrality and tends to be strategically important in terms of the macroscopic structure of a new work. Those nods with high betweenness centrality are called pivotal points or turning points; if a node has a red ring; it means the node has burst in one of its attributes, notably citations.

Therefore, researchers can easily analyze the trends, patterns, and critical changes by studying the size, color of nodes, and links of colorful network. The current version of the CiteSpace II 3.0.R5 is a freely available Java WebStart application, and it was used in this paper to analyze the references. For CiteSpace, the primary input data come from the Web of Science and the PubMed.

2.2 Data Collection and Analysis

In this study, we followed the instruction of CiteSpace II presented by Chen [5], the knowledge domain, crop growth model, was first identified. The input data for CiteSpace II, based on searching the topic of crop growth model from 1995 to 2011, were downloaded from the Web of Science, including Science Citation Index Expanded and Conference Proceedings Citation Index-Science database. The resultant dataset consists of 6,079 records, including authors, titles, abstracts, and cited references. After that, the research front terms were extracted from titles, abstracts, descriptors, and identifiers of cited articles in the dataset. Furthermore, on the CiteSpace interface, the time scaling number was changed to one-year, meaning that the data from 1995 to 2011 were separated into 16 one-year slices. In each slice, the top most cited/specific items were figured out to build the visual network. Pathfinder network scaling was chosen for network pruning. The node types were changed for different objects. Other parameters were taken by CiteSpace II default values. At last, we ran CiteSpace to show and analyze the map. There are three algorithms to produce cluster labels, including tf*idf, log-likelihood ratio tests, and mutual information. The tf*idf is chosen to mark the most significant content of each interesting cluster.

3 Results and Discussion

3.1 Subject Category

Categories on the interface of CiteSpace were selected as the network nodes for the analysis of subject category. The time scaling value was set to 1, and the top 30 most cited or occurred items were chosen to build the network in each slice.

Table 1 The top-15 subject categories according to frequency

No.	Frequency	Category
1	655	Agronomy
2	308	Plant Sciences
3	218	Agriculture
4	171	Environmental Sciences
5	167	Horticulture
6	157	Soil Science
7	149	Ecology
8	136	Meteorology and Atmospheric Sciences
9	122	Water Resources
10	110	Forestry
11	106	Agricultural Engineering
12	70	Geosciences
13	62	Remote Sensing
14	62	Engineering
15	53	Computer Science

The high ranked study field with the most papers is clearly shown in Table 1; Figure 1 depicts the relation network of the subject category related to crop growth model between 1995 and 2011, including 46 nodes and 82 links. It can be seen that Agronomy was the key subject category of research in crop growth model with 655 papers, followed by Plant Science, Agriculture, and Environmental Sciences, contributing 308, 218, and 171 papers, respectively. The frequencies of other subjects varied from 53 to 171, including Environmental Sciences, Horticulture, Soil Science, Ecology, Meteorology and Atmosphere Sciences, Water Resources, Forestry, Agricultural Engineering and Computer Science, indicating that crop growth models have been applied in various research areas.

3.2 Country and Institution

In CiteSpace II, country and institution were selected as nodes to analyze research work forces. A comprehensive network consisting of nodes representing collaborating countries and institutes is presented in Fig. 2, and the top 15 countries and institutes on Crop Growth Model from 1995 to 2011 are listed in Table 2. For example, the USA was the largest contributor with 1,270 papers published, followed by France, China, and Australia, with 433, 424, 422 papers, respectively. Other notable countries were the Netherlands, the United Kingdom, Germany, India, and Canada. These countries developed many useful crop growth models, such as Decision Support Systems in Agrotechnology Transfer (DSSAT) [11], EPIC [10, 20] and CropSyst [18] in the USA, STICS model in France [3], APSIM [12] in Australia.

Fig. 1 The subject category network with 46 nodes and 82 links on the Crop Growth Model researches from 1995 to 2011

Fig. 2 Country and institute related to Crop Growth Model research from 1995 to 2011

Table 2 The top-15 most productive countries and institutes related to Crop Growth Model research from 1995 to 2011

No.	Frequency	Country	Frequency	Institute
1	1270	USA	242	Institute National de la Recherche Agronomique (INRA)
2	433	France	164	United States Department of Agriculture (USDA) ARS
3	424	P. R. China	122	Wageningen University & Research Centre
4	422	Australia	103	University of Florida
5	361	Netherlands	98	Chinese Academy of Sciences
6	351	England	77	Commonwealth Scientific and Industrial Research Organisation (CSIRO)
7	291	Germany	60	University of California, Davis
8	250	India	59	China Agricultural University
9	197	Canada	58	University of Georgia
10	185	Spain	57	Agricultural Research Service (ARS)
11	184	Italy	55	Agriculture & Agri-Food Canada
12	134	Japan	52	University Nebraska
13	132	Brazil	48	Wageningen University
14	102	Denmark	43	Rothamsted Research
15	102	Argentian	42	Indian Agricultural Research Institute

The straight lines linked the institutes in circular nodes with their countries (Fig. 2). The important institutes in the USA included the United States Department of Agriculture (USDA) ARS, University of Florida, University of California, University of Davis, University of Georgia. At the same time, Institute National de la Recherche Agronomique (INRA) in France, Wageningen University and Research Centre in the Netherlands, Commonwealth Scientific and Industrial Research Organisation (CSIRO) in Australia also played important roles on the research and development of plant growth models. The representative institutes in China included the Chinese Academy of Sciences and the China Agricultural University.

3.3 Author Co-citation Network Analysis

Co-citation analyses, mainly including journal co-citation, author co-citation, and document co-citation, were used to evaluate the document likeness for the scientific specialty mapping [21]. For author co-citation analysis, author's impact can be found in terms of citations, and used to evaluate an author's research contribution in a specific field. Basically, authors with most citations can be considered as those scholars who produced a significant and basic effect on the development and evolution in a specific research area.

Figure 3 shows an author co-citation picture with 143 authors and 288 co-citation links, and Table 3 lists the key cited authors in the research area of crop growth model. The prominent authors in Fig. 3 were Ritchie JT, Monteith JL, and Sinclair TR, with 541, 511, 387 citations, separately (Table 3).

3.4 The Most Cited Paper

The highly cited and most influential research references in a specific research domain can be found by document co-citation analysis, and then researchers can easily obtain the reference, especially for new researchers. Each node is surrounded with citation rings, and the ring thickness corresponds to the cited times in the time slice.

The representative author of the cited documents is labeled with year and source of publication (Fig 4). The authors with highly cited papers included Jones CA (1986), Monteith [14], Allen et al. [1], Ritchie [17], Jones et al. [11]. The details of the top-15 most cited papers are listed in Table 4. Based on citation frequency, the highly cited papers are as follows. Jones et al. [9] publication took the first position with the 455 time citations from 1995 to 2011, indicating the most significant publication in the research area of crop growth model. His publication was a book entitled "CERES-Maize: A simulation Model of Maize Growth and Development" with co-editor, Kiniry JR, and with contributions by Dyke PT,

Fig. 3 The author co-citation map related to Crop Growth Model research from 1995 to 2011

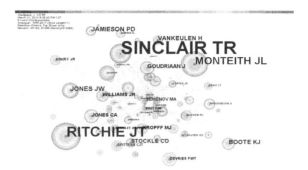

Table 3 The top-15 key cited authors in the author co-citation network related to Crop Growth Model research from 1995 to 2011

No.	Frequency	Author
1	541	Ritchie JT
2	511	Monteith JL
3	387	Sinclair TR
4	323	Jones JW
5	316	Allen RG
6	287	Jones CA
7	281	Goudriaan J
8	278	Devries FWT
9	255	Boote KJ
10	251	*FAO
11	248	Williams JR
12	234	Hoogenboom G
13	228	Doorenbos J
14	222	Spitters CJT
15	217	Kropff MJ

Fig. 4 Co-citation map related to Crop Growth Model research from 1995 to 2011

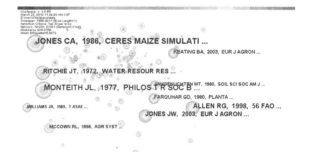

Farmer DB, Godwin DC, Parker JT, Ritchie JT and Spanel DA. In this book, biological and physical relationships used in CERES-Maize model are described, the structure of the input and output files are discussed, the various subroutines and

Table 4 The top-15 highly cited papers with co-citation frequency

No.	Frequency	Cited reference
1	191	Jones CA, 1986, Ceres MAIZE SIMULATI
2	182	Monteith JL, 1977, PHILOS T R SOC B
3	157	Allen RG, 1998, 56 FAO
4	154	Ritchie JT, 1972, WATER RESOUR RES
5	148	Jones JW, 2003, EUR J AGRON
6	124	Priestley CHB, 1972, MON WEATHER REV
7	119	Keating BA, 2003, EUR J AGRON
8	118	Farquhar GD, 1980, PLANTA
9	114	Zadoks JC, 1974, WEED RES
10	111	Mccown RL, 1996, AGR SYST,
11	110	Vangenuchten MT, 1980, SOIL SCI SOC AM J
12	94	Williams JR, 1989, T ASAE
13	81	Stöckle CO, 2003, EUR J AGRON
14	79	Penman HL, 1948, P ROY SOC LOND A MAT
15	79	Tanner CB, 1983, LIMITATIONS EFFICIEN

testing and validation of the model were carried out using the dataset collected internationally.

The second publication was Monteith paper [14] with 182 citations from 1995 to 2011 in the journal of Philosophical Transactions of the Royal Society of London series B-biological sciences. The paper's title is "Climate and the efficiency of crop production in Britain" which analyzed effects of water, solar radiation, temperature on crop growth, and built a basis of crop growth model. The third most cited paper is Allen et al. [1]'s "Crop evaporation—Guidelines for computing crop water requirements". This publication illustrates an evapotranspiration model considering effects of solar radiation, leaf area index, daily air temperature, atmospheric pressure, wind speed, etc., which gave a reference for developing crop growth model. The fourth paper is Ritchie' [17] "model for predicting evaporation from a row crop with incomplete cover" in Water Resources Research with 154 citations. This paper describes a model that calculates the daily evaporation rate from a crop surface under no soil water limited conditions.

The fifth paper was written by Jones et al. [11] with 148 citations, entitled "The DSSAT cropping system model" in *European Journal of Agronomy*. The Decision Support System for Agrotechnology Transfer (DSSAT) was one of the most influential crop growth models used worldwide by many scientists and researchers. Jones's paper describes the DSSAT V4.0 design, as well as approaches to model the primary scientific components, such as soil, crop, weather, and management, and reviews hundreds of published studies with DSSAT various application.

Through detailed analysis of highly cited papers by CiteSpace, the results can assist researchers to quickly collect the main publications on crop growth model for their research.

4 Conclusion

In this paper, using mapping knowledge domains software CiteSpace II, visualization analysis on 6,079 papers in Crop Growth Model field were studied to analyze research category, region, institution, as well as the representative of typical research works. It is supposed to support some information related to crop growth model for the new researchers to easily find important references and application situation. Using the CiteSpace, further study is needed to focus on the research front in the crop growth model field. Research front analysis can provide researchers the latest information, and then quickly provide valuable information or references in their potential research area.

Acknowledgments This research was supported by National High Technology Research and Development Program of China (2013AA102305).

References

1. Allen RG, Pereira LS, Raes D, Smith M (1998) Crop evaporation: guidelines for computing crop water requirements. FAO Irrigation and Drainage paper 56
2. Borgatti SP (2002) Netdraw network visualization. Analytic Technologies, Cambridge
3. Brisson N, Gary C, Justes E, Roche R, Mary B, Ripoche D, Zimmer D, Sierra J, Bertuzzi P, Burger P, Bussiere F, Cabidoche YM, Cellier P, Debaeke P, Gaudillere JP, Henault C, Maraux F, Seguin B, Sinoquet H (2003) An overview of the crop model STICS. Eur J Agron 18(3–4):309–332
4. Chen C (2004) Searching for intellectual turning points: progressive knowledge domain visualization. In: Proceedings of National Academy of Sciences, 101(Suppl.):5303-5310
5. Chen C (2006) CiteSpace II: detecting and visualizing emerging trends and transient patterns in scientific literature. J Am Soc Inform Sci Technol 57(3):359–377
6. Chen CM, Ibekwe-SanJuan F, Hou JH (2010) The structure and dynamics of cocitation clusters: a multiple-perspective cocitation analysis. J Am Soc Inform Sci Technol 61(7):1386–1409
7. Chen C, Hu Z, Liu S, Tseng H (2012) Emerging trends in regenerative medicine: a scientometric analysis in CiteSpace. Expert Opin Biol Ther 12:593–608
8. Jahangirian M, Eldabi T, Garg L, Jun GT, Naseer A, Patel B, Stergioulas L, Young T (2011) A rapid review method for extremely large corpora of literature: applications to the domains of modelling, simulation, and management. Int J Inf Manage 31(3):234–243
9. Jones CA, Ritchie JT, Kiniry Godwin, D.C (1986) Subroutine structure. In: Jones CA, Kiniry JR (eds) CERES-Maize: a simulation model of maize growth and development. Texas A&M University Press, TX
10. Jones CA, Dyke PT, Williams JR, Kiniry JR, Benson VW, Griggs RH (1991) EPIC: an operational model for evaluation of agricultural sustainability. Agric Syst 37:341–350
11. Jones JW, Hoogenboom G, Porter CH, Boote KJ, Batchelor WD, Hunt LA, Wilkens PW, Singh U, Gijsman AJ, Ritchie JT (2003) The DSSAT cropping system model. Eur J Agron 18(3–4):235–265
12. Keating BA, Carberry PS, Hammer GL, Probert ME, Robertson MJ, Holzworth D, Huth NI, Hargreaves JNG, Meinke H, Hochman Z, McLean G, Verburg K, Snow V, Dimes JP, Silburn M, Wang E, Brown S, Bristow KL, Asseng S, Chapman S, McCown RL, Freebairn DM,

Smith CJ (2003) An overview of APSIM, a model designed for farming systems simulation. Eur J Agron 18:267–288

13. Liu ZY, Chen Y, Hong HY (2008) Mapping knowledge domain: methods and application. People Publishing House, Beijing (In Chinese)

14. Monteith JL (1977) Climate and the efficiency of crop production in Britain. Philos Trans R Soc B 281:277–294

15. Niazi M, Hussain A (2011) Agent-based computing from multi-agent systems to agent-based models: a visual survey. Scientometrics 89(2):479–499

16. Persson, OD, Danell R, Wiborg Schneider J (2009) How to use Bibexcel for various types of bibliometric analysis. In: Åström FR, Larsen DB, Schneider J (eds) Celebrating scholarly communication studies: a Festschrift for Persson at his 60th Birthday. International Society for Scientometrics and Informetrics, Leuven, pp 9–24

17. Ritchie JT (1972) Model for predicting evaporation from a row crop with incomplete cover. Water Recour Res 8(5):1204–1213

18. Stöckle CO, Donatelli M, Nelson R (2003) CropSyst, a cropping systems simulation model. Eur J Agron 18:289–307

19. van Eck NJ, Waltman L (2010) Software survey: VOSviewer, a computer program for bibliometric mapping. Scientometrics 84(2):523–538

20. Williams JR (1995) The EPIC model. In: Singh VP (ed) Computer models of watershed hydrology. Water Resources Publications, Highlands Ranch, pp 909–1000

21. Zhao R, Wang J (2011) Visualizing the research on pervasive and ubiquitous computing. Scientometrics 86(3):593–612

22. Zhao LM, Zhang QP (2011) Mapping knowledge domains of Chinese digital library research output, 1994–2010. Scientometrics 89(1):51–87

23. Zhou Jinxia (2011) Documents visibilization analysis of information visibilization based on the Citespace II. Inf Sci 29(1):98–112 (In Chinese)

Lexical Multicriteria-Based Quality Evaluation Model for Web Service Composition

Dunbo Cai and Sheng Xu

Abstract Providing quality aware services to customers is an important problem in the composition of Web services. Often, the quality of services is considered from different aspects that we call criterions. Based on the criterions, customers compare and choose better (composed) Web services. Previous methods compare Web services by comparing their aggregate values, which are weighted sum of each quality (the *WSum model*). However, the WSum model is not proper to support qualitative preferences over criterions. Herein, we proposed a new evaluation model that lexically compares the qualities of Web services, which is called the *Lexical model*. In contrast with WSum model, our proposed Lexical model is more powerful in modeling users' preferences both quantitatively and qualitatively. We will prove that it is in $O(k^2)$ for a WSum model to simulate the Lexical model, where k is the number of criterions. We may note that though the idea of Lexical models has been utilized in other areas, few researchers in the field of Web service composition noticed the way of using the Lexical model introduced here.

Keywords Web service composition · Quality aware service · Lexical order

1 Introduction

Service oriented architecture (SOA) is a popular method in the development of softwares [3], especially for those process oriented softwares. SOA based softwares are open and flexible. In terms of openness, the softwares are able to use

---

D. Cai (✉) · S. Xu
Hubei Province Key Laboratory of Intelligent Robot, Wuhan Institute of Technology,
Wuhan 430205, China
e-mail: dunbocai@gmail.com

S. Xu
e-mail: xusheng20120531@gmail.com

Z. Wen and T. Li (eds.), *Knowledge Engineering and Management*,
Advances in Intelligent Systems and Computing 278,
DOI: 10.1007/978-3-642-54930-4_25, © Springer-Verlag Berlin Heidelberg 2014

new services. And in terms of flexibility, they usually have many options to choose between different services. As the notion of "service" become popular in the area of Information Technology, more and more services are available. The most exciting services are those that can be invoked through the Internet, which is called Web services. In this way, SOA-based softwares could use a single Web service to support customers' simple requests, or composes a few Web services to support complex requests [11]. As there are often many compositions of Web services, the corresponding evaluation model for them is an important topic [9].

An evaluation model defines better compositions in terms of some kind comparison method. For example, when services have many criteria, the quality of each criterion is aggregated in a weighted sum manner [7, 12], which we called the WSum model. However, WSum model is unable to model qualitative preferences of customers. We will show this with an example. Suppose that we have two Web services ws_a and ws_b, and we have two criteria c_1 and c_2 on them. For ws_a, its qualities are 3 and 8 in terms of c_1 and c_2, respectively. And for ws_b, its qualities are 5 and 5. Suppose that c_1 and c_2 are associated with weights 0.6 and 0.4, respectively. Therefore, the aggregated values of ws_a and ws_b are 5 and 5, respectively. The WSum model then may not be able to tell whether ws_a or ws_b is better. However, looking insight the weights of each criterion, we could derive that c_1 is more important than c_2, from the customers' point of view. This finding on qualitative preference can also be verified by consulting customers.

In order to model both the quantitative and qualitative preferences of customers, we proposed a *Lexical model*. Specifically, we consider the qualities of a service as a *quality vector*. The quality vectors of different Web services are compared lexically. To be compatible with the WSum model, for each service we associate a pseudo criteria whose quality is the aggregate value defined by the WSum model. Analysis in Sect. 3 will show that our Lexical model has more powerful modeling power than that of the WSum model.

The following of this paper are organized as follows. In Sect. 2, we introduce some abstract notions in Web services composition and a brief introduction to the WSum model. Then, we introduce our Lexical model and its strength. In Sect. 4, we make conclusions and discuss future work.

2 Background

In the following, we introduce notions of Web services and Web service compositions in a very abstract way. For details on how Web services are described in WSDL language, please refer to manuals in the W3C website.[1]

---

[1] http://www.w3.org/TR/wsdl.

Definition 1 Given a finite set of variables $X = \{x_1, \ldots, x_n\}$, each variable $x_i (i = 1 \ldots n)$ with a finite domain D_{xi}, a universe of variable-value pairs are $U = \{(x, d) | x \in X, d \in D_x\}$. A Web services ws is of the form $<$in(ws), out(ws)$>$, where in(ws) $\subseteq U$ and out(ws) $\subseteq U$. We denote all available Web services, WS.

Definition 2 A customer request R is a subset of X, i.e. $R \subseteq X$. In another words, R is the set of variables whose values a customer concerns.

Definition 3 The situation before a request is proposed is a subset of U, which is called the initial situation, and is denoted by I.

Definition 4 A Web services composition (WSP) problem is $\Gamma = <U$, WS, R, $I>$, where WS is a finite set of Web services.

Definition 5 An execution of a Web service ws on a situation $s \subseteq U$ is defined as ws$(s) =$ out(ws) $\cup \{ <x, d> \in s | x$ does not appear in out(ws)$\}$.

Definition 6 The solution of $\Gamma = <U$, WS, R, $I>$ is a sequence of Web services $\pi = <\text{ws}_{i1}, \ldots, \text{ws}_{im}>$ that each variable in R appears in $\text{ws}_{im}(\ldots \text{ws}_{i1}(s) \ldots)$.

Definition 7 A criteria cr is of the form $<f_{cr}, A_{cr}>$, where f_{cr}: WS $\rightarrow$ **R**, and A_{cr}: $2^{WS} \rightarrow$ **R**.

Note that, for ws, $f_{cr}(\text{ws})$ is the quality of ws with respect to cr, and for π, $A_{cr}(\pi)$ is an aggregate value of qualities of Web services in π. In the notation $A_{cr}(\pi)$, we don't distinguish a sequence of Web services and a set of them. It's worthy to mention some practical criterion. If cr corresponds to "execution cost" of services, then $A_{cr}(\pi) = \sum_{\text{ws} \in \pi} f_{cr}(\text{ws})$. If cr corresponds to "probability of success," then $A_{cr}(\pi) = \prod_{\text{ws} \in \pi} f_{cr}(\text{ws})$.

Definition 8 Given a set of criterion CR $= \{cr_1, \ldots, cr_k\}$, the quality vector of a Web service ws is $\text{ws}_{\{QV,CR\}} = <f_{cr_1}(\text{ws}), \ldots, f_{cr_k}(\text{ws})>$. And the quality vector of a composition of Web services π is $\pi_{\{QV,CR\}} = \{A_{cr_1}(\pi), \ldots, A_{cr_k}(\pi)\}$.

Next, we briefly introduce the weighed sum model (WSum model). In the WSum model, each criterion cr is associated with a weight w_{cr}. The quality of a composition is defined as: WSum$(\pi) = \sum_{cr \in CR} \{w_{cr} * A_{cr}(\pi)\}$.

3 A Lexical Evaluation Model

Lexical order is an important tool for modeling preference in the literature [6], especially in applications of fuzzy logic [1] and constraint satisfaction problems (CSP) [4, 5].

To be specific, we define lexical order over compositions, with respect to a set criterion CR as follows.

Definition 9 Given a set of criterion $CR = \{cr_1, \ldots, cr_k\}$, two compositions of Web services π and π'. π is better than π' (denoted by $\pi \succ_{CR} \pi'$), if

$$\exists i : \wedge_{j=1}^{i-1}\left(A_{cr_j}(\pi) = A_{cr_j}(\pi')\right) \text{ and } A_{cr_i}(\pi) > A_{cr_i}(\pi').$$

For example, if we are considering two criterion, and the quality vectors of compositions π and π' are <4, 5> and <4, 3> respectively, then $\pi \succ_{CR} \pi'$ holds. However, if the quality vectors were <4, 5> and <4, 5>, then $\pi \succ_{CR} \pi'$ doesn't hold. So, lexical order is able to express qualitative preferences. However, if the quality vectors are <4, 5> and <3, 20>, it may be not rational to think that π is better than π'. With this concern, our lexical model is designed as follows.

Definition 10 A WSum extension of a given a set of criterion $CR = \{cr_1, \ldots, cr_k\}$ is $CR = \{cr_{WSum}, cr_1, \ldots, cr_k\}$, with $f_{WSum}(ws) = WSum(<ws>)$, and $A_{WSum}(\pi) = WSum(\pi)$. π is better than π' (denoted as $\pi \succ \pi'$) if $\pi \succ CR_{WSum}\pi'$ holds.

Now, recall our motivating example in the introduction section. According to Definition 10, after the extension of criterion $\{c_1, c_2\}$, as the quality vector of ws_a and ws_b are <5, 3, 8> and <5, 5, 5> respectively, we consider that ws_b be better than ws_a. In the other side, if the quality vector of ws_a is <5.6, 4, 8>, then we consider ws_a the better one. Therefore, our proposed *lexical model* can express both qualitative and quantitative preferences.

Next, we will show that it is more difficult for the Wsum model to simulate the Lexical model while. In terms of "simulate," we expect *optimal solutions* on the one model are also optimal ones on the other. Specifically, we have the following theorem.

Theorem 1 *If there are k criterions, the Lexical model can simulate the Wsum model in time $O(k)$, while the Wsum model can simulate the Lexical model in time $O(k^2)$.*

Proof First, let's simulate the Wsum model with the Lexical model. For a set of criterion $CR = \{cr_1, \ldots, cr_k\}$, suppose the Wsum model uses the set of weights $W = \{w_1, \ldots, w_k\}$ for the respective criterion. We can construct a quality *vector* $< \sum_{cr \in CR} w_{cr} \times A_{cr}(\pi) >$, which is then used to form a Lexicial model. It is easy to see that the optimal solutions on the WSum model are optimal ones on the Lexical model as well. In the other direction, if a Lexical model involves a quality vector $ws_{\{QV, CR\}} = <f_{cr_1}(ws), \ldots, f_{cr_k}(ws)>$ for a Web service composition ws, we provide the following method to find for a WSum model to simulate the Lexical model. Suppose there be m Web service compositions $ws = \{ws_1, \ldots, ws_m\}$, and the optimal one be $ws_j \in ws$ for the Lexical model. To simulate it with a WSum model, the core is to find a set of weights $W = \{w_1, \ldots, w_k\}$ satisfying the following linear equations:

$$w_1 \times f_{cr_1}(ws_1) + \cdots + w_k \times f_{cr_k}(ws_1) = t_1,$$
$$w_1 \times f_{cr_1}(ws_2) + \cdots + w_k \times f_{cr_k}(ws_2) = t_2,$$
$$\cdots$$
$$w_1 \times f_{cr_1}(ws_m) + \cdots + w_k cr_k(ws_m) = t_m,$$

where $t_i(i \in \{1 \ldots m\})$ can be any integer satisfying $t_j > t_i$ for $i \in \{1 \ldots m\}/\{j\}$. An iterative method, such as the Jacobi method [2], can solve the system of linear equations in time $O(k^2)$. So, we finish our proof.

4 Conclusion and Future Work

In this chapter, we show that the WSum model is incapable of expressing qualitative preferences of customers. An evaluation model in the spirit of lexical order in the literature is proposed, which we called Lexical model. It should be noted that our Lexical model does not conflicts with the WSum model, but a more powerful model than it.

The report in this chapter is rather preliminary. However, our proposed evaluation model could be a guide in extracting customers preferences, comparing different compositions, and designing Web service composition algorithm [10, 12] and heuristics [8]. Currently, we are working on the directions.

Acknowledgments This work is supported by National Science Foundation of China (Grant No. 61103136), Educational Commission of Hubei Province of China (Grant No. D20111507), Hubei Province Key Laboratory of Intelligent Robot Open Foundation (Grant No. HBIR200909), and Youths Science Foundation of Wuhan Institute of Technology (Grant No. 12106022).

References

1. Bowen J, Lai R, Bahler D (1992) Lexical imprecision in fuzzy constraint networks. In: Proceedings of the national conference on artificial intelligence, Wiley, pp 616–620
2. Byrne CL (2008) Applied iterative methods. AK Peters Wellesley, Wellesley
3. Erl T (2004) Service-oriented architecture: a field guide to integrating XML and web services. Prentice Hall PTR, London
4. Freuder EC, Wallace RJ, Heffernan R (2003) Ordinal constraint satisfaction. In: Proceedings of the fifth international workshop on soft constraints
5. Freuder EC, Heffernan R, Wallace RJ, Wilson N (2010) Lexicographically-ordered constraint satisfaction problems. Constraints 15(1):1–28
6. Herzberger HG (1973) Ordinal preference and rational choice. Econom: J Econom Soc 41:187–237
7. Kalepu S, Krishnaswamy S, Loke SW (2003) Verity: a qos metric for selecting web services and providers. In: Web information systems engineering workshops, proceedings fourth IEEE international conference on 2003, pp 131–139
8. Liu Y, Ngu AH, Zeng LZ (2004) Qos computation and policing in dynamic web service selection. In: Proceedings of the 13th international world wide web conference on alternate track papers & posters, ACM, pp 66–73

9. Manikrao US, Prabhakar T (2005) Dynamic selection of web services with recommendation system. In: Proceedings of the international conference on next generation web services practices (NWESP), IEEE, pp 117
10. Nie K, Wang H, He J (2013) Semantic web service automatic composition based on discrete event calculus. In: Proceedings of the 2012 international conference on information technology and software engineering, Springer, pp 477–485
11. Rao J, Su X (2005) A survey of automated web service composition methods. In: Semantic web services and web process composition, Springer, pp 43–54
12. Zou G, Lu Q, Chen Y, Huang R, Xu Y, Xiang Y (2012) Qos-aware dynamic composition of web services using numerical temporal planning. IEEE Trans Serv Comput 99(PrePrints):1. doi: http://doi.ieeecomputersociety.org/10.1109/TSC.2012.27

Design and Implementation of Enterprise Resources Content Management System

Qian Mo, Feng Xiao and Da-Zhuang Su

Abstract This paper introduces a content management system (CMS) for enterprises, which combines the enterprise resources with the information-issuing website, and which integrates content collection, content creation, content editing, template production, content approval, content issuing, and content browsing applications into a whole. This system can make users easily and efficiently issue, manage, and exchange information via websites according to users' different demands, indicating that the efficiency of website daily information processing could be improved. This paper elaborates the architecture of the enterprise content management system, presents its major functions and key technologies, and gives the website applications of this system as the example.

Keywords Content management system · Information portal · Enterprise resource

1 Introduction

With the advent of the new medium of Internet, human society has entered into the big-data era with information explosion and overloaded data. A wide variety of Internet websites have constantly sprung up and tens of thousands of new websites emerge every day. There are also a great number of websites with lower page views overwhelmed in this big ocean of data. If new personal websites lack required technical assistance, it is extremely difficult for them to survive and develop themselves in the net-volution. Faced with the numerous Web information resources, people are in urgent demand for a means which can clearly and accurately absorb and manage the content on the websites, as well as the resources,

Q. Mo · F. Xiao (✉) · D.-Z. Su
Computer and Information Engineering Institute,
Beijing Technology and Business University, Beijing, China
e-mail: btbuxf@126.com

Z. Wen and T. Li (eds.), *Knowledge Engineering and Management*,
Advances in Intelligent Systems and Computing 278,
DOI: 10.1007/978-3-642-54930-4_26, © Springer-Verlag Berlin Heidelberg 2014

which can fast and effectively update or publish personal websites, which can share various information in real time, and which has user-friendly and personalized interactive operating interface. What is more important is that users need not have the complicated knowledge of computer programming and they can easily and efficiently maintain and manage personal websites [1–3].

In recent years, the content management system undergoes a rapid development. Many foreign universities conduct in-depth researches on this field. For example, the University of Utah [4] has established a database-driven content management system, which can obtain the latest information from any Web browser. The system can also present the presettings of the system in the Internet browser in the form of a schedule or save it as a calendar.*ical* file. Not only it achieves the precise positioning on the map, but also it provides multiversion layered access control module of text library [5]. Mara established another research direction, i.e., proposing the design prototype of virtual learning PBL content management system, and adding it into forums and chat rooms. By using this system to evaluate the predicted report on students' achievement and then to improve curriculum design, it greatly reduces the communication cost between teachers and students [6]. Seung Hyun Jeon proposed user-centered open IPTV service provider CMS. The system uses the network TV engine, the contents mediator, the media server, and the IPTV configuration file server to construct an open IPTV platform, with the purpose of realizing the integration of social network TV with user-centered CMS and SNS.

2 Related Work

Content Management System (CMS), refers to the relationship integration of people, content, applications, and processes. Content management is involved in making the organization, classification, and management of inner enterprise information resources with various format and media type (which is usually called as information assets) become a more orderly process. Content management aims at solving such issues as information analysis, information filtering, read permission, content security, etc., and then realizes the full integration of the industry value chain of content collection, content creation, content transmission, etc. Although a variety of content after having been digitized are transformed to be electronic data, they can still be processed by the means of digital management, but as for the level of application, the issue becomes more complicated.

This paper proposes an enterprise resource and content management system, providing personalized services by combining content management with information portal technology. The first part is an introduction, the second part is the overview of the research background, the third part describes the technology involved in this system, the fourth section describes the architecture of the enterprise resources content management system, the fifth section presents the main functions of each module, and gives the core algorithm, and the last part is the concluding remarks.

3 System Architecture

3.1 Network Topology

The recommended configuration of the proposed CMS includes cache server, Web server, content management server, and database server. For the application of some small networks, it is feasible to merge several servers together into one server, in order to reduce the cost. While for the large projects, it is advisable to adopt the distributed server technology (i.e., different servers have different functions and then they collaboratively complete some tasks), which will improve the processing speed, load balancing, and reduce the losses caused by system crash. Figure 1 shows the network topology used in CMS.

The CMS introduced in this paper uses Tomcat server and MySQL database. Tomcat server is a free open-source-code Web application server. Because of its advanced technology, stable performance, and free-charge, it is popular among Java lovers and it has been also recognized by some software developers. Its runtime occupies few system resources, its scalability is excellent, and it supports commonly used functions for developing applications, such as load balance, mail services, etc. Tomcat, as a small, lightweight application server, is widely used in medium-sized system and in the application of websites with a small number of page views. Tomcat has become a very popular Web application server.

For the individual users and small/medium-sized enterprises, the functionality provided by MySQL is more than sufficient, because MySQL is open source code software, so it can greatly reduce the time and labor costs of operation and maintenance. The website structure popularized on the Internet is LAMP (Linux + Apache + MySQL + PHP), i.e., using Linux as the operating system, Apache as the Web server, MySQL as the database, PHP server-side scripting interpreter. MySQL has the following advantages:

- Providing multioperating systems support;
- Saving 50,000,000 records;
- Having flexible security permissions and password system;
- Having powerful query capabilities;
- Realizing multithreaded programming to save resources.

3.2 Architecture Diagram

The architecture of enterprise resource content management system is shown in Fig. 2.

The main function of enterprise resources CMS is to quickly and easily create and publish content-rich and user-friendly enterprise information portal home page, according to user-defined theme needs. This system makes it possible to

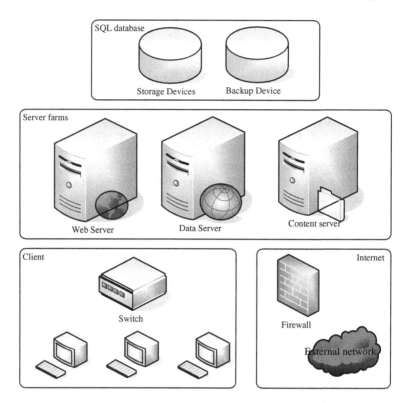

Fig. 1 Enterprise resources content management system network topology

conveniently conduct the remote management of the published content and information. Users only need to master the basic knowledge of computers rather than the complex knowledge of computer programming. They are able to operate the system by themselves right away as soon as they finish reading the operating manual.

4 System Architecture

Enterprise resources content management system adopts commonly used Java Web application development framework as its base, focusing on Ajax and MVC. The main functional modules can be divided into four parts: content management, user management, system management, and database. Under the four main functional modules, there are many relevant issues, such as the structure of back-end server (which is slightly different from the traditional Web application), programming model, security, debugging, demand-analytical methods, development management, operation management, etc.

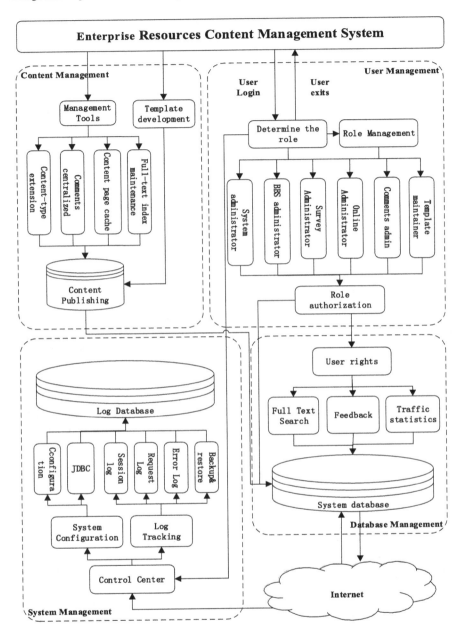

Fig. 2 Architecture diagram of enterprise resources content management system

4.1 Content Management Module

In the main page of content management module, user can choose to create a user-defined website. After the creation, the system will automatically generate the detailed management chart of web resources, based on users' demand. Besides, content management module also embraces two major parts: module management and management tools (see Fig. 3).

The main function of module management is to manage templates, including template images, CSS files, and JSP pages. Users can create, edit, delete, move, and copy the templates, and they can also upload the webpages or image resources which have been created, and files in the templates can be compressed and decompressed by automatically using ZIP. In addition, users can also conduct the operations of deletion, moving, copying, etc.

The three submodules of management tools are as follows:

- content-type extension.
- comment-centralized management.
- full-text index maintenance.

Content-type extension mainly manages the definition of domain types, and the width of entry control and the entry types. It allows users to conveniently manage the design of website controller, and improves the overall publishing rate, then seizing the initiative. Furthermore, between the additional content page cache options, users can choose to "clean all cache" or "clean up the last-7-day no-access cache" to solve the cache deficiency of system and alleviate the situation in which the page cannot be refreshed. By using index maintenance, CMS index can be rebuild, which enables the system to have a quick recovery when encountering baleful attacks and then enhances security of the system.

4.2 User Management Module

The main function of user management module is to manage users' personal information and allocable permissions. Users can freely switch different roles, and the permission allocation is also very flexible. Therefore, a user can be authorized multiple permissions according to his/her quality and responsibility; as well, the correspondent permission could be cancelled with the change of the user's role. This module is subdivided into three parts: role management, department management, and permission management.

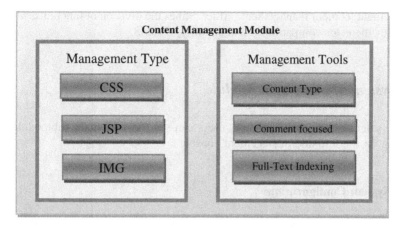

Fig. 3 Content management module of enterprise resources content management system

4.2.1 Sectorial Allocation

Users can create a new department, and then to authorize the department, including adding new users and new roles (i.e., department managers and their responsibility), making this system keeping with the real situation of companies and enterprises, and meeting leaders' operation needs.

4.2.2 Role Assignment

It contains the roles of "system administrator," "BBS administrator," etc. Each role has its own unique ID and role description. The role assignment completely separates website designers with article managers. On the one hand, it makes possible that people with different duties do not interfere with each other. On the other hand, it reflects the confidentiality of enterprises, and protects users' privacy.

4.2.3 Allocation of Permission

The authorized users can enter the interface of the permission management. System resources authorization schedule lists three permissions which could be authorized to users, i.e., role management, user management, and content management, indicating there is a clearer division of labour in this system. Role Management and User Management include the management of roles, department, staff, and permissions, making the staff turnover within the enterprise or between enterprises more clear. It further meets the employers' demands. Content management includes column and content maintenance, content transfer and copy, templates assignment, publication of approval, etc. Complementary to role management, content management further refines the division of functions, while it increases the error-compatibility and stability of CMS.

4.3 System Management Module

In the system management module, users can see the four main submodules by clicking on the left-side management console.

4.3.1 System Configuration

In "Core Set," the sub-option of "System Configuration," users can set the system code, data sources for connection, as well as most of the details of the SQL database management list. The auxiliary test function (e.g., error information minder, situation tracing) can also be activated in this module during the test phase of this system. Users can also add JNDI data source or Jakarta DBCP data source under the suboption of JDBC data source.

4.3.2 Role Assignment

"Running State" displays all the Servlets under running state, and such records administrators could examine as load time, active thread, history thread, initialization error messages, and so on, even administrators are able to enter the users' form to examine the details of each thread. Moreover, administrators can also set data sources used by each program.

4.3.3 Logging

"Logging" can be categorized as the conversation log, the request log, the error log, and the system tracking log. The conversation log will record the IP addresses of all accessed websites, online situation, the type of the used browser, and access time. The request log will record the requester's IP or ID number, time-consuming (ms), the servlets and action, which is further divided into the AutoSQL tracking log and CMS SQL trace log. The error log will record error-reminding information which has appeared in CMS, which could facilitate system debugging. The system tracking log is responsible for recording the operation information of the system, including system changes or running state.

4.3.4 Backup and Recovery

As the name suggests, "Backup and Recovery" is mainly responsible for data backup and recovery. The specific methods of operation will be elaborated in the interface of the inner system.

4.4 Database Management Module

The database module, containing remote management tools, can make a remote execution of generated SQL statements and easily lead them into CMS back-end system. Users can immediately see the generated results. Query Wizard is recommended to realize the agile development. It also allows users to remotely browse the resources and forms which have already stored in the website. These resources and form include the values of columns, primary keys, and indexes. Based on the contents of the database, SQL template statements and Servlet template statements can also be automatically generated in the list view or in the form view. This proposed CMS supports a variety of databases, such as MySQL, SQL Server, Oracle, as well as the users' self-developed database access module.

5 Conclusion

There are two differences between the functions of enterprise resources content management system and those of general resource portal and content classification system: the first is that enterprise resource CMS has concisely and efficiently built a straight path linking the browser interface and back-end database; the second is that it can provide secure solution to user management.

Enterprise resources content management system described in this paper has been completed and passed the project acceptance test. Because of its easy and efficient operation and its compatibility, this system deserves a wide range of application. It makes the multimethods of data management feasible, providing a reliable means to ensure the system stability.

Acknowledgments This paper is support by The Open Research Fund of Key Laboratory of Disaster. Reduction and Emergency Response Engineering of the Ministry of Civil Affairs (DRERE20120105) and The Importation and Development of High-CaliberTalents Project of Beijing Municipal Institutions (IT&TCD201304034).

References

1. Kristensen T (2011) The dynamic content management system. In: International conference on information technology based higher education and training (ITHET), pp 1–8
2. Yong R (2011) A dynamic comprehensive web service for content management. In: IEEE 3rd international conference on communication software and networks (ICCSN), pp 51–54
3. Scott JE (2011) User perceptions of an enterprise Content management system. In: HICSS
4. Taylor R (2012) Web-based content management system for equipment information. In: University/Government/Industry, Micro/Nano Symposium (UGIM), 19th Biennial, vol 1, p 1
5. Abu Kasim NA, Gunawan TS (2012) Virtual-learning content management system for problem- based learning (PBL) courses. In: Computer and communication engineering (ICCCE), international conference on, vol 3, pp 948–952
6. Jeon SH, An S, Choi JK, Yoon C, Lee H-W (2012) User centric content management system for open IPTV over SNS. In: ICT Convergence (ICTC), international conference on, vol 10, pp 258–263

Determination of Soluble Solids Content in Cuiguan Pear by Vis/NIR Diffuse Transmission Spectroscopy and Variable Selection Methods

Wenli Xu, Tong Sun, Wenqiang Wu, Tian Hu, Tao Hu
and Muhua Liu

Abstract The objective of this research was to assess soluble solids content (SSC) of Cuiguan pears by visible/near infrared (Vis/NIR) and variable selection methods. Vis/NIR transmission spectra of Cuiguan pears were taken by a USB4000 spectrometer. A variety of pretreatment methods and three variable selection methods were used to select important variables in this study. After that, Partial least squares (PLS) was used to develop calibration model. The results show that competitive adaptive reweighted sampling (CARS) is an effective variable selection method for SSC of Cuiguan pears. PLS using the selected variables by CARS combined with multiplicative scattering correction (MSC) obtains the best results. Compared with full spectrum PLS, The number of variables in MSC-CARS-PLS model reduces from 1,400 to 84, the correlation coefficient rises from 0.88 to 0.96, and the root mean square error decreases to 0.29 °Brix.

Keywords Visible/near infrared · Diffuse transmission · Competitive adaptive reweighted sampling · Soluble solids content · Cuiguan pear

1 Introduction

Cuiguan pear is grown in southern China, and is well accepted by consumers due to its early ripening, large fruit, thinly skin, juicy crisp, sweet and refreshing, delicate appearance, and edible quality. With the development of people's living standards, consumers pay more attention to the quality of Cuiguan pear than before, simply relying on artificial subjective selection cannot meet the demands of

W. Xu · T. Sun · W. Wu · T. Hu · T. Hu · M. Liu (✉)
College of Engineering, Jiangxi Agricultural University, 1225 Zhimin Street,
Nanchang 330045, People's Republic of China
e-mail: suikelmh@sina.com

Z. Wen and T. Li (eds.), *Knowledge Engineering and Management*,
Advances in Intelligent Systems and Computing 278,
DOI: 10.1007/978-3-642-54930-4_27, © Springer-Verlag Berlin Heidelberg 2014

consumers, so rapid and nondestructive testing of internal quality of pear [such as soluble solids content (SSC)] to enhance its market competitiveness is necessary.

The visible/near infrared (Vis/NIR) spectroscopy is a rapid, nondestructive technique, and it has been widely used in fruit internal quality determination, e.g., sweetness and sourness of apple [1], SSC, total acidity (TA) and firmness (Fi) of apricots [2], total SSC of watermelons [3], astringency in peach fruit [4], SSC and total acidity of oranges [5], SSC, titratable acidity (TA), vitamin C and surface color of mandarin [6], Brix value in intact "Gannan" navel oranges [7], SSC, Fi of pears [8]. And other documents also did relevant research [9–15].

In this paper, Vis/NIR transmission spectroscopy was used to assess SSC of Cuiguan pears. A variety of pretreatment methods and three variable selection methods were used to figure out the most suitable combination for modeling SSC of Cuiguan pears.

2 Materials and Methods

2.1 Samples

Pears were purchased from a local market in Nanchang, Jiangxi province, China. Samples were cleaned by humid cloth, and air-dried. Each Cuiguan pear was treated as a sample, numbered at the peduncle area. They were stored at room temperature for 1 day before the experiment. Samples were sorted by the value of SSC, then a part of the samples with the same SSC values were selected as the calibration set, which makes modeling sample content range cover the prediction set content range, so that the calibration model can well apply to prediction set samples.

The proportion of calibration set and prediction set was 3:1. Total of 147 samples were used as calibration set and 49 samples were used as the prediction set.

2.2 Experimental Setup

A spectral measuring system in Fig. 1 was designed to obtain the diffuse transmission spectra. The system is mainly consisted of a USB4000 spectrometer, a fiber optics probe, microcontroller system, transportation unit, and power supply unit. Some units are not shown in Fig. 1. The USB4000 spectrometer (Ocean Optics, USA) was equipped with a 3,648 pixel CCD detector (TCD1304AP, Toshiba, Japan), and the wavelength range is from 465 to 1,150 nm. Two halogen lights source (15 V, 150 W) were placed at both sides of the fruit, and light irradiation center substantially aligned on samples equatorial area. The code of

Fig. 1 Vis/NIR spectrum
online acquisition device

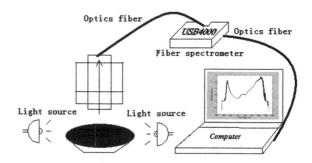

Fiber optics probe is ocean Optics Qp600-2-VIS-NIR. White polytetrafluoroethylene ball with 80 mm diameter was used as a reference. The SSC was obtained by
a PR-101 handheld refractometer (Atago Co. Ltd, Tokyo, Japan).

2.3 Spectral Acquisition

The experiment was carried out at room temperature. The device was preheated for
30 min before the experiment. Self-developed spectrum acquisition software was
used to collect spectra of Cuiguan pears. First, the polytetrafluoroethylene ball was
used to collect the dark (the dark current of the spectrometer) and reference (the
spectral data of the white polytetrafluoroethylene ball) spectra. Then, spectra were
taken at three equidistant positions around the equator (approximately 120°) for
each fruit, and the average spectra was used as the data for subsequent analysis.

2.4 SSC Measurement

The true value of SSC of samples was obtained by the measurement method of
National Standard of China (GB12295-90) after the spectral acquisition. After the
nuclear removed, the fruit was cut into small pieces and squeezed into juice by
juice machine. Samples juice was filtered by filter paper, and then dripped on the
test window of the PR-101 digital refractometer to measure SSC. At last, the same
value of two measurements was selected as the true SSC value of samples.

2.5 Data Processing Method

In this paper, a variety of pretreatment methods were used, such as savitzky–golay
(SG), standard normal variate (SNV), multiplicative scattering correction (MSC),
unit vector normalization (UVN), and wavelet transformation (WT). Three

variable selection methods were used, such as genetic algorithm (GA), successive projections algorithm (SPA), and competitive adaptive reweighted sampling (CARS). The partial least squares regression (PLSR) was used to create a linear model between SSC of pears and its spectrum. The pretreatment methods and PLSR were processed by a chemometrics software called Unscrambler $\times 10.1$ (CAMO AS, Trondheim, Norway), and the variable selection methods were implemented by Matlab 7.5 (The Math Works, Inc, USA).

The root mean square errors of calibration (RMSEC), the root mean square errors of prediction (RMSEP), correlation coefficient of calibration (r_c), correlation coefficient of prediction (r_p) were taken as a model performance evaluation criteria. In order to ensure the performance of the model and avoid over fitting phenomenon, a high correlation coefficient and smaller error model needs to be chosen.

3 Results and Discussion

3.1 Measurement Results of SSC

Table 1 presents the statistical results of SSC of the measured samples in this study. In this paper, the range of SSC in the calibration set is from 7.6 to 13.4 °Brix, and the range of SSC in the prediction set of SSC is from 8.2 to 12.1 °Brix. The range of SSC in the calibration set covers the range in the prediction set, so the model established by calibration set can be better applied to predict the prediction set.

3.2 Spectral Analysis

Figure 2 shows the average spectrum of all Cuiguan pears. Both ends of the spectrum were discarded because of low signal-to-noise ratio. So 616–887 nm was selected in this experiment fort modeling, and this wavelength range contains 1,400 variables.

3.3 Spectral Pretreatment Methods

Table 2 shows the results of PLSR model with different pretreatment methods. MSC is the optimum pretreatment methods, and PLS with MSC obtains the best results. The performance of MSC-PLS model is listed as follows, r_c is 0.98, RMSEC is 0.27 °Brix, r_p is 0.91, PMSEP is 0.43, and the factors is 11.

Table 1 The statistical results of SSC of Cuiguan pears

Parameter	Data set	N	Range	Mean	S.D.
SSC/°Brix	Total samples	196	7.6–13.4	10.27	1.21
	Calibration set	147	7.6–13.4	10.31	1.26
	Prediction set	49	8.2–12.1	10.13	1.03

N number of samples; *S.D.* standard deviation

Fig. 2 Vis/NIR diffuse transmission average spectra of all Cuiguan pears

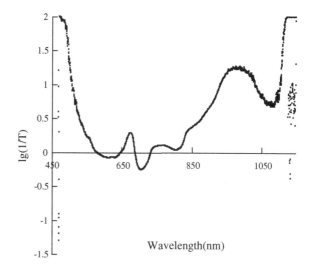

3.4 Variable Optimum

Three variable selection methods were conducted in Matlab 7.5 to select important variables using the MSC spectra, and the numbers of selected variables for GA, SPA, and CARS were 98, 40, and 84, respectively. Table 3 shows the model results of different selected important variables, which are obtained by PLSR. From Table 3, it can be seen that CARS is the best variable selection method. Compared with GA, CARS selects fewer variables, r_c rose from 0.95 to 0.98, r_p rose from 0.93 to 0.96, RMSEC decreased from 0.40 to 0.26, RMSEP decreased from 0.48 to 0.29. Compared with SPA, although the number of variables was increased, the performance of the model was greatly improved. Compared with the original spectrum, the number of variables decreased from 1,400 to 84, r_p rose from 0.88 to 0.96, RMSEC dropped from 0.48 to 0.26, the number of factors dropped from 11 to 9. The results indicate that CARS variable selection method is an effective method that can effectively be applied to the Cuiguan pear spectral variables preferably, to simplify prediction model, and improve the accuracy of the model.

Table 2 The model results of different pretreatment methods

Pretreatment methods	r_c	RMSEC	r_p	RMSEP	Factors
Non	0.96	0.33	0.88	0.48	11
SG	0.96	0.34	0.88	0.48	11
SNV	0.97	0.29	0.92	0.43	11
MSC	**0.98**	**0.27**	**0.91**	**0.43**	**11**
UVN	0.94	0.42	0.85	0.55	11
WT	0.96	0.42	0.88	0.50	11

Table 3 The model results of different select important variable by PLSR

Method (variables)	r_c	RMSEC	r_p	RMSEP	Factors
Non (1,400)	0.96	0.33	0.88	0.48	11
MSC-GA (98)	0.95	0.40	0.93	0.42	8
MSC-SPA (40)	0.95	0.41	0.89	0.51	9
MSC-CARS (84)	**0.98**	**0.26**	**0.96**	**0.29**	**9**

Fig. 3 Predicted results of the calibration model

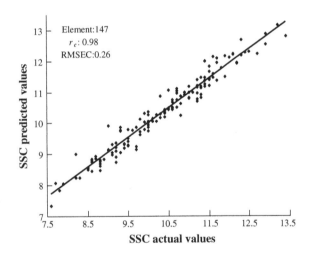

3.5 The Best Model

Total of 196 Cuiguan pear samples were used for the experiment, 147 samples were used as the calibration, and the remaining 49 samples were used as the prediction set. First, a pretreatment using MSC for the spectrum (616–914 nm) of Cuiguan pear was conducted then the variables selected by the CARS were used to establish the best model of Cuiguan pear SSC. The predicted results of calibration samples by MSC-CARS-PLS model with leave-one-out cross-validation are shown in Fig. 3, and the results of prediction samples are shown in Fig. 4.

Fig. 4 Predicted results of the prediction model

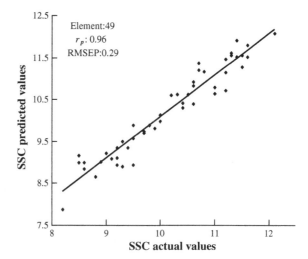

4 Conclusions

Visible/near infrared (Vis/NIR) transmission spectra of Cuiguan pears were taken by a USB4000 spectrometer. A variety of pretreatment methods and three variable selection methods were used in this paper for data analysis. The results show that MSC is the best pretreatment method, and CARS method is an effective wavelength selection method that can select the important variables for SSC of Cuiguan pear efficiently. The correlation coefficients and RMSEs in the calibration and prediction sets in the MSC-CARS-PLS model are 0.98, 0.26 and 0.96, 0.29, respectively. SSC prediction model for Cuiguan pear based on the Vis/NIR diffuse transmittance spectroscopy can be effectively optimized by MSC combined with CARS.

References

1. Jha SN, Garg R (2010) Non-destructive prediction of quality of intact apple using near infrared spectroscopy. J Food Sci Technol 47:207–213
2. Camps C, Christen D (2009) Non-destructive assessment of apricot fruit quality by portable visible-near infrared spectroscopy. J LWT-Food Sci Technol 42:1125–1131
3. Flores K, Sánchez M, Pérez-Marín D et al (2008) Prediction of total soluble solid content in intact and cut melons and watermelons using near infrared spectroscopy. J Near Infrared Spectrosc 16:91–98
4. Takano K, Senoo T, Uno T et al (2007) Distinction of astringency in peach fruit using near-infrared spectroscopy. J Hortic Res (Japan) 6:137–143
5. Cayuela JA (2008) Vis/NIR soluble solids prediction in intact oranges (Citrus sinensis L.) cv. Valencia late by reflectance. J Postharvest Biol Technol 47:75–80

6. Xudong S, Hailiang Z, Yande L (2009) Nondestructive assessment of quality of Nanfeng mandarin fruit by a portable near infrared spectroscopy. Int J Agric Biol Eng 2:65–71

7. Liu Y, Gao R, Hao Y et al (2012) Improvement of near-infrared spectral calibration models for brix prediction in 'Gannan' navel oranges by a portable near-infrared device. J Food Bioprocess Technol 5:1106–1112

8. Paz P, Sánche MT, Pérez-Marín D et al (2009) Instantaneous quantitative and qualitative assessment of pear quality using near infrared spectroscopy. Comput Electron Agric 69:24–32

9. Tsuchikawa S, Sakai E, Inoue K et al (2003) Application of time-of-flight near-infrared spectroscopy to detect sugar and acid content in Satsuma mandarin. J Am Soc Hortic Sci 128:391–396

10. Jha SN, Jaiswal P, Narsaiah K et al (2012) Non-destructive prediction of sweetness of intact mango using near infrared spectroscopy. J Scientia Horticulturae 138:171–175

11. Sun T, Lin H, Xu H et al (2009) Effect of fruit moving speed on predicting soluble solids content of 'Cuiguan' pears (Pomaceae pyrifolia Nakai cv. Cuiguan) using PLS and LS-SVM regression. J Postharvest Biol Technol 51:86–90

12. Wang JH, Qi SY, Tang ZH et al (2012) Temperature compensation for portable Vis/NIR spectrometer measurement of apple fruit soluble solids contents. J Guang pu xue yu guang pu fen xi = Guang pu 32:1431

13. LeiMing Y, HaiNing G, Song L et al (2012) Non-destructive analysis of soluble solids content in apple by VIS/NIR semi-transmittance. J Food Saf Qual 3:448–452

14. Moghimi A, Aghkhani MH, Sazgarnia A et al (2010) Vis/NIR spectroscopy and chemometrics for the prediction of soluble solids content and acidity (pH) of kiwifruit. J Biosyst Eng 106:295–302

15. Paz P, Sánchez MT, Pérez-Marín D et al (2008) Nondestructive determination of total soluble solid content and firmness in plums using near-infrared reflectance spectroscopy. J Agric Food Chem 56:2565–2570

Ontology-Based Design Knowledge Representation for Complex Product

Liu Yang, Linfang Qian, Shengchun Ding and Yadong Xu

Abstract In order to represent the design knowledge of complex product, logical descriptions of different knowledge are conducted by character analysis for the generalized design knowledge. Then a knowledge representation system is proposed in this chapter utilizing the three-layer structure of "design object ontology—design process ontology—data resource" (OPR). And the semantic description frame is defined based on the suggested upper merged ontology (SUMO) to implement the ontology collaborative construction using the ontology web language (OWL). Based on the constructed ontology, the storage structure for generalized knowledge is determined to actualize the generalized knowledge representation. Finally, the feasibility and effectiveness of the proposed technique in this chapter is demonstrated by the design knowledge representation for the self-propelled gun.

Keywords Generalized knowledge · Ontology · Knowledge representation · Self-propelled gun

1 Introduction

The development of complex product is a process of inheriting and reusing the design knowledge, the knowledge reuse plays an important role in product design [1–3]. Researches focused on the product design knowledge representation and

L. Yang (✉) · L. Qian · Y. Xu
School of Mechanical Engineering, Nanjing University of Science and Technology, Nanjing 210094, China
e-mail: ylnjust@126.com

S. Ding
School of Economics and Management, Nanjing University of Science and Technology, Nanjing 210094, China

Z. Wen and T. Li (eds.), *Knowledge Engineering and Management*,
Advances in Intelligent Systems and Computing 278,
DOI: 10.1007/978-3-642-54930-4_28, © Springer-Verlag Berlin Heidelberg 2014

modeling have been carried out in the literature. For example, Li et al. [4] succeeded in the product design knowledge representation, acquisition, and application by using knowledge engineering technique to classify the product design knowledge. Yu et al. [1] utilized the structure of "domain ontology—index knowledge—data resource" to establish the design knowledge representation system. Zhang et al. [5] studied the design knowledge modeling technique based on design knowledge management, and summarized the advantages of ontology-based design knowledge modeling. Zhang et al. [6] studied the ontology-based knowledge representation, and proposed a parallel ontology-based knowledge modeling technique. Wang et al. [7] presented an ontology-based multidesigner collaborative product design knowledge modeling method. As the generalized design knowledge exists in all kinds of knowledge resource, such as books, design handbooks, national standards, patents, experience of designers, and design cases in the format of texts, pictures, videos, and models. The generalized design knowledge for complex product should be classified and described reasonably combined with the design process to achieve the generalized knowledge representation. Thus, designers can reuse the knowledge conveniently in design process. In this chapter, the generalized design knowledge is classified into four kinds, i.e., design object knowledge, design process knowledge, design cases, and design resource. The knowledge characters are described logically based on UML static diagram. The knowledge representation system is proposed in this chapter utilizing the three-layer structure of "design object ontology-design process ontology-data resource" (OPR), and the semantic description framework is established based on the suggested upper merged ontology (SUMO), the ontology web language (OWL) is selected for the modeling of ontology to guarantee the sharing and reusability of the ontology. The implementation steps and knowledge storage structure of multidesigners collaborative ontology construction in the LAN environment are investigated to achieve the generalized knowledge representation of complex products. Finally, the effectiveness of the proposed method in the design knowledge representation of complex products is demonstrated by the design knowledge representation for the self-propelled gun.

2 Character Analysis of Product Design Knowledge

The complex product design process contains several design steps, and the designer should determine the design scheme and design parameters for the design object by analysis and calculation together with his design experience and existed design cases. In this chapter, the generalized design knowledge is classified into four kinds, i.e., object knowledge, process knowledge, design cases, and design resource according to the knowledge utilizing pattern in the complex product design process. The relationships among different types of knowledge are shown in Fig. 1. The design object knowledge is the base of the generalized knowledge, and the design process knowledge and design cases also have corresponding

Fig. 1 Relationships among different types of knowledge

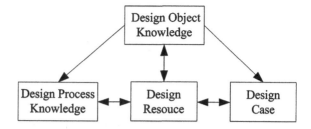

relationships with the object knowledge. The design resource is used to help to describe the design object knowledge and design process knowledge as well as the design cases.

2.1 Design Object Knowledge

The design object knowledge refers to the inherent concepts and their properties in complex product design field, which can be used to describe the component, classification, relationships between different components, and design experience, etc. Complex products often have a variety of different types which may have different components, such as partly different subsystem or completely different subsystems. And this may also exist in subsystems and components. The class, object, and relationship of UML static diagram are used to accomplish the logical description for design object knowledge, parts of which are shown in Fig. 2.

2.2 Design Process knowledge

The design process knowledge contains design steps for overall product and components in different design stages, parameters, the parameter constraints, and the experience in selecting parameters involved in these steps. Complex product design process is usually divided into several stages, including overall design, scheme design, detailed design and process design, etc. The design object and design process of different stage is usually not the same, the design steps of the same object at different stage also may not the same.

And the relationships between the parameter and design step can be constructed by dividing design process of the system, subsystem, component, and part at different design stage into several sequential design steps and extracting the parameters, parameter constrains, and the experience in selecting parameters used in design process. Part of logical description of the design process knowledge is shown in Fig. 3, there are seven parametric properties as shown in Fig. 3, which can be chosen according to practice. And different types of the parametric property can be chosen, including integer type, floating-point type and text type, etc.

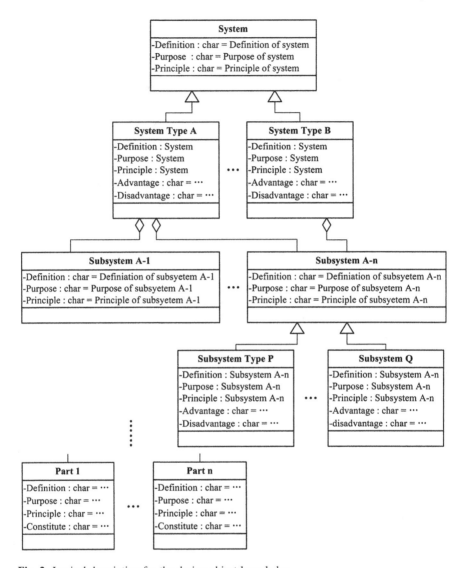

Fig. 2 Logical description for the design object knowledge

2.3 Design Cases

Design cases refer to the successful cases accumulated in the design process, including the case properties indicating the case characteristics and the case parameters containing the design process parameters and design result parameters. These design parameters of a successful case own good reference value to new design tasks. Corresponding to the design object, the description of design case can

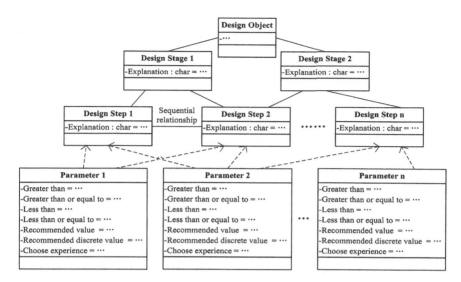

Fig. 3 Logical description for the design process knowledge

Fig. 4 Logical description for the design case

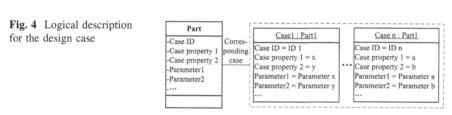

be accomplished by defining values of case property and case parameters when case property and case parameters are added to the design object class. The logical description of design case taking parts for example is shown in Fig. 4.

2.4 Design Resource

Product design resource includes images, videos, models, and calculation programs used in design process. Images include product function diagram, structural diagram, product photos, etc. Videos include product simulation animation and videos recording the test. Models include 2D and 3D models of the design cases, dynamic models, FEM models, and optimization models used in the design process. The reuse of these design resource can help designers to complete product design faster and more intuitive. As shown in Fig. 5, the logical description of design resource is constructed by numbering the resource and establishing mappings between the design resource and knowledge of other types based on analyzing the relationship between design resource and design objects/design process/design cases [8].

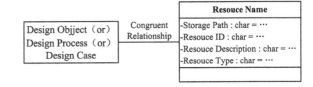

Fig. 5 Logical description for the design resource

3 Ontology-Based Knowledge Representation

The commonly used methods for knowledge representation include object-oriented representation, ontology-based representation, production representation, frame representation, etc. The ontology-based representation of complex product design knowledge is adopted in this chapter by analyzing complex product design characteristics and description. Ontology is a conceptualized description of domain knowledge, the core of the ontology is the concepts and the relationship between concepts [9]. Knowledge representation can be achieved by defining the three elements of ontology, i.e., classes, relationships, and cases.

3.1 Ontology-Based Knowledge Representation System

The structure of design object ontology-design process ontology-data resource (OPD) is adopted to achieve the design knowledge representation.

Definition 1 Design object Ontology
Design object ontology is used to describe the design object knowledge, which can be defined using sextuple as follows:

$$DOO = \{OC, OP, DP, CID, CP, RID\}$$

where DOO is the ontology of design object, OC is the set of design object class, OP is the set of object property, DP is the set of data property, CID is the ID of design case, CP is the set of case property, and RID is the ID of design resource.

Definition 2 Design process ontology
Design process ontology is used to describe the design process knowledge, which can be defined using sextuple as follows,

$$DPO = \{OC, DSA, DST, DPA, OP, DP\}$$

where DPO is the ontology of design process, DSA is the set of design stage class, DST is the set of design step class, and DPA is the set of design parameter class.

Data resource is used to describe case parameters and design resource.

3.2 Ontology-Based Semantic Description Frame

Semantic relations in ontology include two categories, i.e., the data properties and object properties. The data properties are used to describe the characteristics of ontology classes, and the object properties are used to describe semantic relations among ontology classes. In order to guarantee the sharing and reusability of the established ontology, a normative frame for the semantic description should be defined. In this chapter, complex product design knowledge ontology semantic description frame has been established based on the reuse SUMO. SUMO is an upper ontology defined by the W3C organization, and is normative top-level ontology recognized by IEEE organization, including the normative reusable class, data properties and object properties. For complex product design process, pertinent semantic relations can be created for different complex products based on semantic relations inherited from SUMO, such as product classification methods, Table 1 shows some semantic relations inherited from SUMO.

4 Modeling and Storage of Knowledge

4.1 Ontology Modeling Steps

OWL is a newly normative ontology description language proposed by W3C, the construction thought of logical description is fused in its syntax and semantics with well-defined semantics. Thus, the OWL is fit to represent concepts and relationships between concepts [2], and is chosen as ontology modeling language in this chapter. Multidesigners collaborative construction in the LAN environment is achieved by using Protégé modeling tool of Stanford, with the specific steps as follows:

(1) Configure network environment of ontology collaborative construction server and ontology modeling computer terminal. Ontology modeling project is created and all kinds of semantic relationships should be added under ontology description frame on the sever;

(2) Build the design object classes, add the data properties and corresponding values to the design object classes to create the relationships among the design objects;

(3) Build the design process classes, including design stage classes, design step classes and design parameter classes, add the data properties and corresponding values to these classes and create the relationships between the design process classes and design object classes, as well as that among the design process classes;

(4) Add individuals and description properties to object class corresponding to the case, define the values of description properties and the cases ID;

Table 1 Part of semantic relationships from SUMO

Object properties	Data properties
Subclass	lessThan
part	greaterThan
contains	lessThanOrEqualTo
SubAttribute	greaterThanOrEqualTo
equal	hasPurpose

Fig. 6 Knowledge storage structure

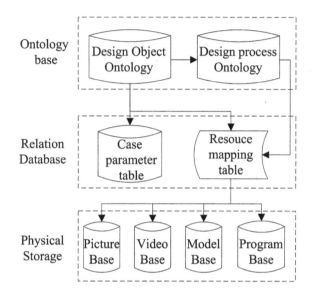

(5) Add resource properties and ID to the classes or individuals corresponding to design resource;

(6) Check the built ontology;

(7) Export the finally finished OWL modeling file.

4.2 Storage of Ontology

Knowledge storage structure is shown in Fig. 6, the generalized knowledge is stored on the base of ontology, and ontology is stored in OWL/RDF Database format. Aiming at the large amount of case parameters, each case should be numbered to create unified case parameter table, including "number," "case ID," "case name," "parameter name," and "parameter value" and others, thus the corresponding relationship between the case parameters and individuals of ontology can be established by the unique case ID. Then, number the design resource, and create resource mapping table, including "number," "resource ID,"

Table 2 User-defined semantic relationship for self-propelled artillery

Object properties		Data properties	
Relation	Explanation	Relation	Explanation
justEarlierThan	Design step prior to others	Has-definition	Has a certain definition
justLaterThan	Design step later than others	hasAdvantage	Has some advantage
byBoreStructure	By bore structure	hasDisAdvantage	Has some disadvantage
byOperationalPurpose	By operational purpose	hasPrinciple	Has a certain principle

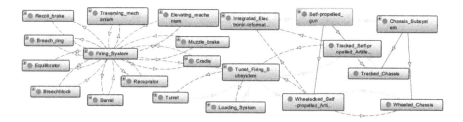

Fig. 7 Graphical view of design object knowledge

"resource name," "resource description," "storage location," and so on, thus the corresponding relationship between resource and class or individual of ontology can be established by the unique resource ID. Finally, the case parameter table and resource mapping table are stored using relationship data, resource in graph base, video base, model base, and calculation program base is physically stored directly.

5 Example Verification

Self-propelled gun is of two kinds, wheeled and tracked, each kind contains turret-firing subsystem, chassis subsystem, and intergraded electronic-information subsystem, but the compositions of different subsystem are not the same. The self-propelled gun is a typical complex mechanical product, its design process include several design stages, such as overall design, scheme design and detailed design, etc [10]. The design process ontology and design object ontology of self-propelled gun can be constructed based on logical description for design process knowledge of self-propelled gun. The established ontology includes more than 860 classes, and more than 350 individuals. Some user-defined semantic relations are created on the basis of inheriting the SUMO, Table 2 shows some user-defined semantic relations for user-propelled gun. And more than 500 design resources have been stored, also the mappings between design resources and ontology have been succeeded, and the design knowledge representation of user-propelled gun is finally accomplished. The Ontograf plugin of Protégé is used to display created ontology graphically, Fig. 7 gives a graphical view of some design object

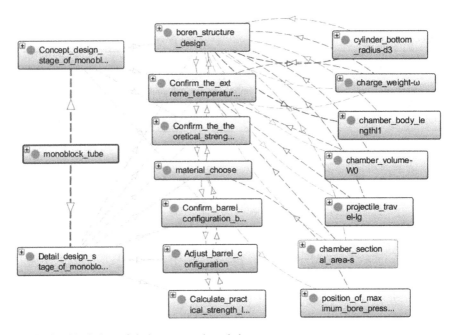

Fig. 8 Graphical view of design process knowledge

knowledge, and Fig. 8 gives a graphical view of design process knowledge taking monocular barrel for example, including some of the design stage classes, design process classes, and the design parameter classes.

6 Conclusions

Aiming at the characters of the generalized design knowledge of complex product and the design knowledge is classified to four kinds combined with the product design process in this chapter. The characteristics of different types of knowledge are analyzed, and logical description of knowledge is carried out using UML static diagram. The ontology-based knowledge representation system utilizing the three-layer structure of "design object ontology-design process ontology-data resource" and the semantic description frame based on the SUMO have been represented to achieve the representation of generalized knowledge. The steps of ontology collaborative construction using the OWL has been determined to guarantee the sharing and reusability of the established ontology. The feasibility and effectiveness of knowledge representation method proposed in this chapter is demonstrated by self-propelled gun design knowledge representation.

The proposed ontology-based design knowledge representation method for complex product can succeed in multidesigner collaborative generalized design knowledge representation, and make the design knowledge contact closely with design process. This makes it convenient for the designer to reuse the existed design knowledge and cases, and thus improve the design efficiency.

Acknowledgments This research is partially supported by National Defense Basic Research Project under Grant No. A2620110003 and A2620133003.

References

1. Yu X, Liu J, He M (2011) Design knowledge retrieval technology based on domain ontology for complex products. Comput Int Manufact Syst 17(2):225–231
2. Rezayat M (2000) Knowledge-based product development using XML and KCs. Comput Aided Des 32(2):299–309
3. Tu L, Zhang S (2010) Knowledge representation and modeling oriented to reusable complex product design. Chin J Mech Eng 21(7):777–781
4. Li Z, Jin X, Jia H et al (2006) The knowledge representation and reuse in product design. J Shanghai Jiaotong Univ 40(7):1183–1186
5. Zhang D, Liao W, Hu J et al (2005) Design knowledge modeling based on ontology. J South China Univ Technol (Natural Science Edition) 35(5):26–31
6. Zhang M, Hao J, Yan Y et al (2010) Ontology-Based Knowledge Modeling. Trans Beijing Inst Technol 30(12):1405–1408, 1431
7. Wang Y, Wang F, Le C et al (2012) Ontology-based knowledge modeling of collaborative product design for multi-design teams. Chin J Mech Eng 23(22):2720–2725
8. Attene M, Robbiano F, Spagnuolo M et al (2009) Characterization of 3D shape parts for semantic annotation. Comput Aided Des 46:756–763
9. Zheng J, He L, Ye X (2010) Ontology-based knowledge representation for computer-aided fixture design. J Comput Res Dev 47(7):1276–1285
10. Zhang X, Zheng J, Yang J (2005) The theory of gun design. Beijing Institute of Technology Press, Beijing

Research on Dynamic Ontology Construction Method for Knowledge Fusion in Group Corporation

Jihong Liu, Wenting Xu and Hao Jiang

Abstract In Group Corporation, the large number of distributed and heterogeneous knowledge resources makes it difficult to give full play to their integrated benefit. Knowledge Fusion is proposed to solve this problem. Aiming at the specific characters of Group Corporation, this paper presents a method for Multiple Domain Ontology Construction. Dynamic ontology is a temporary ontology built on the base of Multiple Domain Ontology to compact the knowledge requirements and form the target knowledge resources. A Dynamic Ontology Construction Method is also proposed by analyzing knowledge requirements for more effective Knowledge Fusion. At the end of the paper, an application example of Dynamic Ontology Construction is presented to prove the feasibility of the method.

Keywords Dynamic ontology · Multiple domain ontology · Knowledge Fusion · Group Corporation

1 Introduction

As the most important intelligent resource, knowledge contributes to the product design and manufacture, in this knowledge-intensive world. However, there are a large number of distributed and heterogeneous knowledge resources scattered around the inside of Group Corporation. Due to the lack of unified organization and management, the knowledge resources can neither give full play to the integrated benefit, nor support the business processes effectively [1]. With the development of semantic web and text mining technology, ontology-based Knowledge Fusion has been widely researched and applied. Aimed at specific

J. Liu (✉) · W. Xu · H. Jiang
School of Mechanical Engineering and Automation, Beihang University, Beijing, China
e-mail: ryukeiko@buaa.edu.cn

Z. Wen and T. Li (eds.), *Knowledge Engineering and Management*,
Advances in Intelligent Systems and Computing 278,
DOI: 10.1007/978-3-642-54930-4_29, © Springer-Verlag Berlin Heidelberg 2014

knowledge requirements, Knowledge Fusion can integrate knowledge resources effectively at the semantic level and provide more targeted services.

Knowledge Fusion is an interdiscipline of knowledge science and information science. Its basic concepts and fundamental techniques are derived from the information fusion and knowledge integration [2]. Knowledge engineers fuse knowledge with different types, contents, features, and positions to give full play to the integrated benefit of on-demand knowledge supply, thus achieve the goal of knowledge innovation. As an explicit specification of a conceptualization [3], ontology could effectively support knowledge acquisition, organization, management, and application by the uniform semantic description of knowledge resources [4, 5].

However, the Knowledge Fusion in Group Corporation is quite complicated, which contains sorting, modeling, acquisition, and processing of the multi-source and heterogeneous knowledge. Therefore, adjustments for specific situations are necessary to organize knowledge resources by ontology in Group Corporation. The Group Corporations focus on the dynamic matching between requirements and services. Hence, two features should be considered in the Knowledge Fusion of practical business processes.

- As the variety of knowledge requirements, the knowledge resources of Group Corporation are updating continuously. An effective service matching between knowledge requirements and resources should be built.
- Although the Domain Ontology has constructed extensive knowledge networks, knowledge requirements may be concentrated on specific nodes or segments.

Therefore, based on the knowledge requirements and the Multiple Domain Ontology, this paper constructs the Dynamic Ontology and the connection framework between knowledge requirements and resources as the basis of the extraction and fusion of knowledge segments. By analyzing knowledge requirements, the relevant ontology segments are extracted from various Domain Ontologies to construct the Dynamic Ontology.

2 Multiple Domain Ontology Construction in Group Corporation

The knowledge structure of Group Corporation is complex. It involves diverse domains and applies to various objects. Their knowledge may come from different sources and correlate with each other. Since it is complicated to construct a unified ontology, different ontologies are encouraged to be built up in the organizations by defining the hierarchy structure of ontology. Guarino, Uschold [6], and Jayaram [7] divided ontology into three categories: Upper Ontology, Domain Ontology, and Application Ontology. The hierarchical ontology organization and construction idea, and the standardized ontology construction may lay a relatively stable knowledge base for the organization and management of knowledge resources in Group Corporation. When constructing ontology in Group Corporation, the actual situation of the large number of the subordinate businesses and departments should be considered. In the unified upper concept frame structure of ontology, concepts and relationships from each divided domain should be collected to form the Domain Ontologies. The mapping relations between the Domain Ontologies should be also built to construct the Multiple Domain Ontology across the enterprise.

2.1 Domain Ontology and Multiple Domain Ontology

Domain Ontology is a kind of specific ontology to describe the knowledge concepts in the domain. It contains a series of concepts and the relationships between them. Based on Domain Ontology, Multiple Domain Ontology (MDO) contains the ontologies of various domains and their mapping relationships. The definition is as follows.

$$MDO = \ <DO_1, DO_2, \cdots, DO_i, \cdots, MR> \qquad (1)$$

DO_i is the ontology of one domain. MR stands for the Mapping Relation Set of the Multiple Domain Ontology. Through the ontology mapping, the concept relationships between different Domain Ontologies can be built to achieve the integration of different ontologies.

Semantic relations include two-way Symmetry and Transitivity and one-way Reciprocal. The characteristics of these relationships can reduce the complexity of the relationship construction in the process of the ontology construction. Namely, it can form semantic relations automatically by making use of the reasoning function of the ontology.

2.2 Multiple Domain Ontology Organization and Construction Method

According to the hierarchical idea of Multiple Domain Ontology and the arrangement of existing organizations and relevant disciplines, the organization hierarchy of Multiple Domain Ontology is built. Meanwhile, by sorting each discipline separately, the relevant terms and relations are sorted out.

The Upper Ontology is utilized to organize Domain Ontologies totally.

- The Upper Ontology is used to describe the hierarchy structure and the mapping relations between the Domain Ontologies. It is related to various Domain Ontologies by its structure which is divided referring to the relevant Upper Ontologies. Meanwhile, as the Concept Set and Relation Set, the Mapping Relations between all the Domain Ontologies are stored in the Upper Ontology.
- To get the complete Multiple Domain Ontology, the concept hierarchies, terms, and relations in the professional domain are subdivided based on the Upper Ontology.

The ontology construction mainly refers to the description language, construction method, and construction tool.

In this paper, the most popular Web Ontology Language (OWL) [8] is taken as the ontology description language. Protégé, which is developed by the ontology research team of Stanford University, is chosen as the ontology construction tool.

The research groups have come up with a number of ontology construction methods, including skeleton method, enterprise modeling method, Methontology, IDEF-5, etc. However, special methods should be established to adapt to the different backgrounds and applications. Combining with the characteristics of Multiple Domain Ontology and the application requirements of professional domain, the basic process of Multiple Domain Ontology construction is proposed as shown in Fig. 1.

- MDO system construction
  In the organization hierarchy of Multiple Domain Ontology and the ontology mapping model, the basic requirements of Multiple Domain Ontology modeling and mapping are analyzed. The information will be used to build the Multiple Domain Ontology system which includes hierarchy, concept structure, and concept relations.
- Concept collection and relation division of Domain Ontology
  Based on the subdivided ontology hierarchy and the concept structure, the standard specification and the relevant professional information are analyzed. As the term concepts are extracted, the semantic relations between concepts are built.
- Mapping relation construction
  After all the Domain Ontologies are constructed, each domain classification and concept should be analyzed to build cross-domain classifications and concept mappings. Then the mapping of Multiple Domain Ontology can be achieved. To make sure that the related work can match the condition of the Group Corporation, this process requires relevant professional department personnel to participate in.
- MDO evaluation
  The evaluation mainly focuses on the effectiveness of the ontology construction. On the one hand, in the preceding work process, relevant professional department personnel should be invited to participate in. This measure can ensure the

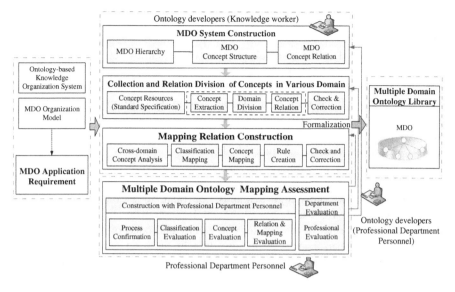

Fig. 1 Basic process of multiple domain ontology construction

quality of the related work. On the other hand, relevant professional department personnel are also asked to review the results which are expected to be improved in the using process.

Finally, through the ontology description language, the ontology can be expressed in a formal representation and stored.

3 Dynamic Ontology Construction Based on Knowledge Requirements

Multiple Domain Ontologies form the knowledge network of Group Corporation by mapping. According to knowledge requirements, Multiple Domain Ontologies are analyzed and processed by Jena to construct the Dynamic Ontology and provide conceptual framework for Knowledge Fusion.

According to the knowledge requirements, Dynamic Ontology is a temporary ontology built on the base of Multiple Domain Ontology. It results from the dynamic reorganization of the partial segments of MDO.

Dynamic Ontology inherits the existing semantic relation of MDO, and form new semantic relation aiming at the knowledge requirements. The main goal of Dynamic Ontology is to compact the knowledge requirements and form the target knowledge resources.

3.1 Process of Dynamic Ontology Construction

Jena was originated in the research program of semantics network done by HP laboratory, and was later transferred to Apache software foundation to manage. Jena has an open-source software framework and provides large number of tools and Java libraries. It has been being widely used in the development of ontology-based application system. The core of Jena is the adoption of processing based on triple (S (subject), P (predicate), O (Object)) to achieve the operation for all models, resources, classes, and individuals. This paper mainly adopts OntModels API (API of OWL) and storage mechanism to retrieve, expand, and store the ontologies.

According to the knowledge demands, the main idea of dynamic ontology construction is to retrieve and expand the ontology segments in Multiple Domain Ontologies and reorganize the ontology hierarchy and concept relations.

Based on the term segmentation, the domain term set can be acquired by analyzing the knowledge demands described by users. By IKAnalyzer segmentation, the domain term dictionary formed by MDO is used in term segmentation [9]. To build and expand the semantic relations between the ontology segments, the users should add the term relations manually.

Considering specific term relations, ontology segments can be acquired by Jena-based expansion, retrieval, and storage of ontology. And the term relation addition is used to reorganize the ontology segments.

- Multiple Ontology Expansion includes the Upper Ontology Expansion and every Domain Ontology Expansion.

  (1) Upper Ontology Expansion. By retrieving in Upper Ontology with terms, this method could get the professional domain the concept belongs to, as well as the mapping concept and corresponding domain (if the mapping concept exists).
  (2) Domain Ontology Expansion. Based on the domain derived from the expansion of Upper Ontology, the following information could be obtained from the domain which each concept belongs to. Ontology Hierarchy (Hierarchy) is the class which the concepts belong to. The concepts include the concept itself, the mapping concept, and the concept associated with semantics. Concept and Relation include the concepts directly related to the input concepts, as well as the concepts directly related to mapping concepts.

- Reorganization of ontology segments includes reorganization of dynamic ontology hierarchy and reorganization of ontology concept and relation.

  (1) Reorganization of Dynamic Ontology Hierarchy removes the repetitive classes in different expanded ontology segments and connects the concepts to form new conceptual hierarchy.

(2) Reorganization of Concept and Relation removes the reiteration of expanded concepts and relationships. And it estimates the relationships added by users between the segmentations.

Both expansion and reorganization are based on related port provided by Jena, the next section details the process to expand and reorganize the construction of Dynamic Ontology.

3.2 Semantic-Based Multiple Ontology Expansion

Ontology expansion proceeds in Upper Ontology and Domain Ontologies by obtaining ontology word set.

- In the process of the Upper Ontology Extension, the Upper Ontology model is obtained by using the OntModel of Jena firstly. And then the concept of word set in model hierarchy is also obtained. If there is a mapping, the concept-belong domain is also acquired. The mapping concept of concept is retrieved to build two-way associative retrieval with the use of above-mentioned triple(S, P, O). The corresponding relations and concepts and the domain which the concept belongs to will be obtained by retrieving the triples.
- In the process of Domain Ontology Expansion, the concept hierarchy and related concepts are obtained by retrieving the domains of the acquired concepts and the mapping expanded concepts. And these have the relations, such as Is_Sibling and Is_Synonym, with the segmentation word set should be aimed at particularly to get their classes.

By iteration, the concept-centric ontology segments can be obtained.

3.3 Reorganization of Expansion Ontology Segments

After getting ontology segments, the Dynamic Ontology can be constructed by building concept, relation, and ontology concept hierarchy. As a container to receive ontology segments, ontology model is built with the support of Jena.

By traversing ontology segments of all classes and removing repetitive classes, the classes in ontologies and their hierarchy relation are built. In the hierarchy relations of Dynamic Ontologies, the segmentation term is the top class and the expanded classes are the SubClasses. The concepts are subjected to the corresponding classes after traversing all the ontology segments and removing the repetitive concepts. And the relations between concepts in segments are traversed to build semantic relations. Finally, after sorting all relationships, the Dynamic Ontology with completed concept and structure is formed by reasoning. The above-mentioned rule of Symmetry and Transitivity is used in this process.

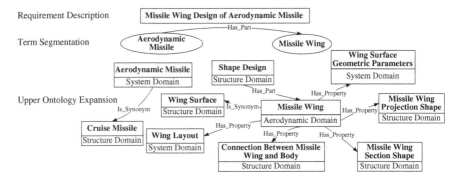

Fig. 2 Term requirement analysis and upper ontology expansion

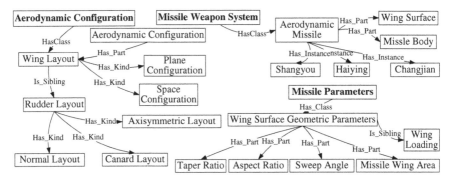

Fig. 3 System domain expansion

4 An Application Example of Dynamic Ontology Construction

As an example, one Group Corporation that develops aerodynamic weapons has a lot of subunits, departments, and domains. The Multiple Domain Ontology including the domains of system, control, structure, and aerodynamics can be constructed by arrangement of the knowledge resources from different domains.

As shown in Fig. 2, the key terms, "aerodynamic missile" and "missile wing," can be acquired by the term segmentation of the requirement "missile wing design of aerodynamic missile." On the basis of adding the semantic relation "aerodynamic missile Has_Part missile wing," the terms and domains which have mapping relations with the two terms can be retrieved in the Upper Ontology by making use of Jena. It lays the foundation for Ontology Segment acquisition.

Then, the terms that have semantic relations with the domains of system, control, structure, and aerodynamics can be obtained with their class. The system domain expansion is shown in Fig. 3 as an example.

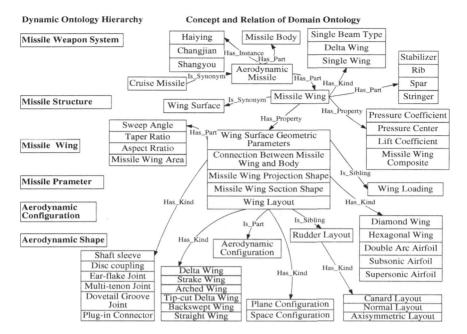

Fig. 4 Result of dynamic ontology construction

The Ontology Segments expanded from the domains should be reorganized to remove the repetitive terms and relations and construct the Dynamic Ontology with new hierarchy and relations. As shown in Fig. 4, every concept belongs to the relevant hierarchy in Dynamic Ontology. For instance, lift coefficient, missile wing layout, and pressure center are all in the hierarchy of aerodynamic shape. On the basis of existing knowledge requirements and MDO, the Dynamic Ontology which includes all the knowledge resources of the Group Corporation can be constructed to refine the knowledge demands and adapt to the dynamic updating of knowledge resources in Group Corporation.

5 Conclusion

On the basis of organizing knowledge resources by Multiple Domain Ontology, this paper analyzes the knowledge requirements in Group Corporation. By searching the ontology segments related to requirements and reorganizing these segments, the Dynamic Ontology can be constructed on-demand as the foundation of Knowledge Fusion. A Dynamic Ontology Construction Method is proposed in this paper, as well as an application example. The Dynamic Ontology formed by Multiple Domain Ontology reorganization can support the accurate and comprehensive matching between knowledge requirements and knowledge resources to provide more targeted knowledge services.

References

1. Li X, Yang H, Jing S, Yan Y, Wang X, Feng Y (2012) Knowledge service modeling approach for group enterprise cloud manufacturing. Comput Integr Manuf Syst 18(08):1869–1880
2. Hou J (2005) Research on several key techniques in knowledge fusion. Zhejiang University, Hangzhou
3. Gruber TR (1995) Toward principles for the design of ontologies used for knowledge sharing? Int J Hum Comput Stud 43(5):907–928. doi:10.1006/ijhc.1995.1081
4. Tempich C (2003) SWAP: ontology-based knowledge management with peer-to-peer. In: Workshop ontologiebasiertes Wissensmanagement, WM
5. Matsokis A, Kiritsis D (2010) An ontology-based approach for product lifecycle management. Comput Ind 61(8):787–797
6. Provine R, Schlenoff C, Balakirsky S, Smith S, Uschold M (2004) Ontology-based methods for enhancing autonomous vehicle path planning. Robot Auton Syst 49(1):123–133
7. Jayaram U, Kim O, Zhu L (2010) Knowledge representation and ontology mapping methods for product data in engineering applications. J Comput Inf Sci Eng 10:021004-1
8. McGuinness DL, Van Harmelen F (2004) OWL web ontology language overview. W3C Recommendation 10(2004–03):10
9. Yu X, Liu JH, He M (2011) Design knowledge retrieval technology based on domain ontology for complex products. Comput Int Manuf Syst 17(2):225–231

Shanghai Component Stock Index Forecasting Model Based on Data Mining

Wei Shen, Xin Wu and Tiyong Zhang

Abstract As it is known to all, many factors may have influence on the movement of stock index. In stock index forecasting, how many quantitative indicators should be introduced in order to obtain the best forecasting result? And is it true that more indicators translate into higher forecasting accuracy? These issues have long been puzzling to researchers of stock index forecasting. In this paper, we carried out data mining on some quantitative indicators with influence on the movement of stock index, then we had short-term forecasting of Shanghai Component Stock Index with BP+GA model. Results of our research are as follows: forecasting with combination of indicators has better result than forecasting with single indicators; combinations of indicators through selection and optimization have the best result; more indicators introduced into forecasting model do not translate into higher accuracy. The results of our research in this paper demonstrate the necessity and significance of data mining in stock index forecasting.

Keywords BP neural network · Genetic algorithm · Forecast method · Stock index forecast · Data mining

1 Introduction

There are many quantitative factors having influence on the movement of stock index. To solve the above problem, we first carried out data mining on the quantitative indicators with influence on the movement of stock index. Procedures

W. Shen (✉) · X. Wu (✉) · T. Zhang
School of Business and Administration, North China Electric Power University, Beijing, China
e-mail: shenwei_1965@126.com

X. Wu
e-mail: shenwei_1965@126.com

Z. Wen and T. Li (eds.), *Knowledge Engineering and Management,*
Advances in Intelligent Systems and Computing 278,
DOI: 10.1007/978-3-642-54930-4_30, © Springer-Verlag Berlin Heidelberg 2014

of data mining are as follows: first, select a few frequently used technical indicators with higher correlation to the movement of stock index and started single-indicator forecasting; next, optimize indicators according to their performance in single-indicator forecasting and obtained an optimized combination for forecasting. In the process of data mining, we tried to observe if forecasting accuracy changes with the number of technical indicators.

The next step is to select forecasting model. Statistical models (GARCH,SV) require complete samples and more information about their distribution; however, in China's stock market, numbers of samples of listed companies change fast and the related information is not complete; therefore, the forecasting result would not be good if statistical models are used. Intelligent forecasting models, by comparison with statistical models, are more data-tolerant. GRAY model and SVM have their advantages in small sample forecasting but not suitable for large samples.

Many researchers, when doing stock index forecasting, chose different indexes and different forecasting models. For example, Tsai and Hsiao [1] and Chu et al. [2] chose some quantitative indicators correlate with stock index fluctuation and introduced them into forecasting model in order to increase forecasting accuracy. Neural-network forecasting models have the capacity of parallel processing large amount of quantitative factors and suitable for short-term forecasting, Hill uses this forecasting model [3]. Bildirici chooses single closing price when they do stock index forecasting [4]. Chang [5] chose some quantitative indicators correlate with stock index fluctuation. Koopman et al. [6], Huang [7], and Chakravarty and Dash [8] also chose single closing price when they do stock index forecasting. Asadi et al. [9] also chose some quantitative indicators correlate with stock index fluctuation and introduced them into forecasting model in order to increase forecasting accuracy, his research is same with. Hsieh et al. [10] selected 14 input indicators with higher correlation to stock index movement through stepwise regression (SRCS) and introduced them into a cycle neural-network forecasting model based on wavelet transformation and artificial bee colony algorithm before carrying out forecasting to Dow Jones Industrial Average, FTSE 100 indices. Forecasting showed this model had relatively good result. But the above-mentioned paper failed to have further discussion whether the 14 indicators is the best choice [10]. Roh and Chen use neural-network forecasting models, like Hill, to solve the problem [11].

In consideration of data features of Shanghai component stock indices and the pros and cons of various intelligent models, we choose BP+GA forecasting model which is theoretically advanced and has higher forecasting accuracy.

Basic thinking of this paper: first, a brief introduction of theory; next, data mining on technical indicators with influence on the movement of stock index; and then, short-term forecasting of Shanghai Component Stock Index with BP+GA model, testing and verifying of the effectiveness data mining in enhancing forecasting accuracy; finally, we reach conclusion.

2 Fundamentals

2.1 Back-Propagation Neural Network

The back-propagation neural network (hereinafter referred to as BP) is an error back-propagation network, which is composed of input, hidden, and output layers. During the forward propagation, input signals are dealt with from the input layer across the hidden layer and transferred to the output layer. The state of neurons in one layer only affects the neurons in the next layer. If the expected output cannot be obtained in the output layer, the error value is then propagated backward through the network, and changes are made to the weights in each layer. The weight changes are calculated to reduce the error signal for the case in question. The cycle is repeated until the overall value of error drops below a predetermined threshold.

Figure 1 shows the neural network with one hidden layer.

With strong nonlinear mapping ability, BP learning algorithm is extensively applied in neural network. It can distribute characters and rules hidden in samples on the link weights of neural network. Another feature of BP learning algorithm is its fault tolerance. When some links are damaged, ineffective or altered, the overall performance of the network is reduced very little, because the storage of information is distributed in the whole system. Like all algorithms, BP network has its own shortcomings. First, the convergent speed of the classic BP learning algorithm is slow, and it is easy to get in local minimum. In training process, if the weight value is adjusted higher, the weight sum of all or most nerve cells will be larger, hence almost put the adjustment process of network weight value to a stop.

2.2 Genetic Algorithm

The genetic algorithm is a self-adaptive optimized search algorithm based on natural selection and genetics. It was first put forward by Professor John Holland in Michigan University. Its idea comes from natural laws such as biological genetics and survival of the fittest. It takes all individuals in the colony to participate in the process of selection, exchange and variance to produce a new generation of colony with ideal result. The search can be implemented in different areas of parameter space in the colony generation subrogation toward the optimal direction and will not get in local minimization. In general, the genetic algorithm includes four parts: encoding mechanism, fitness function, genetic operators, and control parameters (see Fig. 2).

The selection operation means random distribution and survival of the fittest. The cross operation is to randomly select pair individuals in the group to implement part genes exchange according to certain cross probability Pc and selected cross mode, and the formed filial generation is the new chromosome containing

Fig. 1 BPNN sketch map
with one hidden layer

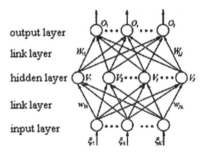

output layer

link layer

hidden layer

link layer

input layer

Fig. 2 Work principle of
genetic algorithm

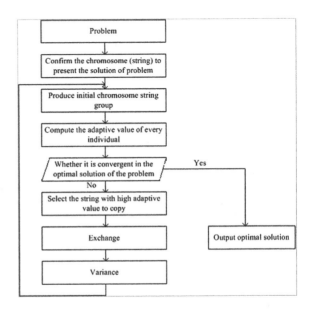

parent gene. Finally, the variance operation is used to simulate the genetic mutation
induced by various occasional factors of biology, and it could change the genetic
value of original chromosome by appointed variance rate Pm, produce the chromosome that the parent generation does not possess, increase the diversification of
group gene, extend the search space, and avoid search to get in local solution.

2.3 BP Algorithm Optimized with GA

When many local minimum points exist, the BP algorithm is easy to get in the local
minimum point. To address this problem, researchers have put forward many
improved algorithms. The genetic algorithm possesses essential parallel computation ability, and it can adopt several arithmetic operators abstracted from natural
selection mechanism to operate the parameter coding character string. This operation
is implemented aiming at the colony composed by multiple feasible solutions, so it

Table 1 Single-indicator forecasting with BP+GA

Date	Real value	MACD		BIAS6		KDJ		PSY	
		Forecasting value	Error rate	Forecasting value	Error rate	Forecasting value	Error rate	Forecasting value	Error rate
27-05	2709.95	2673.37	1.35	2652.50	2.12	2747.62	1.39	2690.44	0.72
30-05	2706.36	2653.86	1.94	2666.58	1.47	2728.28	0.81	2731.53	0.93
31-05	2743.47	2770.08	0.97	2689.97	1.95	2702.32	0.15	2703.69	1.45
1-06	2743.57	2794.60	1.86	2775.40	1.16	2699.95	1.59	2812.43	2.51
2-06	2705.18	2731.69	0.98	2731.69	0.98	2734.94	1.10	2724.39	0.71
3-06	2728.02	2686.01	1.54	2751.48	0.86	2679.73	1.77	2760.21	1.18
7-06	2744.3	2712.19	1.17	2708.62	1.30	2766.80	0.82	2674.87	2.53
8-06	2750.29	2798.42	1.75	2786.59	1.32	2812.45	2.26	2800.35	1.82
9-06	2703.35	2660.37	1.59	2658.74	1.65	2677.94	0.94	2753.09	1.84
10-06	2705.14	2670.24	1.29	2669.70	1.31	2746.26	1.52	2674.84	1.12
Average error rate (%)			1.44		1.41		1.37		1.48

can search different areas of parameter space simultaneously in generation subrogation, and make the search toward the optimal direction. The training of forward neural network is in fact a process of optimization, i.e., looking for optimal link weight value to make the difference between the outputs of NN with objective function the minimum. So we adopt the genetic algorithm to train the forward neural network, i.e., to establish the neural network based on genetic algorithm and BP algorithm. When BP is optimized with GA in a new algorithm, we can overcome blindness of seeking superior and avoid the condition of local convergence.

3 Stock Index Forecasting Based on Data Mining

The basic method of data mining: first, introduce single indicators into forecasting model and compare forecasting results; then, put single indicators into different groups and find out groups with better performance.

We choose four popular indicators, namely MACD, BIAS6, KDJ, and PSY4, as input quantitative indicators to the forecasting model. Shanghai Component Stock Index data from April 25, 2007 to May 26, 2011 are taken as training sample, and forecasting is carried out to closing prices of 10 days of Shanghai Component Stock Indices from May 27, 2011 to June 10.

3.1 Forecasting with Single Indicators

Introduce the above four indicators to BP+GA model one by one, for single-indicator forecasting. The results are shown in Table 1.

From the above table, we can see that when using single indicators, there is little difference in the performance of the four indicators, of which, KDJ has the smallest error rate.

Table 2 Stock index forecasting by BP+GA with multi-indicator group

Date	Real value	MACD+KDJ		BIAS6+PSY		MACD+KDJ+PSY		MACD+KDJ+PSY+BIAS6	
		Forecasting value	Error rate	Forecasting value	Error rate	Forecasting value	Error rate	Forecasting value	Error rate
27-05	2709.95	2764.69	2.02	2681.39	1.05	2679.08	1.14	2673.37	1.35
30-05	2706.36	2691.07	0.57	2690.36	0.59	2693.83	0.46	2679.57	0.99
31-05	2743.47	2700.53	1.57	2738.88	0.16	2721.35	0.81	2722.37	0.77
1-06	2743.57	2766.48	0.84	2802.98	2.17	2793.84	1.83	2798.65	2.01
2-06	2705.18	2733.31	1.04	2728.26	0.85	2733.38	1.04	2739.28	1.26
3-06	2728.02	2682.87	1.66	2720.88	0.26	2689.29	1.42	2705.32	0.83
7-06	2744.3	2739.50	0.18	2694.65	1.81	2729.26	0.55	2724.53	0.72
8-06	2750.29	2805.43	2.01	2799.32	1.78	2804.53	1.97	2787.42	1.35
9-06	2703.35	2669.15	1.27	2747.17	1.62	2737.36	1.26	2674.28	1.08
10-06	2705.14	2708.25	0.12	2672.41	1.21	2709.72	0.17	2685.54	0.72
Average error rate (%)			1.12		1.15		1.07		1.11

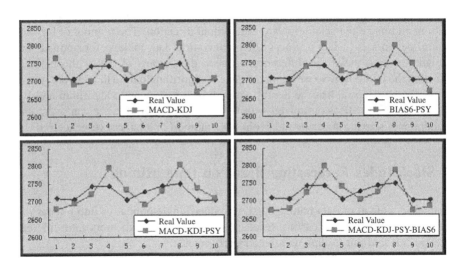

Fig. 3 Forecasting results of BP+GA model with multi-indicator combination

3.2 Forecasting with Multiple Indicators

Put indicators into various combinations. First, forecasting with combination of two indicators, and find out those with better results, and then add other indicators into them respectively till the optimal indicator group is obtained. Forecasting results are shown in Table 2 and Fig. 3.

From the above table, we can see that indicator groups selected after data mining outperform single indicators in forecasting. The combination of MACD+KDJ+ PSY has better result with average error rate at 1.07 %. Still, we believe there is room for increasing its forecasting accuracy.

4 Conclusion

On the basis of data mining on certain technical indicators with influence on the movement of stock index, we used BP+GA model in short-term forecasting of Shanghai Component Stock Index. Our conclusion is as follows.

Group indicators selected through data mining have higher forecasting accuracy than single indicator; combination of indicators through selection and optimization has the best forecasting result; more input indicators do not translate into higher accuracy. In this paper, MACD+KDJ+PSY combination has relatively better result and the average forecasting error is 1.07 %. Our research fully demonstrates the necessity and significance of data mining in stock index forecasting.

Acknowledgments This work is supported by Beijing Natural Science Foundation (9132011).

References

1. Tsai C-F, Hsiao Y-C (2010) Combining multiple feature selection methods for stock prediction: union, intersection, and multi-intersection approaches. Decis Support Syst 50(1):258–269
2. Chu H-H, Chen T-L, Cheng C-H, Huang C-C (2009) Fuzzy dual-factor time-series for stock index forecasting. Expert Syst Appl 36(1):165–171
3. Hill T, O'Connor M, Remus W (1996) Neural network models for time series forecasts. Manage Sci 42(7):1082–1092
4. Bildirici M, Ersin OO (2009) Improving forecasts of GARCH family models with the artificial neural networks: an application to the daily returns in Istanbul Stock Exchange. Expert Syst Appl 36(4):7355–7362
5. Chang P-C, Wang D, Zhou C (2012) A novel model by evolving partially connected neural network for stock price trend forecasting. Expert Syst Appl 39(1):611–620
6. Koopman SJ, Jungbacker B, Hol E (2005) Forecasting daily variability of the S&P 100 stock index using historical, realised and implied volatility measurements. J Empir Finance 12(3):445–475
7. Huang S-C (2011) Forecasting stock indices with wavelet domain kernel partial least square regressions. Appl Soft Comput 11(8):5433–5443
8. Chakravarty S, Dash PK (2012) A PSO based integrated functional link net and interval type-2 fuzzy logic system for predicting stock market indices. Appl Soft Comput 12(2):931–941
9. Asadi S, Hadavandi E, Mehmanpazir F, Nakhostin MM (2012) Hybridization of evolutionary Levenberg–Marquardt neural networks and data pre-processing for stock market prediction. Knowl-Based Syst 35(11):245–258
10. Hsieh T-J, Hsiao H-F, Yeh W-C (2011) Forecasting stock markets using wavelet transforms and recurrent neural networks: an integrated system based on artificial bee colony algorithm. Appl Soft Comput 11(2):2510–2525
11. Roh TH (2007) Forecasting the volatility of stock price index. Expert Syst Appl 33(4):916–922

An Improved Design Rationale Reuse Method Based on Semantic Information

Jihong Liu and Kuan Wang

Abstract Design rationale (DR) explains why an artifact is designed the way it is, which is a kind of important trait knowledge. Successful reusing of DR knowledge plays an important role in guiding design thinking and could raise design efficiency and design quality significantly. In order to satisfy the rising requirements of DR reuse, a formal expression of DR is described and a semantics-based design rationale reuse model based on the expression is proposed. Designers' knowledge needs are analyzed based on semantics, and isolated DR elements retrieved are extended into DR segments by reasoning. After putting any isolated DR nodes in the context of design, the DR segment is easier to understand. Finally, a prototype system is developed and several examples are given to prove the validity of proposed methods.

Keywords Design rationale · Knowledge reuse · Semantic retrieval · Knowledge reasoning

1 Introduction

Engineering design is a creative process that relies heavily upon associations to previous design experiences and similar products. Generally, the outcomes of design processes are standard technical documents such as drawings, reports, and specifications. The reusing methods of these outcomes are like black-box methods [1], in which very important information, such as what design intent is, why certain option is selected or neglected, how to evaluate the solutions, etc., is rarely captured [2]. Nevertheless, the missing information listed above can be valuable,

J. Liu (✉) · K. Wang
School of Mechanical Engineering and Automation, Beihang University,
Beijing 100191, China
e-mail: ryukeiko@buaa.edu.cn

Z. Wen and T. Li (eds.), *Knowledge Engineering and Management*,
Advances in Intelligent Systems and Computing 278,
DOI: 10.1007/978-3-642-54930-4_31, © Springer-Verlag Berlin Heidelberg 2014

even critical, for the successor designer to understand, modify or recreate. For example, when there is a system defect or a need for design modification, such information can help designers analyze the root cause and diagnose change impacts. Design rationale (DR) is an efficient way of capturing such information and can be viewed as a valuable intellectual asset of an enterprise. It includes the reasons behind a design decision, the justification for it, the other alternatives considered, the tradeoffs made, and the argumentation that led to the decision [3]. Therefore, DR should be captured and retained in an efficient and effective manner.

Much progress has been made on the research of DR since the early 1980s [4], and there are several prominent representation schemas including argument-based representation, functional representation, and intent-driven representation. However, the attention of the DR research is mainly on the representation and capture stage, research on the reuse of DR is only in the beginning, and reuse methods are not sufficient to meet industrial demands [5]. In this paper, we developed a new DR reuse model based on semantics. Designers' knowledge needs are semantically extended and the DR elements retrieved are extended into DR segments. DR segments are complete record of corresponding design problem-solving process rather than isolated DR elements or complete DR case. The remainder of the paper is organized as follows: The related work in the area of DR reuse is reviewed briefly in Sects. 2 and 3 describes the intent-driven representation of DR and the formal expression representing and reasoning it; Sect. 4 describes the reuse model mainly by utilizing the two sources of semantic information. After that, the implementation of a prototype system and the evaluation of the methods proposed are given in Sect. 5. Finally, Sect. 6 concludes the paper and discusses the future work.

2 Related Works

The reuse of DR is realized by its successful retrieval. The main retrieval strategies used by existing DR systems include navigating by designers, retrieval by queries, and automatic triggering during design processes. In this paper, we concentrate on the retrieval-by-queries strategy, which is the most common and efficient way of DR retrieval [6]. Based on the Dred system, Kim et al. [4] proposed a method using NLP (Nature Language Processing) techniques to evaluate the similarity of rationale records which could better understand designers' retrieval requirement. Wang et al. [5] utilized the dependencies between two DR elements so as to develop methods and tools to facilitate the provision of useful and reusable design knowledge in new design projects. Based on ISAL, Liang et al. [7] proposed retrieval plug-in supporting retrieving relevant DR from multiple aspects. By applying analysis functions on retrieved DR, visualization is able to suggest some useful DR insights, e.g., key components of certain issues and technology development trend, etc. REMAP/MM [6] uses a deductive query language to define various types of adhoc queries, and provides a graphical interface for displaying queries and retrieving desired information. In addition, a number of DR systems including Kuaba, SEURAT, and Compendium utilize keyword-based retrieval.

The main problems of the analyzed retrieval methods given above include: (1) Discarding the structured information contained in DR model during reusing, however, the structural information is a vital component of DR knowledge. Since the structural information is discarded, retrieved results are isolated DR elements or complete DR case. For the isolated element, it makes little sense to use, because the isolated element only has detailed information of itself. For example, an issue node only describes the problem to solve, but cannot convey how the problem is solved. For the complete DR case, since the whole design process is usually complicated, it is a time-consuming and difficult work for designers to find out the useful part in it. (2) Retrieval queries tend to be ambiguous, and it is hard for keyword-based method to correctly interpret the underlying meanings of terms contained. Moreover, when the cases are created by individuals, the useful knowledge is seldom retrieved, since the word searchers use are not the same as those by which the knowledge they seek has been indexed.

3 Design Thinking Process Model

Before developing the detailed reuse method of DR, a good DR model is needed. The role of this model is to organize the enormous amount of materials that are included in design procedures and design thinking processes. Representation of this model was built in our previous work, which could represent design intent, design operation, design decision, and design justification. In order to achieve the understanding and reasoning of DR knowledge both by human and computers, a formal expression of the model was proposed based on OWL.

3.1 Representation of the Model

The elements of the DR model introduced in this paper include "design intent," "design option," "design decision," "design operation," and "design justification (divided into evidence and criteria)". Relations between the above elements include "decomposed-into," "achieved-by," "decided-by," "refer-to," "realized-by," and "initiate and return-to."

Design intent (I) describes the purposes or plans for a designer to solve design problems. Design option (O) is the potentially alternative solution to realize the design intent proposed by designers. Design decision (D) is the ultimate decision designers make after comparing against several design options. Design operation (Op) is the action whose execution can help realize the design intent. Design Justification(C or E) explains how the option is generated and why the decision is made. The detailed definition of these concepts can be found in reference [2].

Decomposed-into <intent, intent> is the relationship between an intent I1 and two or more intents (I2, I3). Achieved-by <intent, option> means that the option is an alternative that is proposed to achieve the design intent. Decided-by <option, decision> means the option is selected by the designer to be the final design decision. Realized-by <decision, operation> is the relationship between decision and operation, meaning that the operation turns the decision into actual activities. Follows <option/decision, justification> is the relationship that the option/decision is proposed by conforming to the justification. The relationships are shown in Fig. 1.

To achieve the reasoning and processing of the DR model with computers, the OWL (Web Ontology Language) is utilized as the basic expressive form for DR model, because OWL could better support the representation of semantic information. Furthermore, the exchangeable capacity of OWL allows us to record and transfer DR files in different systems. The elements and relationships between elements mentioned above are, respectively, defined as OWL language classes and object properties, besides; some other descriptions of elements are also treated as object properties, such as the version attribute of design intent.

3.2 Example

For illustrative purposes, here is a simple example describing how bill processing mechanism for ATM is designed. Figure 2 shows part of the process in formal graph structure. The original intent in the case is "Design a bill processing mechanism for an ATM" (I1). To achieve this intent, the designer may decompose I1 into four sub-intents, e.g., "to Transport bill" (SI1), "Separate bill" (SI2), "Identity bill" (SI3), and "Store bill" (SI4). Only after all of the four sub-intents are achieved, I1 could be achieved. In order to achieve SI2, several options such as "Friction-type mechanism" (O1) and "Vacuum-type mechanism" (O2) were proposed by designers. The evidences for these options are separately "principal of vacuum" (E1) or "principal of friction" (E2). It is impossible to allow designers to propose as many plausible options as possible in design processes, since they can come back to other options when a selected option is proved to be infeasible. In this case, the final decision was vacuum-type (D1), the criteria may be "simple structure" (C1), "low cost" (C2) and "no damaging to bills" (C3). Then the designer draws the sketch of the decision (Op1).

4 A Model for DR Reuse Based on Semantic Information

Three major steps in the DR knowledge reuse model are proposed in this paper, and they are: (1) representation of design thinking processes, which is introduced in Sect. 1; (2) semantics-based DR elements retrieval; and (3) semantics-based DR segments extraction. The reusing model is shown in Fig. 3.

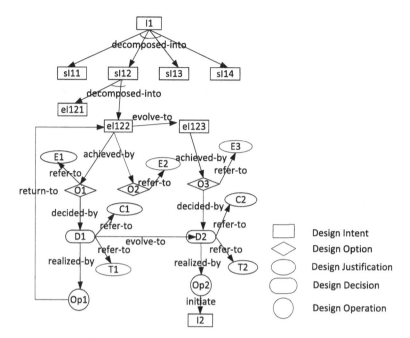

Fig. 1 DR elements and relationships

4.1 Semantics-Based DR Elements Retrieval

The semantics-based retrieval mainly consists of three steps: (1) query statement segmentation, (2) semantic extension of segmented terms to extended-wordset, and (3) retrieve DR elements. Considering that the retrieval requirements are only related to the field of product design, ambiguity processing and new words recognition problems are not too acute. Therefore, dictionary-based word segmentation method is selected. Since common dictionary does not contain professional terminology composed of multiple phrases, it is necessary to build the domain dictionary. For semantic extension, we utilized domain ontology, WordNet, and HowNet. Domain ontology is responsible for semantic extension of domain terminologies, and WordNet/HowNet is responsible for the synonym extension of common terms. During the extension process, the extension weight for each extended term and the part-of-speech weight for original words are both taken into consideration. The weight indicates the importance of the corresponding term in the query statement.

Keywords can be classified into four different categories: domain terminology (W_D), which is from original query and perfectly match domain terminology; extended domain terminology (W_{De}), which is received through ontology-based semantic reasoning of W_{De}; common term (W_N), which is from original query and could not match any domain terminologies; extended common term (W_{Ne}), which is received through reasoning of W_N. Therefore, the extension weight (w_e) and part-of-speech weight (w_c) comprised the final weight of keywords.

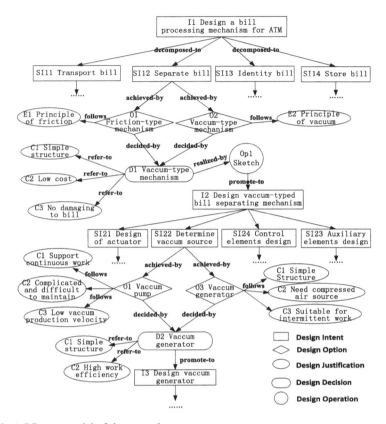

Fig. 2 A DR case model of the example

4.2 Semantics-Based DR Segment Extraction

Since a single DR element only contains a sentence or two to introduce itself, it does not include any context information of design process and thus is hard for designers to understand. Therefore, in this paper, the elements retrieved through semantic retrieval should be extended into a group of interrelated elements, i.e., a sub-graph which refers to as "DR segment." From the perspective of both easy to understand and reuse, taking into account human cognitive characteristics, the DR segment should meet the following three principles: (1) Completeness, which means DR segments should represent design process completely; (2) Independence, which means DR segments should be independent each other as much as possible; (3) Moderate complexity, which means DR segments should not be too complexity or conciseness.

Based on the above-mentioned analysis, the DR segment proposed in this paper is the record of the realization process of design intent. First, we define the min-length DR segment (MLS) as the realization of element-Intent, which starts from

Fig. 3 A DR case model of the example

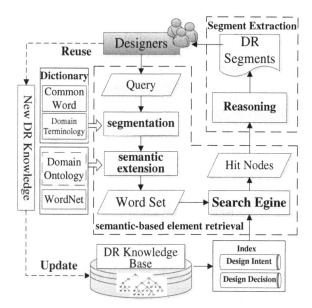

the element-Intent node and ends with the Operation node. Since there is a one-to-one correspondence between a MLS and the element-Intent, we can use intent element as the representative of the corresponding MLS.

Obviously, the MLS perfectly satisfied the three principles described above. Therefore, we define MLS as the basic unit of DR segment. However, if the retrieved result belongs to the original-Intent or sub-Intent, single MLS could not cover entire design process, because the original-Intent or sub-Intent is achieved by more than one MLS. Therefore, the segment extraction method is divided into two cases: if the retrieved results do not belong to the original-Intent or sub-Intent, extracted segments are their corresponding MLS; and if the retrieved results belong to the original-Intent or sub-Intent, extracted segments includes both Intent elements and their following MLSs.

The DR segment extraction method is achieved by structure reasoning. First, we define the reasoning direction (RD) as $<E, PT>$, where E refers to one of the DR elements, which is the starting point of reasoning. And PT is a set of relationships and directions as $<P, n>$, where P refers to a kind of relation and n refers to the reasoning depth, the value of n can be -1, ∞ or 1. If $n = \infty$, the reasoning contains the transitive closure of E under the relationship P. The stepwise procedure of DR segment extraction is as follows:

Step 1: Initial intent identify
Considering that segments start with intent elements and end when the intents achieved. Therefore, the first step of segment extraction is to identify the intent.

Step 2: Intent elements extract

For the DR model described above, the Intent element with a higher level of abstraction are usually decomposed into sub-intents and then into e-intents. Therefore, intent nodes in a DR model are expressed in the form of an intent-tree. Besides, whether the intent exacted in this step has an evolve-to relationship must be considered. The strategy is to extract the version input and all of its later versions.

Step 3: Option element extract

After the element-intents are obtained, corresponding option elements can be identified through the archived-by relation reasoning with $RD1 = <I, PT1>$, $PT1 = \{<p, n>, <achieved\_by, 1>\}$.

Step 4: Decision element extract

The decision elements are extracted by the decided-by relation reasoning with $RD2 = <I, PT2>$, $PT2 = \{<p, n>, <decided \_by, 1>\}$. The design decision also has possible evolution. The strategy to deal with it is the same as intent element described in step 2, i.e., extract the version input and all of its later versions.

Step 5: Operation element extract

The last step of DR segment extraction is to get each decision's operation through the realized-by relation reasoning with $RD3 = <I, PT3>$, $PT3 = \{<p, n>, <realized\_by, 1>\}$.

5 The Prototype System

A prototype system based on the knowledge representation is developed. The system helps designers to record their design rationale and achieves the semantic retrieval of DR knowledge (Fig. 4).

5.1 Graphical Modeling Interface

A graphical modeling interface is developed to help designers to record DR knowledge. The graphical user interface and the case in Sect. 3.2 are graphically built as shown in Fig. 5. The DR elements and their relationships will be represented as nodes and links in the system, and the system provides several icons to represent different types of elements and relationships.

Fig. 4 Graphical modeling
interfaces

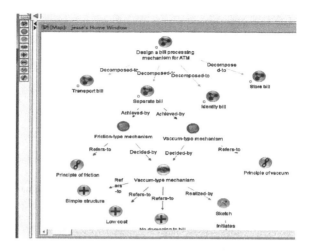

5.2 Semantic Retrieval of DR Knowledge

The test cases used in the validation process include: ATM bill processing mechanism, mechanical system design for pogo stick, bending die design, and the design for four different kinds of gear. The test is carried out from two aspects: the semantic retrieval as is shown in Fig. 5a and the segment extraction method as is shown in Fig. 5b.

For the case in Fig. 5a, five different segments of transmission mechanisms are given as retrieval results while the query is "transmission mechanism design." Compared with the semantic retrieval method, the traditional keyword-based method could not return any of these results.

In another case, the designer wants to retrieve any existed designs of vacuum generator. After executing the query of "vacuum generator," the retrieved DR elements include "design vacuum generator" (I3) and "absolute vacuum" (D3). Based on the elements, segment extraction is processed, first, the initial Intent is identified as I3, then one Intent element (SI32:"calculate flow capacity") and three MLSs ("ascertain vacuum," "ascertain pumped air volume," and "ascertain working frequency") are identified through reasoning. These Intents and MLSs comprise the final segment the designer needs, as shown in Fig. 5b.

The segment in Fig. 5b shows one of the vacuum generators' design processes which is part of the bill processing mechanism design. The experience point that the segment comprises 18 nodes is neither complicated nor simple, and it could completely convey the vacuum generator design process. What's more, designers could understand how to design a vacuum generator quickly and effectively.

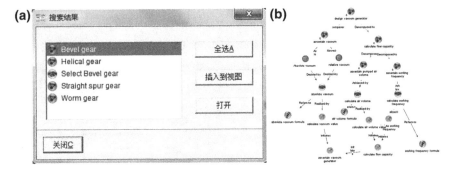

Fig. 5 Example of semantic retrieval. **a** Example of semantic retrieval result. **b** Example of segment extracted

6 Conclusions and Future Work

An improved DR knowledge reusing model is proposed based on semantics in two ways: The semantics-based DR elements retrieval helps to understand the underlying meanings of designers' knowledge needs; The DR segments extraction method put any retrieved elements in contexts. To support the methods, formal expression of DR knowledge is proposed. A graphical DR modeling and reusing system is developed and the methods proposed are validated.

The proposed reusing model is a preliminary research. Future work will be focused on the following issues: First, the reuse strategy of "recommend relevant pieces of DR knowledge within a design process automatically" should be studied. Second, the reasoning ability of DR model should be strengthened, and the application of reasoning should be extended.

References

1. Nomaguchi Y, Tomiyama T (2002) Design knowledge management based on a model of synthesis. In: IFIP TC5/WG5.2 Fifth Workshop on Knowledge Intensive CAD. Kluwer Academic Publishers, Malta, pp 131–149
2. Liu J, Sun Z (2008) Representing design intents for design thinking process modeling. J Mech Eng 45:182–189
3. Dorst K (2011) The core of 'design thinking' and its application. Des Stud 32(6):521–532
4. Kim S, Bracewell RH, Wallace KM (2011) A framework for design rationale retrieval. In: International conference of engineering design, Melbourne, p 8
5. Wang H, Aymler LJ, Bracewell RH (2012) The retrieval of structured design rationale for the re-use of design knowledge with an integrated representation. Adv Eng Inform 26(2):251–266

6. Regli WC, Hu X et al (2000) A survey of design rationale systems: approaches, representation, capture and retrieval. Eng Comput 16(3–4):209–235
7. Liang Y, Wen F et al (2010) Interactive interface design for design rationale search and retrieval. In: Proceedings of the ASME2010 design engineering technology conference and information in engineering conference, pp 308–318

A New Study of Representation of Spatio-temporal Data with Information Granules

Mingli Song, Wenqian Shang and Witold Pedrycz

Abstract This paper proposes an algorithm to represent them in a granular way—information granules. Information granules can be regarded as a collection of conceptual landmarks using which people can view the data and describe them in a semantic way. The key objective of this paper is to introduce a new granular way of data analysis through their granulation. Several experiments are done with synthetic data and the results show a clear way how our algorithm performs.

Keywords Spatio-temporal data · Information granules · FCM · Granular descriptor

1 Introduction

Spatio-temporal data are visible in various application fields in which weather conditions and their observation stations are a typical example. Spati-temporal system can be effectively analyzed through separating data into spatial and temporal or using data as a whole entity. In either way, we can conclude that both the spatial data and the temporal data will affect the final result (the system).

M. Song (✉) · W. Shang
School of Computer Science, Communication University of China, Beijing, China
e-mail: songmingli@cuc.edu.cn

W. Shang
e-mail: shangwenqian@163.com

W. Pedrycz
Electrical and Computer Engineering, University of Alberta, Alberta, AB, Canada
e-mail: wpedrycz@ualberta.ca

Z. Wen and T. Li (eds.), *Knowledge Engineering and Management*,
Advances in Intelligent Systems and Computing 278,
DOI: 10.1007/978-3-642-54930-4_32, © Springer-Verlag Berlin Heidelberg 2014

Fig. 1 A collection of
stations along with associated
time series

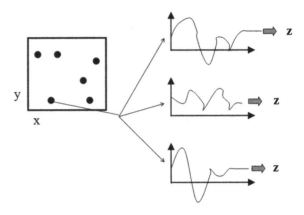

For example, in the following Fig. 1, we encounter a collection of spatial data—stations characterized by x-y coordinates. For each station, there is an associated suite of time series.

In this paper, we concentrate on the study of representation of spatio-temporal data with information granules, which will help the subsequent research of spatio-temporal data. The spatial and temporal data can be described at various levels of abstraction. In fact, this level of abstraction helps to realize a user-friendly approach to data analysis. In essence, the user can establish the most suitable perspective to view the data; in other words, the user can identify a proper level of abstraction. We here name this level of abstraction the granularity. This view inherently brings the concept of information granules into picture. Information granules are regarded as a collection of conceptual landmarks using which one can look at the data. Information granules can be modeled in many formats, such as sets (intervals), fuzzy sets, rough sets, etc. From data points to information granules, we show two ways: adopt clustering mechanisms; first describe data points (time series) in a shorter format (we get some coefficients) and then form information granules on those coefficients. The proper level of granularity can be optimized in an evolutionary way in which an objective function with one or two criteria is needed.

The material is organized as follows. Section 2 introduces the basic concepts of information granules and a typical clustering method—Fuzzy C-Means. In Sects. 3 and 3.1, spatial data and temporal data representation are discussed. The conclusions are in Sect. 4.

2 Formats of Information Granules and Fuzzy C-Means

There are some different formalisms and concepts of information granules, for example, sets (intervals), fuzzy sets, rough sets, etc., cf. [1, 2]. We will first review some basic concepts and operations on these information granules.

Set theory occupies an important and unique place in modern mathematics since it can be shown that it can be used as a starting point for the derivation of all other branches of mathematics. In set theory, interval arithmetic which first appeared in [3–5] offered an important generalization of arithmetic defined on real numbers. If information granules are represented in the form of intervals X, assume that X = [a, b], in which a and b are numeric values.

Fuzzy sets provide a possibility to formally express concepts of continuous boundaries. These concepts are everywhere. When expressing ideas, describing concepts, and communicating with people, we always use terms to which the yes-no does barely apply. Fuzzy set A is generally described by a membership function which maps the universe of discourse X in which A is defined into a unit interval:

$$A : X \rightarrow [0, 1]$$

Formally, $A(x)$ denotes a degree of membership that describes an extent to which x belongs to A.

The description of information granules completed with the aid of some vocabulary is usually imprecise. Intuitively, such description may lead to some approximations called lower and upper bounds. This is the essence of rough sets introduced by [6, 7]; refer also to [10].

Clustering and fuzzy clustering have been regarded as a synonym of structure discovery in data. The result, no matter what technique has been used, comes as a collection of information granules which serve as a quantification of concepts serving as descriptors of the phenomenon behind the data. Fuzzy C-Means is one of the commonly used mechanisms of fuzzy clustering. With FCM, one is concerned with discovering the prototypes and partition matrix U by a minimization of the following objective function Q being regarded as a sum of the squared distances:

$$Q = \sum_{i=1}^{c} \sum_{k=1}^{N} u_{ik}^{m} \|\mathbf{x}_k - \mathbf{v}_i\|^2 \tag{1}$$

where v_i's are n-dimensional prototypes of the clusters, $i = 1, 2,\ldots, c$, and $U = [u_{ik}]$ stands for a partition matrix; u_{ik} is the membership degree of data x_k in the ith cluster.

3 Spatial Data Representation

A position on the earth can be described by a combination of x-coordinate and y-coordinate. The information granules in the space x-y form regions by bringing together points being in close vicinity to each other. From the algorithm perspective, one can consider to use a clustering mechanism. It should be noted that a distance function used to quantify the closeness of points needs to be carefully defined. From the user perspective, the formed information granules provide a better way to look at the regions on earth, their location, and size, and reflect the inner homogeneity of the points.

We define a collection of spatial information granules in the x-coordinate and y-coordinate, respectively,

$$A_i : R \rightarrow [0, 1], \ i \ = 1, 2, \ldots, \ n$$
$$B_j : R \rightarrow [0, 1], \ j \ = 1, 2, \ldots, \ m$$

The selected forms of fuzzy sets can be: intervals, triangular membership functions, and Gaussian membership functions, etc. In this case, they are described by the corresponding membership functions: $A_i(x)$ and $B_j(y)$. Thus, we have n fuzzy sets on the x-coordinate and m fuzzy sets on the y coordinate. These fuzzy sets come with well-defined semantics. For instance, we may have the linguistic descriptors such as,

A_i—east region of x-coordinate, B_j—north region of x-coordinate.

If we realize Cartesian products of these fuzzy sets, we easily find that the resulted granular descriptors are also meaningful and serve as a description in the region of x-y space located in the northeast of the plane. We represent this by $A_i \times B_j$ meaningful and serve as a description in the region of x-y space located in the northeast of the plane. We represent this by $A_i \times B_j$ in Fig. 2.

After the development of information granules on spatial data, we obtain not only fuzzy sets but also some clusters characterized by their closeness, please refer to Fig. 3.

Here we give another example (in Fig. 4), in which we have 200 stations with x-y coordinates. For each coordinate, we define four triangular functions.

As to Gaussian functions, there are two parameters needed to be carefully dealt—average value and standard deviation. It is a headache problem to decide the value of standard deviation and an evolutionary method may be a choice.

3.1 Temporal Data (Time Series) Representation

A time series is a collection of observations made chronologically. The nature of time series data includes: large in data size, high dimensionality, and necessary to update continuously. Moreover time series data, which is characterized by its numerical and continuous nature, is always considered as a whole instead of individual numerical field. There are many ways for their representation which are popular in research. One of the major reasons for time series representation is to reduce the dimension (i.e., the number of data point) of the original data.

The simplest method perhaps is sampling [8]. In this method, a rate of m/n is used, where m is the length of a time series P and n is the dimension after dimensionality reduction. Thus, the coefficients are the n sampled values.

Fig. 2 Spatial granular descriptors A_i and B_j

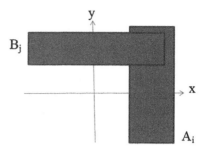

Fig. 3 From a collection of stations to spatial information granules

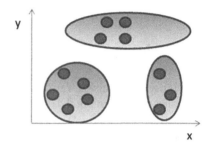

Fig. 4 Four triangular functions in A and four triangular functions in B

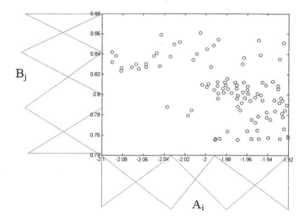

However, the sampling method has the drawback of distorting the shape of sampled/compressed time series, if the sampling rate is too low. An enhanced method is to use the average (mean) value of each segment to represent the corresponding set of data points. First, we divide the m points of times series into n parts and calculate the average of each segment as our coefficients. Some other ways of representing time series are in [9]: piece wise linear approximation (PLA), parameters of the autoregressive time series model (AR model), coefficients of the Fourier expansion, cepstral coefficients, components of Discrete Wavelet Transforms (DWT), parameters of fuzzy rule-based system (in case of fuzzy modeling of the series) and others.

Assume that we have a time series T with t_1 points and we cut it into t_2 segments with equal length (have same number of time points, for easy analysis). In general, we may have $t_i = t_2$ coefficients or $t_i = 2*t_2$ coefficients according to the method selected. As to the sampling method and average value method, we have $t_i = t_2$ coefficients; as to piecewise linear approximation method and Fourier expansion method, we have $t_i = 2*t_2$ coefficients, etc. Normalize those coefficients by a certain method since we have to unify and compare the method for both spatial data and temporal data. Adopt a clustering mechanism like FCM on the normalized coefficients and we obtain a granular way of description—membership degrees (U in FCM) and prototypes.

4 Conclusions

In this paper, we show a way how to represent spatio-temporal data with information granules. Spatio-temporal data are split into spatial and temporal parts and analyzed, respectively. Spatial data are described by fuzzy sets like triangular membership functions, Gaussian functions, and so on and their Cartesian products whereas temporal data particularly time series are dealt with in the way of representation methods and clustering mechanisms. A series of experimental results done by synthetic data shows a clear outstanding performance of our idea. It is a future research topic that how to combine these two parts into one granular descriptor.

Acknowledgments This study is supported by special youth Engineering Project of Communication University of China numbered 3132013XNG1320, XNG1246 and XNG1319.

References

1. Bargiela A, Pedrycz W (2003) Granular computing: an introduction. Kluwer Academic Publishers, Dordrecht
2. Pedrycz W, Gomide F (2007) Fuzzy systems engineering: toward human-centric computing. Wiley, New York
3. Warmus M (1961) Approximations of inequalities in the calculus of approximations: classification of approximate numbers. Bull Acad Polonaise Sci 9(4):241–245
4. Warmus M (1956) Calculus of approximations. Bull Acad Polonaise Sci 4(5):253–259
5. Sunaga T (1958) Theory of interval algebra and its applications to numerical analysis. Gaukutsu Bunken Fukeyu-kai, Tokyo
6. Pawlak Z (1982) Rough sets. Int J Comput Inf Sci 11:341–356
7. Pawlak Z (1991) Rough sets: theoretical aspects of reasoning about data. Kluwer Academic Publishers, Dordrecht
8. Astrom KJ (1969) On the choice of sampling rates in parametric identification of time series. Inf Sci 1(3):273–278
9. Fu T (2011) A review on time series data mining. Eng Appl Artif Intell 24:164–181
10. Skowron A (1989) The relationship between the rough set theory and evidence theory. Bull Pol Acad Sci Tech 37:87–90

Does Control Matter? Exploring the Effects of Formal Control and Social Control on Knowledge Transfer in Service Supply Chain

Adejiang Dawuti

Abstract This paper presents the effect of control on knowledge transfer in service supply chain with an empirical data from China Mobile. Formal control and Social control are mechanisms usually taken to promote knowledge transfer by buyer in service supply chain. So, the role of formal control, social control and their complement to knowledge transfer are studied in this paper. Using the data collected from China Mobile, we found formal control is positively related to knowledge transfer, and social control is also positively related to knowledge transfer. In addition, there is a complement effect between formal control and social control to knowledge transfer. Theoretical and managerial implications are discussed.

Keywords Control · Knowledge transfer · Service supply chain

1 Introduction

Knowledge transfer from partners is critical to any firm to acquire competitive advantage [1]. Studies discussed knowledge transfer between two cooperation parties with control mechanism [2, 3]. There are two kinds of control: formal control and social control [4]. And the roles of formal control and social control to knowledge transfer are all empirically tested [3]. However, no studies paid attention to the role of control mechanism complement to knowledge transfer.

In this study, we aim to contribute to the literature about the above gap by examining the role of formal control, social control and their complement to knowledge transfer. Specifically, two research questions will be addressed: the

A. Dawuti (✉)
School of Management, University of Chinese Academy of Sciences, Beijing, China
e-mail: adejiang2013@126.com

Z. Wen and T. Li (eds.), *Knowledge Engineering and Management*,
Advances in Intelligent Systems and Computing 278,
DOI: 10.1007/978-3-642-54930-4_33, © Springer-Verlag Berlin Heidelberg 2014

relationship between control and knowledge transfer will be discussed; and the role of control complement will be discussed. Empirically, we will test these relationships with the data collected from China Mobile Group, which is the largest mobile service company in China with near 70 % market share in China Mainland. And the following contents of this paper are organized as followed: We develop our theory and hypotheses after this introduction; then, we describe our measurement and test results; finally, we discuss our results and their implications.

2 Theory and Hypotheses

There are two kinds of control in buyer-supplier dyad: formal control and social control [3, 5], Formal control indicates buyer-supplier relationship is managed through stipulated contractual agreements [3]. And social control indicates buyer-supplier relationship depend on the belief of the other's competence, reliability or goodwill [3, 4]. Applying the approach of relational view that effective governance can enhance knowledge sharing, both formal control and social control are contributed to knowledge transfer. The rights, duties, and responsibilities will be specified with the contracts which facilitate the exchange of explicit knowledge [6, 7]. Moreover, specific contracts will constitute a cooperation framework for communication and information exchange, which also facilitates the knowledge transfer [3]. Different from formal control, social control will enhance social interaction which increases the opportunity to exchange resources and knowledge [8]. Meanwhile, social control increases the trust, vision and willingness that contribute to valuable knowledge sharing [3, 9].

With the extension of the relational view, we will discuss the role of control mechanism complement to the knowledge transfer. Social control encourages the exchange of unspecific but valuable knowledge, which overcomes the limitation of contracts [3]. Therefore, stronger social control is beneficial for the positive relationship between formal control and knowledge transfer. Formal control establishes a formal framework which reduces the cognitive barrier to knowledge transfer and also assures the mutual trust to share valuable information [3]. Therefore, stronger formal control is also beneficial for the positive relationship between social control and knowledge transfer.

2.1 Control and Knowledge Transfer

Formal control indicates to manage the relationship with the explicit and clear contracts. Behavioral boundaries between the buyer and supplier are pre-specified with contracts before the exchange. These contracts include the formal operation procedures related to the communication of explicit knowledge, such as training program and technology communication and so on [10]. With the contracts, the explicit information and knowledge will be transferred. In addition, specific contracts will constitute a cooperation framework for communication and information exchange. Lane and Lubatkin argued the knowledge learning through passive and active methods [11]. Passive learning occurs when firms acquire knowledge from seminars and consultants, and active learning occurs when firms acquire the knowledge through arm's length. Both of the learning is based on the relationship of two parties. And specific contracts provide a relationship framework to communicate and interact, which will contribute to the knowledge sharing. Therefore, we propose:

H1: formal control is positively related to knowledge transfer from buyer's perspective in supply chain

Social control are always with the interaction, trust between buyer and supplier [4, 12]. There are two channels contributed to knowledge transfer. Firstly, interaction between buyer and supplier enhances the incentive to share knowledge and the ability to recognize and assimilate the knowledge. Interaction makes two parties comfortable in the exchange [9], and the intense to exchange information will be strengthened [13]. Secondly, once trust is established, supplier will provide buyer with more production technology [14]. And trust overcomes the willingness obstacle to share valuable knowledge, and more valuable information can be transmitted [15]. Therefore, we propose the following hypothesis:

H2: social control is positively related to knowledge transfer from buyer's perspective in supply chain

2.2 The Complement Between Formal Control and Social Control

Social control can improve tacit knowledge sharing through complementing the limitation of formal control [3]. Explicit knowledge and tacit knowledge are two kinds of knowledge [16]. And explicit knowledge can be transferred without loss of integrity, but tacit knowledge is complex and hard to be transferred [3]. Formal control can contribute to the explicit knowledge sharing, but it is hard to transfer tacit knowledge through formal control. Social control can encourage the exchange of unspecific but valuable knowledge by reducing conflict and fostering mutual understanding. Therefore, we propose:

H3a: the positive relationship between formal control and knowledge transfer will be stronger when social control is stronger

Formal control establishes a formal framework which reduces the cognitive barrier to knowledge transfer. The acquisition of new knowledge is based on the prior related knowledge [17]. If the firm does not possess the basis of related knowledge, knowledge absorption capability is limited. Contracts can provide a formal framework with related knowledge by specifying the roles, rules and procedures. Therefore, stronger formal control is beneficial for the positive relationship between social control and knowledge transfer. In addition, formal control assures the mutual trust to share valuable information [3]. Whereas social control establishes norms and expectations of appropriate behavior to reduce the risk perception, it can not avoid the opportunism risk [5]. Contracts provide a formal assurance which complements the social control. With safe boundaries of knowledge flows, contracts promote the role of trust to knowledge transfer [3]. Therefore, we propose the following hypothesis:

H3b: the positive relationship between social control and knowledge transfer will be stronger when formal control is stronger

3 Data and Methods

The data was collected in China with the cooperation of the Research Institution of China Mobile which is the largest mobile communication research institution in China. We studied the relationship between control and knowledge transfer between the service firm and its suppliers, and the unit of analysis was the individual vertical dyad. We collected data over a ten-month period from September 2009 to June 2010. China Mobile Group is the leading mobile communication company in China and it owns thirty-one 100 % interest subsidiaries. Each of the thirty-one subsidiaries was asked to evaluate fifteen different relationship units with its main suppliers. Therefore, 465 questionnaires were sent out. We received 411 questionnaires, but 57 were excluded from the analysis unit because of excessive missing values. The effective response rate is high to 88.4 %. In order to enhance the data quality received by the questionnaires, we set cheater questions in the questionnaire. Cheater questions were embedded in the questionnaire to identify the informant who may not be following the required procedures. Through the cheater questions 83 questionnaires were excluded without following the required procedures (either deliberately or through negligence). Thus, we ended up with a final sample of 271 buyer-supplier relationship units.

Multi-item constructs in this study were measured through seven-point Likert scales. Where possible, the scales from the literature were used or adapted in this study. Knowledge transfer was measured with four items reflecting the technology knowledge and valuable information transferring from the supplier to the buyer [9]. Formal control reflects to manage the buyer-supplier relationship through the contract that specifies the obligations and roles of two parties [3]. Formal control was measured by three items [15]. Social control means that the buyer-supplier relationship depends on the belief of the supplier's competence, goodwill and trust [3, 4]. Social control mechanism was measured by four items [15].

We took three control variables in this study. The three control variables are length of cooperation, product life stage, importance of product. Length of cooperation was operationalized as the years the two sides have cooperated. Product life stage is another control variable because the young stage product is a rare resource to the buyer. We measured this variable using a 5-point item: the stage of the product from the supplier. Importance of product from the supplier is also a key determinant to the cooperation relationship. And we measured this variable using a 5-point scale (1 = not important, 5 = important) and the following item: the importance of the product purchased from the supplier.

We used previously validated measurement items to help ensure the validity and reliability. Then, we translated the English version of the items into Chinese version and then back-translated to English to check against the original English version. We conducted a confirmatory factor analysis to assess the unidimensionality, reliability, convergent validity and discriminant validity of the multi-item scales with AMOS 7.0. The overall measurement model fit index shows ($\chi^2(41) = 103.166$, $P < 0.01$; comparative fit index (CFI) = 0.969, incremental fit index (IFI) = 0.969, normed fit index (NFI) = 0.950; root mean square error of approximation (RMSEA) = 0.075) the model provides a reasonable fit with the observed data. All the factor loadings are significant ($P < 0.001$) and related to their respective constructs. Therefore, the constructs are the unidimensionality.

Next, we assessed the reliability of the measures using composite reliability (CR). The composite reliabilities in the Table 1 are all above 0.70, which indicate that all the constructs demonstrate adequate reliability [18]. We also examined the average variance extracted (AVE) for each construct. The AVEs are all above the recommended 0.50 level. Finally, we used methods recommended by Fornell and Larcker to test discriminant validity. Table 2 shows that the square root of AVE for each individual construct is more than the correlation between each pair of construct, thus supporting the discriminant validity.

4 Results

We performed multiple regressions to test the hypotheses, and Table 3 reports the results of regression models. The results in model 4 show that formal control has a significantly positive effect on the knowledge transfer (b = 0.172, $P < 0.05$). Hypothesis 1, which indicates formal control is positively related to knowledge transfer from buyer's perspective in supply chain, is supported. Also, the results in Model 4 show that social control has a significantly positive effect on the knowledge transfer (b = 0.341, $P < 0.05$). Hypothesis 2, which indicates that social control is positively related to knowledge transfer from buyer's perspective in supply chain, is supported.

Hypothesis 3a means the relationship between formal control and knowledge transfer will be strengthened when social control is stronger. And Hypothesis 3b indicates the relationship between social control and knowledge transfer will be strengthened when formal control is stronger. The results in model 1 show that formal control is positively related to social control. And the results in model 2

Table 1 The result of confirmatory factor analysis, reliability and validity

Factors	Items	Standardized weight	Standardized error	Critical ratio	Composite reliability	AVE
Knowledge	KT4	0.701			0.879	0.647
acquisition	KT3	0.795	0.089	11.991		
	KT2	0.904	0.084	13.238		
	KT1	0.805	0.091	12.134		
Formal control	FC3	0.863			0.884	0.720
	FC2	0.912	0.059	18.327		
	FC1	0.763	0.063	14.736		
Social control	SC4	0.792			0.912	0.723
	SC3	0.827	0.069	15.289		
	SC2	0.941	0.055	17.902		
	SC1	0.834	0.060	15.458		

$\chi 2$ (41) = 103.166, P < 0.01; comparative fit index (CFI) = 0.969, incremental fit index (IFI) = 0.969, normed fit index (NFI) = 0.950; root mean square error of approximation (RMSEA) = 0.075

Table 2 Descriptive statistics, pearson correlation matrix

	1	2	3	4	5
1. Knowledge acquisition	0.804				
2. Formal control mechanism	0.449[**]	0.849			
3. Social control mechanism	0.376[**]	0.578[**]	0.850		
4. Length of cooperation	−0.022	−0.027	−0.040	N/A	
5. Product life stage	−0.058	0.004	−0.003	0.034	N/A
6. Importance of the product	0.045	0.064	0.056	0.202[**]	0.087

N = 271; [*] P < 0.05; [**] P < 0.01; [***] P < 0.001. The diagonal is square root of the average variance extracted

show that social control is positively related to formal control. The positive coefficients suggest a complementary relationship. The greater formal control is, the greater social control is, and vice versa.

And the results in model 5 show that the interaction between formal control and social control is positively related to knowledge transfer (0.109, P < 0.05). Therefore, Hypothesis 3a, the positive relationship between knowledge transfer and formal control is positively moderated by social control, is supported. In addition, Hypothesis 3b, the positive relationship between knowledge transfer and social control is positively moderated by formal control, is supported.

In order to use figure to interpret the moderated effect, all the variables except formal control and social control constrained to mean, and formal control and social control took the value of one standard deviation below and above mean. Then, we plotted the Fig. 1, which indicates that when social control is stronger, the positive relationship between formal control and knowledge transfer is strengthened. As shown in Fig. 1, the relationship between formal control and

Table 3 Regression result

	Social control	Formal control	Knowledge transfer		
	Model 1	Model 2	Model 3	Model 4	Model 5
(Constant)	2.436***	2.157***	4.925***	2.182***	1.832***
Control variables					
Length of cooperation	−0.011	−0.014	−0.026	−0.009	−0.014
Product life stage	0.009	−0.006	−0.042	−0.047	−0.052
Importance of the product	0.027	0.024	0.051	0.018	0.017
Independent variables					
Formal control (FC)	0.547***			0.172*	0.190**
Social control (SC)		0.597***		0.341***	0.381***
Interaction					
FC*SC					0.109***
F value	31.986***	31.953***	0.547	13.161***	14.120***
R squared	0.332	0.331	0.006	0.207	0.252
Adjusted R squared	0.321	0.321	0.005	0.252	0.234

Significance level * P < 0.10; ** P < 0.05; *** P < 0.01; **** P < 0.001

Fig. 1 The effect of social control

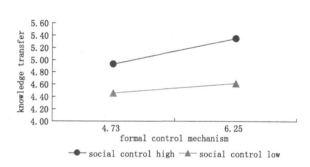

knowledge transfer will be strengthened when social control is stronger. Also, we plotted the Fig. 2, which indicates that when formal control is stronger, the positive relationship between social control and knowledge transfer is strengthened.

5 Discussion and Conclusion

Contributions to the research of relational view. Our findings contribute to the theory of the relational view. We used the logic in the relational view that effective governance can enhance knowledge sharing to argue the role of control to knowledge transfer. In addition, with extending the relational view we discussed the complement of control to the knowledge transfer. Our results extend the relational view and its implication in supply chain.

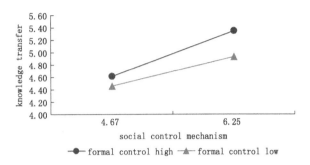

Fig. 2 The effect of formal control

Contributions to the research of governance in supply chain. Our findings contribute the governance literatures in supply chain to analyze the role of control to knowledge transfer. Control mechanism has been attracting the scholars' concerns for the positive effects on knowledge transfer [3]. However, few study argued the role of their complements to knowledge transfer. Our results indicate that two formal control and social control have complementary effect on knowledge transfer. Those enriched the literatures related to governance in supply chain.

This study has limitations that suggest the future research direction. First, the same as most survey data researches, our results relied on the issues related to common method variance. We believe that common method variance problem should not be serious in our data analysis with our test. However, we only use the one side data collected from China Mobile Group to test our theoretical model. In the future two side paired data used to test our model will be better to reduce the common method variance. Second, our sample is only collected from one industry, the specific context of China mobile telecommunication in which technological change and market change very fast. The environments contingency is important in this context. Therefore, further work is needed to verify the impacts of environment factors.

References

1. Cheung M, Myers M, Mentzer J (2010) Does relationship learning lead to relationship value? a cross-national supply chain investigation. J Oper Manage 28(6):472–487
2. Mesquita L-F, Anand J, Brush T-H (2008) Comparing the resource-based and relational views: knowledge transfer and spillover in vertical alliances. Strateg Manag J 29(9):913–941
3. Li J-J, Poppo L, Zhou K-Z (2010) Relational mechanisms, formal contracts, and local knowledge acquisition by international subsidiaries. Strateg Manag J 31(4):349–370
4. Poppo L, Zenger T (2002) Do formal contracts and relational governance function as substitutes or complements. Strateg Manag J 23:707–725
5. Dyer J-H, Singh H (1998) The relational view: cooperative strategy and sources of interorganizational competitive advantage. Acad Manag Rev 23(4):660–679
6. Reuer J-J, Arino A (2007) Strategic alliance contracts: dimensions and determinants of contractual complexity. Strateg Manag J 28:313–330
7. Moriizumi S, Chu B, Cao H, Matsukawa H (2011) Supply chain risk driver extraction using text mining technique. Inf Int Interdisc J 14:1935–1946

8. Inkpen A-C (2008) Knowledge transfer and international joint ventures: the case of NUMMI and general motors. Strateg Manag J 29:447–453
9. Yli-Renko H, Autio E, Sapienza H-J (2001) Social capital, knowledge acquisition, and knowledge exploitation in young technology-based firms. Strateg Manag J 22:587–613
10. Yakubovich V (2005) Weak ties, information, and influence: how workers find jobs in a local Russian labor market. Am Sociol Rev 70(3):408–421
11. Lane P-J, Lubatkin M-H (1998) Relative absorptive capacity and interorganizational learning. Strateg Manag J 19:461–477
12. Inkpen A-C, Tsang E-W (2005) Social capital, networks, and knowledge transfer. Acad Manag Rev 30:146–165
13. Larson A (1992) Network dyads in entrepreneurial settings: a study of the governance of exchange relationships. Adm Sci Q 37:76–104
14. Liu Y, Luo Y, Liu T (2009) Governing buyer-supplier relationships through transactional and relational mechanisms: evidence from China. J Oper Manage 27:294–309
15. Koza K-L, Dant R-P (2007) Effects of relationship climate, control mechanism, and communications on conflict resolution behavior and performance outcomes. J Retail 83(3):279–296
16. Hahn M-H, Lee K-C, Seo Y-W, Lee D-S (2011) The impact of organizational culture and knowledge-sharing on individual creativity. Inf Int Interdisc J 14:3081–3088
17. Cohen W-M, Levinthal D-A (1990) Absorptive capacity: a new perspective on learning and innovation. Adm Sci Q 35:128–152
18. Fornell C, Larcker D-F (1981) Evaluating structural equation models with unobservable variables and measurement error. J Mark Res 18:39–50

A Performance Measurement Method to Production Efficiency

Ci Chen and Junhong Chen

Abstract This paper provides an alternative way of performance measurement on production efficiency using optimization method. This paper provides an alternative way of performance measurement on production efficiency using optimization method. It is believed that any network production can be transformed into a series system where each stage in the series has a parallel structure. Thus, we first present models for simple series structure and parallel structure production under evaluation, separately. Then the models for general network structure production that consists of both the series and parallel structures are given. Performance efficiency including the relationship between the whole production and its divisions in series structure, parallel structure and general network structure are discussed.

Keywords Efficiency estimate · Performance evaluation

1 Introduction

With the development of industrial specialization and global supply chain, it is rare to see that a manufacturing process is completed by a single production unit nowadays. Actually, in almost all industrial sections, including manufacturing, servicing and retailing, the productions are operated with multiple stages in a

This paper is supported by KJCX201104013, the technology innovation construction program in BAAFS.

C. Chen (✉) · J. Chen
Institute of Agricultural Integrated Development, Beijing Academy of Agricultural and Forestry Sciences, Beijing, China
e-mail: sweetchenci@163.com

Z. Wen and T. Li (eds.), *Knowledge Engineering and Management*,
Advances in Intelligent Systems and Computing 278,
DOI: 10.1007/978-3-642-54930-4_34, © Springer-Verlag Berlin Heidelberg 2014

network form. Generally, the larger and more complex the production is, the more it becomes to be measured effectively [1].

Among many performance evaluation methods, data envelopment analysis (DEA) has been widely used to evaluate the relative performance of a set of production processes called Decision Making Units (DMUs) [2]. Classical DEA models make no assumptions concerning the internal operations of a DMU. Rather, DEA treats each DMU as a "black box" by considering only the inputs consumed and the outputs produced by it. Nowadays, several authors abandon the "black box" perspective in the assumption that, in some particularly contexts, the knowledge of the internal structure of DMUs can give further insights for the DMU performance evaluation [3]. Those models are called "Network DEA models", which are first introduced in the innovative book [4, 5] by Fare and Grosskopf (see also [6]). For the DMU observed, they first built the whole company's production possibility set in the sense that each company or DMU exhibits constant returns to scale and strong disposability of inputs and outputs since each division or subunit does the same. Then the whole company's efficiency is calculated by the Farrel radial projection on the frontier of company's production possibility set.

This paper constructs an alternative network DEA model for performance evaluation on network structure production. It is believed that any network production can be transformed into a series system where each stage in the series has a parallel structure. Thus, by constructing network DEA models for series structure production and parallel structure, separately, a network DEA model for a network structure production is given. Performance efficiency including the whole production and its divisions in series structure, parallel structure and general network structure are discussed. It is proved that, in series structure production, the whole production efficiency is less than or equal to the product of all of its divisions'. And in parallel structure production, the whole production is equal to the maximum efficiencies of its divisions'.

The remainder of this paper unfolds as follows. Section 2 develops network DEA model for series production, the network DEA model for parallel production is given in Sect. 3. In Sect. 4, a general network DEA model is constructed. Conclusions are given in Sect. 5.

2 Series Structure

For convenience, consider n companies or DMUs, each of which consists of two divisions connected in series. The inner structure is depicted in Fig. 1. Let X_j and Z_j be defined as the input and output vectors of the company, respectively. Denote Y_j as the intermediate product vector or $DMU_j, j = 1, \ldots n$.

For intermediate product, we only need to ensure that the level that taken as the output of division S_2 should not be larger than that taken as the input of division S_1.

Fig. 1 Series structure
production

With regard to the initial input X and final output Z, we assume that they all satisfy the strong free disposability, as it is common in conventional DEA models.

When the operations of all divisions are taken into account, the whole company efficiency of DMU_{j0}, called DMU_0 in short, is calculated from the following network DEA model:

$$
(D_{\text{series}})
\begin{cases}
\min \quad \theta \\[4pt]
s.t. \quad \displaystyle\sum_{j=1}^{n} X_j \lambda_j^1 \leq \theta X_0 \\[10pt]
\quad \displaystyle\sum_{j=1}^{n} Y_j \lambda_j^1 \geq \sum_{j=1}^{n} Y_j \lambda_j^2 \\[10pt]
\quad \displaystyle\sum_{j=1}^{n} Y_j \lambda_j^2 \leq Y_0 \\[10pt]
\quad \displaystyle\sum_{j=1}^{n} Z_j \lambda_j^2 \geq Z_0 \\[10pt]
\quad \lambda_j^1, \lambda_j^2 \geq 0, \quad j = 1, \ldots n
\end{cases}
$$

The efficiencies for division S_1 and S_2 are solved by conventional DEA models:

$$
(D_1)
\begin{cases}
\min \quad \theta_1 \\[4pt]
s.t. \quad \displaystyle\sum_{j=1}^{n} X_j \lambda_j^1 \leq \theta_1 X_0 \\[10pt]
\quad \displaystyle\sum_{j=1}^{n} Y_j \lambda_j^1 \geq \sum_{j=1}^{n} Y_0 \\[10pt]
\quad \lambda_j^1 \geq 0, \quad j = 1, \ldots n
\end{cases}
\qquad
(D_2)
\begin{cases}
\min \quad \theta_2 \\[4pt]
s.t. \quad \displaystyle\sum_{j=1}^{n} Y_j \lambda_j^2 \leq \theta_2 Y_0 \\[10pt]
\quad \displaystyle\sum_{j=1}^{n} Z_j \lambda_j^2 \geq Z_0 \\[10pt]
\quad \lambda_j^2 \geq 0, \quad j = 1, \ldots n
\end{cases}
$$

Let $\theta^*, \theta_1^*, \theta_2^*$ be the optimal values of model (D_{series}), (D_1), (D_2) respectively.

Theorem 1 *The Whole production efficiency is less than or equal to the products of all its divisional efficiencies, viz.* $\theta^* \leq \theta_1^* \theta_2^*$.

Proof Suppose $\lambda_j^{1*}, \theta_1^*, \quad j = 1, \ldots n$ is the optimal pair of solutions of (D_1). $\lambda_j^{2*}, \theta_2^*, \quad j = 1, \ldots n$ is the optimal pair of solutions of (D_2), thus we have:

$$\sum_{j=1}^{n} X_j \lambda_j^{1*} \leq \theta_1^* X_0 \tag{2.1}$$

$$\sum_{j=1}^{n} Y_j \lambda_j^{1*} \geq Y_0 \tag{2.2}$$

$$\sum_{j=1}^{n} Y_j \lambda_j^{2*} \leq \theta_2^* Y_0 \tag{2.3}$$

$$\sum_{j=1}^{n} Z_j \lambda_j^{2*} \geq Z_0 \tag{2.4}$$

By multiplying θ_2^* on both sides of (2.1) and (2.2), we obtain $\theta_2^* \sum_{j=1}^{n} X_j \lambda_j^{1*} \leq \theta_2^* \theta_1^* X_0$, $\theta_2^* \sum_{j=1}^{n} Y_j \lambda_j^{1*} \geq \theta_2^* Y_0$.

Define $\lambda_j^{1'} = \theta_2^* \lambda_j^{1*}$, $j = 1, \ldots n$, then the above two in-equations change into:

$$\sum_{j=1}^{n} X_j \lambda_j^{1'} \leq \theta_2^* \theta_1^* X_0 \tag{2.5}$$

$$\sum_{j=1}^{n} Y_j \lambda_j^{1'} \geq \theta_2^* Y_0 \tag{2.6}$$

Take (2.6) and (2.3) together, we have

$$\sum_{j=1}^{n} Y_j \lambda_j^{1'} \geq \theta_2^* Y_0 \geq \sum_{j=1}^{n} Y_j \lambda_j^2 \tag{2.7}$$

(2.5), (2.7), (2.3) and (2.4) together imply that $\theta_2^* \theta_1^*, \lambda_j^{1*}, \lambda_j^{2*}$, $j = 1, \ldots n$ is an feasible solution of (D_{series}), thereby we have $\theta^* \leq \theta_1^* \theta_2^*$.

3 Parallel Structure

For simplicity, consider a company consists of two independently divisions connected in parallel as shown in Fig. 2. All inputs and outputs of the two divisions constitute the overall input and output index system of the whole company production. Suppose there are n companies, each of which has the same structure.

Fig. 2 Parallel structure
production

In parallel structure production, with regard to the initial input X and final output Y, we assume that they all satisfy the strong free disposability, as it is common in conventional DEA models. Thus, the whole company efficiency of DMU_0 in short, is solved by:

$$
(D_{\text{parallel}})
\begin{cases}
\min \quad \theta \\
s.t. \quad \displaystyle\sum_{j=1}^{n} X_j^1 \lambda_j^1 \leq \theta X_0^1 \\
\displaystyle\sum_{j=1}^{n} X_j^2 \lambda_j^2 \leq \theta X_0^2 \\
\displaystyle\sum_{j=1}^{n} Y_j^1 \lambda_j^1 \geq Y_0^1 \\
\displaystyle\sum_{j=1}^{n} Y_j^2 \lambda_j^2 \geq Y_0^2 \\
\lambda_j^1, \lambda_j^2 \geq 0, \quad j = 1,\ldots n
\end{cases}
$$

Consider the DEA model for efficiency evaluation on division S_{11} and S_{12} as follows:

$$
(D_{11})
\begin{cases}
\min \quad \theta_{11} \\
s.t. \quad \displaystyle\sum_{j=1}^{n} X_j^1 \lambda_j^1 \leq \theta_{11} X_0^1 \\
\displaystyle\sum_{j=1}^{n} Y_j^1 \lambda_j^1 \geq Y_0^1 \\
\lambda_j^1 \geq 0, \quad j = 1,\ldots n
\end{cases}
\qquad
(D_{12})
\begin{cases}
\min \quad \theta_{12} \\
s.t. \quad \displaystyle\sum_{j=1}^{n} X_j^2 \lambda_j^2 \leq \theta_{12} X_0^2 \\
\displaystyle\sum_{j=1}^{n} Y_j^2 \lambda_j^2 \geq Y_0^2 \\
\lambda_j^2 \geq 0, \quad j = 1,\ldots n
\end{cases}
$$

Let $\theta^*, \theta_{11}^*, \theta_{12}^*$ be the optimal values corresponding to models (D_{parallel}), (D_{11}), (D_{12}), respectively.

Theorem 2 *The whole company efficiency is equal to the maximum efficiencies of its divisions', viz.* $\theta^* = \max\{\theta_{11}^*, \theta_{12}^*\}$.

Proof First, suppose $\theta^*, \lambda_j^{1*}, \lambda_j^{2*}, \quad j = 1, \ldots n$ is an optimal pair of solutions of model (D_{parallel}). Obviously, $\theta^*, \lambda_j^{1*}, \quad j = 1, \ldots n$ is feasible for model (D_{11}), thus we have $\theta_{11}^* \leq \theta^*$. In the same vein, we get $\theta_{12}^* \leq \theta^*$. Hence, we obtain, $\max\{\theta_{11}^*, \theta_{12}^*\} \leq \theta^*$.

Let $\theta_{11}^*, \lambda_j^{1*}, \quad j = 1, \ldots n$ be the optimal solution of model $(D_{11}), \theta_{12}^*, \lambda_j^{2*}, \quad j = 1, \ldots n$ of model (D_{12}). So we have:

$$\sum_{j=1}^{n} X_j^1 \lambda_j^{1*} \leq \theta_{11} X_0^1 \tag{3.1}$$

$$\sum_{j=1}^{n} Y_j^1 \lambda_j^{*1} \geq Y_0^1 \tag{3.2}$$

$$\sum_{j=1}^{n} X_j^2 \lambda_j^{2*} \leq \theta_{12} X_0^2 \tag{3.3}$$

$$\sum_{j=1}^{n} Y_j^2 \lambda_j^{2*} \geq Y_0^2 \tag{3.4}$$

Without loss of generality, suppose $\theta_{11}^* = \max\{\theta_{11}^*, \theta_{12}^*\}$. By referring to (3.3), we have

$$\sum_{j=1}^{n} X_j^2 \lambda_j^{2*} \leq \theta_{11}^* X_0^2 \tag{3.5}$$

Note that (3.1), (3.5), (3.2) and (3.4) together imply that $\theta_{11}^*, \lambda_j^{1*}, \lambda_j^{2*}, \quad j = 1, \ldots n$ is feasible for model (D_{parallel}). Thus, we derive $\theta^* \leq \theta_{11}^* = \max\{\theta_{11}^*, \theta_{12}^*\}$.

4 General Structure

In this section, we discuss a more general case that contains both of the series and parallel structure. For convenience, consider the structure depicted in Fig. 3.

Suppose there are n companies or DMUs denoted as $DMU_1, DMU_2, \ldots DMU_n$, each of which has exactly the same structure. As discussed in the Sects. 2 and 3, when the operations of intermediate products are taken into account, the whole company efficiency is assessed by:

Fig. 3 General structure production

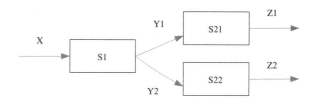

$$(D)\begin{cases} \min \quad \theta \\ s.t. \quad \sum_{j=1}^{n} X_j \lambda_j^1 \leq \theta X_0 \\ \sum_{j=1}^{n} Y_j^1 \lambda_j^1 \geq \sum_{j=1}^{n} Y_j^1 \lambda_j^2 \quad \sum_{j=1}^{n} Z_j^1 \lambda_j^2 \geq Z_0^1 \\ \sum_{j=1}^{n} Y_j^2 \lambda_j^1 \geq \sum_{j=1}^{n} Y_j^2 \lambda_j^3 \quad \sum_{j=1}^{n} Y_j^2 \lambda_j^3 \leq Y_0^2 \\ \sum_{j=1}^{n} Y_j^1 \lambda_j^2 \leq Y_0^1 \qquad \sum_{j=1}^{n} Z_j^2 \lambda_j^3 \geq Z_0^2 \\ \lambda_j^1, \lambda_j^2, \lambda_j^3 \geq 0, \quad j=1,\ldots n \end{cases}$$

The efficiency evaluations for division S_1, S_{21}, and S_{22} are solved by classic DEA models, where we denote (D_1), (D_{21}) and (D_{22}).

Theorem 3 *For the structure depicted in* Fig. 3, *the relationship between the whole efficiency and divisional efficiencies is:* $\theta^* \leq \theta_1^* \max\{\theta_{21}^*, \theta_{22}^*\}$.

Proof Denote $\theta_1^*, \lambda_1^{1*}, \theta_{21}^*, \lambda_j^{21*}$, $j=1,\ldots n$ as the optimal pair of solutions of model (D), (D_1), (D_{21}) and (D_{22}), respectively. Thus, we have the following:

$$\sum_{j=1}^{n} Y_j^1 \lambda_j^{21*} \leq \theta_{21}^* Y_0^1 \leq Y_0^1 \qquad (4.1)$$

$$\sum_{j=1}^{n} Z_j^1 \lambda_j^{21*} \geq Z_0^1 \qquad (4.2)$$

$$\sum_{j=1}^{n} Y_j^2 \lambda_j^{23*} \leq \theta_{22}^* Y_0^2 \leq Y_0^2 \qquad (4.3)$$

$$\sum_{j=1}^{n} Z_j^2 \lambda_j^{22*} \geq Z_0^2 \qquad (4.4)$$

and $\sum_{j=1}^{n} X_j^1 \lambda_j^{1*} \le \theta_1^* X_0^1$, $\sum_{j=1}^{n} Y_j^1 \lambda_j^{1*} \ge Y_0^1$, $\sum_{j=1}^{n} Y_j^2 \lambda_j^{2*} 0^2$. Without loss of generality, suppose $\theta_{21}^* \ge \theta_{22}^*$. By multiplying θ_{21}^* on both sides of the above three in-equations, we get

$$\theta_{21}^* \sum_{j=1}^{n} X_j^1 \lambda_j^{1*} \le \theta_{21}^* \theta_1^* X_0^1 \tag{4.5}$$

$$\theta_{21}^* \sum_{j=1}^{n} Y_j^1 \lambda_j^{1*} \ge \theta_{21}^* Y_0^1 \tag{4.6}$$

$$\theta_{21}^* \sum_{j=1}^{n} Y_j^2 \lambda_j^{1*} \ge \theta_{21}^* Y_0^2 \tag{4.7}$$

(4.1) and (4.6) can be represented as $\theta_{21}^* \sum_{j=1}^{n} Y_j^1 \lambda_j^{1*} \ge \theta_{21}^* Y_0^1 \ge \sum_{j=1}^{n} Y_j^1 \lambda_j^{21*}$. (4.3) and (4.7) can be represented as $\theta_{21}^* \sum_{j=1}^{n} Y_j^2 \lambda_j^{1*} \ge \theta_{21}^* Y_0^2 \ge \sum_{j=1}^{n} Y_j^2 \lambda_j^{22*}$.

Let $\lambda_j^{1'*} = \theta_{21}^* \lambda_j^{1*}$, $j = 1, \ldots n$, then (4.5), (4.6) and (4.7) can be expressed in the following way:

$$\sum_{j=1}^{n} X_j^1 \lambda_j^{1'*} \le \theta_{21}^* \theta_1^* X_0^1 \tag{4.8}$$

$$\sum_{j=1}^{n} Y_j^1 \lambda_j^{1'*} \ge \sum_{j=1}^{n} Y_j^1 \lambda_j^{21*} \tag{4.9}$$

$$\sum_{j=1}^{n} Y_j^2 \lambda_j^{1'*} \ge \sum_{j=1}^{n} Y_j^2 \lambda_j^{22*} \tag{4.10}$$

(4.1), (4.2), (4.3), (4.4), (4.8) and (4.9) together imply that $\lambda_j^{1'*}, \theta_1^* \theta_{21}^*, \lambda_j^{21*}, \lambda_j^{22*}, j = 1, \ldots n$ is feasible for model (D). Thus, we have $\theta^* \le \theta_{21}^* \theta_1^*$, viz. $\theta^* = \theta_1^* \max\{\theta_{21}^*, \theta_{22}^*\}$.

5 Conclusions

This paper provides an alternative way of performance evaluation on network structure production using data envelopment analysis (DEA). We present models for simple series structure and parallel structure production under evaluation, separately. Then the models for general network structure production that consists of both the series and parallel structures are given. Performance efficiency including the relationship between the whole production and its divisions in series structure, parallel structure and general network structure are discussed.

References

1. Beamon B (1999) Measuring supply chain performance. Int J Oper Prod Manage 19(3):275–292
2. Charnes A, Cooper WW, Rhodes E (1978) Measuring the efficiency of decision making units. Eur J Oper Res 2(6):429–444
3. Castelli L, Psesnti R, Ukovich W (2008) A classification of DEA models when the internal structure of the decision making units is considered. Ann Oper Res. doi:10.1007/s10479-008-0414-2
4. Fare R, Grosskopf S (1996) Intertemporal production frontiers: with dynamic DEA. Kluwer Academic Publisher, Boston
5. Fare R, Grosskopf S (2000) Network DEA. Socio Econ Plan Sci 34:35–49
6. Farrel MJ (1957) The measurement of productive efficiency. J Roy Stat Soc A 120:253–290

Teaching Strategies Reasoning Based on Dynamical Uncertainty Causality Graph in e-Learning System

Wansen Wang and Guizhen Wang

Abstract This article is based on the emotional cognition and learning style of the ECLS model, giving a reasoning tools-Dynamical Uncertainty Causality Graph (DUCG) that the teaching strategies reasoning needs, then it uses DUCG for network teaching reasoning based on the emotional cognitive interaction.

Keywords Emotional cognitive interaction · Bayesian network · DUCG · Teaching strategies reasoning

1 Introduction

Dealing with knowledge representation and reasoning of uncertain causal is one of the key issues that must be addressed in artificial intelligence system. So far a number of theoretical models have been proposed, such as certainty factor [1], evidence reasoning [2], PROSPECTOR [3], fuzzy logic [4], BN [5–9], etc. Among them, BN is increasingly popular [10–12]. The concise of knowledge representation manner and efficient inference algorithm is one of the core research questions in BN.

However BN has some shortcomings, there are two different conditions in the subvariables of BN: single-valued and multivalued, but the simplicity of

Supported by the National Natural Science Foundation of China under Grant No. 60970052, Beijing Natural Science Foundation (The Study of Personalized e-learning Community Education based on Emotional Psychology. 4112014).

W. Wang · G. Wang (✉)
Information Engineering College Capital Normal University, Beijing, China
e-mail: 1039463205@qq.com

Z. Wen and T. Li (eds.), *Knowledge Engineering and Management*,
Advances in Intelligent Systems and Computing 278,
DOI: 10.1007/978-3-642-54930-4_35, © Springer-Verlag Berlin Heidelberg 2014

knowledge representation and reasoning methods in the single-valued case is not applied in the multivalued case. Dynamical uncertainty causality graph (DUCG) is the very tool to overcome the shortcomings and other defects of BN. It graphically expresses the uncertain causality relationship of any cases simply; meanwhile, it simplifies graphics and unfolds events based on the evidence in order to obtain the concerned assumption and its state probability expressions. In addition, DUCG allows the incomplete representation of knowledge, so it beyonds the theoretical framework of BN [3]. The teaching strategy reasoning mentioned in this article belongs to the uncertainty reasoning category, so we use DUCG for uncertainty reasoning between the emotional cognition and the teaching strategies.

2 The Basic Introduction of DUCG

First, please look at Table 1, before the introduction of DUCG, we define a new set of symbol systems. Capital letters represent variables or events, lowercase letters represent the probability distribution of the variables or the probability of occurred events; the first subscript of variables or events is used to distinguish variables and the second subscript is used to distinguish the state of variables (that is event). In other words, the first and second subscript is determined a state of a variables in an event. For example, X_n represents the variable with numeral n, X_{nk} represents the K-th state of variable X_n, which is also an event, that means the variable X_n is in the state of K. Correspondingly, $x_{nk} = P_r\{X_n = x_{nk}\}, P_{nk;ij}$ is the probability of occurrence of the connection event, in DCD $P_{nk;ij} = P_r\{X_n = x_{nk}/X_i = x_{ij}\}$ represents the event $X_n = x_{nk}$, which is caused by the event $X_i = x_{ij}$.

This article conducts uncertainty reasoning between emotional cognitive and teaching strategies by multivalued DUCG model (M-DUCG), because the subvariable analysis in this paper is multivalued, so we only describe the M-DUCG. For single-valued DUCG (S-DUCG), it adds two tools of condition: connection and default events on the basis of DCD [13] which has been specifically expressed in the literature [14].

The M-DUCG is a theoretical model to handle multivalued conditions.

First, we give the formula (1)

$$X_{nk} = \sum_i (r_{n;i}/r_n) \sum_{j_i} A_{nk;ij_i} V_{ij_i} \tag{1}$$

j_i marks the status of the parent variables $V_{ij_i}, r_{n;i}$ is defined as the degree of association between X_n and $V_i, r_n = \sum_i r_{n;i}$.similar to $P_{nk;ij}, A_{nk;ij}$ is defined as the random event X_{nk} which is caused by the occurred conditions V_{ij_i} independently. $A_{nk;ij}$ refers to the role of events between the parent event and subevent. (Note the formula (1) is an expression which is banded by the sum of products, it is composed by the event of the coefficient $(r_{n;i}/r_n)$).

Table 1 The symbol systems

Variable	Meaning
X_n	The variable with numeral n
X_{nk}	The variable X_n is in the state of K
$P_{nk;ij}$	The probability of occurrence of the connection event (the event $X_n = x_{nk}$ is cause by the event $X_i = x_{ij}$)
j_i	The status of the parent variables V_{ij_i}
V_{ij_i}	The parent variable
$r_{n;i}$	The causal relationship between V_i and X_n
$A_{nk;ij}$	The random event X_{nk} which is caused by the occurred conditions V_{ij_i} independently (similar to $P_{nk;ij}$)

The following figure is the interpretation of this assumption (Fig. 1):
$\rightarrow$ represents the role of event variables after weights. $F_{n;i} \equiv (r_{n;i}/r_n)A_{n;i}, A_{n;i}$ is an event matrix with the element of $A_{nk;ij}$. k marks the matrix rows, j_i marks the matrix columns. $F_{n;i} \equiv (r_{n;i}/r_n)A_{n;i}$ represents the role of event variables with full total weight, it is an event matrix, whose elements are $F_{n;i} \equiv (r_{n;i}/r_n)A_{n;i}$, for simplicity, j_i abbreviated as j in the case not to be confused.

Similar to $P_{nk;ij}, A_{n;i}$ represents the uncertainty mechanism that V_{ij} really caused by V_{ij} under the condition that V_{ij} has occurred, if this mechanism works, then there is $X_{nk;ij}$, then X_{nk} occurs; if this mechanism does not work, then $X_{nk;ij} = 0$ (empty set). Since this mechanism is independent of any other factors, $A_{n;i}$ is an independent random variable. Of course, for $K \neq K', A_{nk;ij}$ and $A_{nk';ij}$ are mutually exclusive, for X_{nk} and $X_{nk'}$ are mutually exclusive. We define $a_{nk;ij} \equiv P_r\{A_{nk;ij}\}, a_{nk;ij}$ means that the original parameters given by experts.

$r_{n;i}$ indicates the causal relationship between V_i and X_n, which is given by experts. The scope of $r_{n;i}$ is: $r_{n;i} > 0$, the size of $r_{n;i}$ represents the strength of the association between V_i and X_n, actually, what really works is $(r_{n;i}/r_n)$, it is the contribution weights of the probability distribution, that means when $A_{nk;ij}$ which with different k or j but the same n and i causes $X_{nk;ij}$, there by X_{nk} bounds to the same weight coefficient $(r_{n;i}/r_n)$.

In accordance with formula (1), we can easily obtain the following results:

$$P_r\{X_{nk}|\bigcap_i V_{ij_i}\} = \sum_i (r_{n;i}/r_n)a_{nk;ij_i} \tag{2}$$

$$x_{nk} = P_r\{X_{nk}\} = \sum_i (r_{n;i}/r_n) \sum_i a_{nk;ij}P_r\{V_{ij}\}$$
$$= \sum_i (r_{n;i}/r_n) \sum_i a_{nk;ij}v_{ij} \text{ (where } v_{ij} = P_r\{V_{ij}\}) \tag{3}$$

Fig. 1 The M-DUCG model

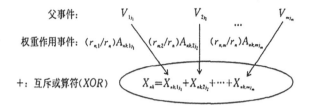

3 The Teaching Strategies Reasoning Based on DUCG

In Fig. 2 [15], since the learning style is the inherent characteristics of students, it will not change in the learning process. When students start learning, the system begins to adjust the teaching content and navigation strategy according to the student's characteristic tends of his learning style. Thus, we just need the cognitive state dimension and emotional state dimensions to reasoning teaching strategies, the calculation of the level of cognitive and emotional state can be seen in papers [15], but it did not give the calculation method of the master status and the state of readiness.

Master status which means the situation that students have grasped, it can be got through the examination test, the test scores is on a 100-establish, we establish the appropriate test content based on what students have learned. If the test results are between 80 and 100, then the master status is good; if the test results are between 60 and 80, then the master status is general; if the test results are below 60, the master status is poor. Our determination is based on the number of tests and test scores, it grasps the credibility of the master status according to the number of tests and the test scores.

Preparation status which means students' store of knowledge for restructuring pilot study before learning the knowledge, preparation status has two levels: good level and general level (Only good and general preparation status can enter the current knowledge learning), the calculation of the preparation status is based on the state of his mastery during each part of his study.

According to ECLS model, each variable in DUCG is expressed as follows:

B_1 means pleasure degree:

$B_{11} \equiv \{happiness\}$ $B_{12} \equiv \{sadness\}$

B_2 means interest degree:

$B_{21} \equiv \{interesting\}$ $B_{22} \equiv \{reluctant\}$

B_3 means wake-up degree:

$B_{31} \equiv \{excitement\}$ $B_{32} \equiv \{fatigue\}$

X_4 means emotion:

$X_{41} \equiv \{positive\ emotions\}$ $X_{42} \equiv \{negative\ emotions\}$

B_5 means cognitive level:

Fig. 2 The ECLS model

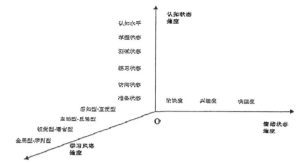

$B_{51} \equiv \{$the high level of cognitive$\}$ $B_{52} \equiv \{$the general level of cognitive$\}$
$B_{53} \equiv \{$the low level of cognitive$\}$

B_6 means master status:

$B_{61} \equiv \{$good master status$\}$ $B_{62} \equiv \{$general master status$\}$
$B_{63} \equiv \{$badmaster status$\}$

B_7 means preparation status:

$B_{71} \equiv \{$good preparation status$\}$ $B_{72} \equiv \{$general preparation status$\}$

X_8 means teaching strategies:

$X_{81} \equiv \{$learning the next knowledge$\}$ $X_{82} \equiv \{$review the current knowledge$\}$
$X_{83} \equiv \{$learning the basic knowledge$\}$ $X_{84} \equiv \{$take a break$\}$
$X_{85} \equiv \{$encourage to learn$\}$ $X_{86} \equiv \{$warning to learn$\}$

and so on.

The corresponding DUCG model of DAG structure has been showed in the following figure (Fig. 3):

The DUCG consists of two modules: $\{B_1, B_2, B_3, X_4\}$, $\{X_4, B_5, B_6, B_7, X_8\}$ and the internal directed arc of both two modules.

For example, the system gets his learning personalization features in the learning process are as follows:

Happiness: credibility CF 0.89; Interesting: credibility CF 0.91;

Excitement: credibility CF 0.85; High cognitive level: credibility CF 0.92

Fig. 3 The DAG structure

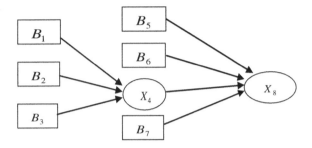

High grasp status: credibility CF 0.88; High preparation status: credibility CF 0.89

The matching rules:

Rule 1: IF happiness and interest and excitement THEN positive emotion.

Rule 2: IF positive emotion AND high level of cognitive and good master status and good preparation status THEN learning the next knowledge.

For the first module, we gave the parameters as follows:

$$a_{4;1} \equiv \begin{bmatrix} a_{41;11} & a_{41;12} \\ a_{42;11} & a_{42;12} \end{bmatrix} = \begin{bmatrix} 0.83 & - \\ - & 0.83 \end{bmatrix} \quad a_{4;2} \equiv \begin{bmatrix} a_{41;21} & a_{41;22} \\ a_{42;21} & a_{42;22} \end{bmatrix} = \begin{bmatrix} 0.92 & - \\ - & 0.92 \end{bmatrix}$$

$$a_{4;3} \equiv \begin{bmatrix} a_{41;31} & a_{41;32} \\ a_{42;31} & a_{42;32} \end{bmatrix} = \begin{bmatrix} 0.80 & - \\ - & 0.80 \end{bmatrix} \quad b_1 \equiv [b_{11} \quad b_{12}]^T = [0.89 \quad 0.11]^T$$

$$b_2 \equiv [b_{21} \quad b_{22}]^T = [0.91 \quad 0.09]^T \qquad b_3 \equiv [b_{31} \quad b_{32}]^T = [0.85 \quad 0.15]^T$$

$$r_{4;1} = 1, r_{4;2} = 2, r_{4;3} = 1 \Rightarrow r_4 = \sum_i r_{4;i} = 4$$

$$\Rightarrow (r_{4;1}/r_4) = 0.25 (r_{4;2}/r_4) = 0.5 (r_{4;3}/r_4) = 0.25$$

According to formula (3)

$$P_r\{X_{41}\} = \sum_i (r_{4;i}/r_4) \sum_{j_i} a_{41;ij_i} v_{ij_i}$$

$$= 0.25 \times (0.83 \times 0.89) + 0.5 \times (0.92 \times 0.91) + 0.25 \times (0.80 \times 0.85)$$

$$= 0.763257$$

The threshold value of emotional is 0.72; $P_r\{X_{41}\} \geq 0.72$

Rule 1 can be determined as positive emotion.

For the second module, we the given parameters are as follows:

$$a_{8;4} \equiv \begin{bmatrix} a_{81;41} & a_{81;42} \\ a_{82;41} & a_{82;42} \\ a_{83;41} & a_{83;42} \\ a_{84;41} & a_{84;42} \\ a_{85;41} & a_{85;42} \\ a_{86;41} & a_{86;42} \\ \cdots & \cdots \end{bmatrix} = \begin{bmatrix} 0.90 & - \\ 0.90 & - \\ 0.90 & - \\ - & 0.90 \\ 0.80 & 0.90 \\ - & 0.90 \\ \cdots & \cdots \end{bmatrix}$$

$$a_{8;7} \equiv \begin{bmatrix} a_{81;71} & a_{81;72} \\ a_{82;71} & a_{82;72} \\ a_{83;71} & a_{83;72} \\ a_{84;71} & a_{84;72} \\ a_{85;71} & a_{85;72} \\ a_{86;71} & a_{86;72} \\ \cdots & \cdots \end{bmatrix} = \begin{bmatrix} 0.90 & 0.70 \\ - & 0.85 \\ - & 0.75 \\ - & 0.90 \\ 0.90 & 0.75 \\ - & 0.80 \\ \cdots & \cdots \end{bmatrix}$$

$$a_{8;5} \equiv \begin{bmatrix} a_{81;51} & a_{81;52} & a_{81;53} \\ a_{82;51} & a_{82;52} & a_{82;53} \\ a_{83;51} & a_{83;52} & a_{83;53} \\ a_{84;51} & a_{84;52} & a_{84;53} \\ a_{85;51} & a_{85;52} & a_{85;53} \\ a_{86;51} & a_{86;52} & a_{86;53} \\ \cdots & \cdots & \cdots \end{bmatrix} = \begin{bmatrix} 0.90 & 0.75 & - \\ 0.75 & 0.90 & - \\ - & - & 0.90 \\ - & 0.75 & 0.90 \\ - & 0.75 & 0.90 \\ - & 0.90 & 0.75 \\ \cdots & \cdots & \cdots \end{bmatrix}$$

$$a_{8;6} \equiv \begin{bmatrix} a_{81;61} & a_{81;62} & a_{81;63} \\ a_{82;61} & a_{82;62} & a_{82;63} \\ a_{83;61} & a_{83;62} & a_{83;63} \\ a_{84;61} & a_{84;62} & a_{84;63} \\ a_{85;61} & a_{85;62} & a_{85;63} \\ a_{86;61} & a_{86;62} & a_{86;63} \\ \cdots & \cdots & \cdots \end{bmatrix} = \begin{bmatrix} 0.95 & 0.70 & - \\ 0.70 & 0.95 & - \\ - & - & 0.95 \\ - & - & - \\ - & 0.95 & 0.70 \\ - & 0.70 & 0.95 \\ \cdots & \cdots & \cdots \end{bmatrix}$$

$$b_4 \equiv [b_{41} \quad b_{42}]^T = [0.76 \quad 0.24]^T \qquad b_5 \equiv [b_{51} \quad b_{52} \quad b_{53}]^T = [0.92 \quad 0.08 \quad 0]^T$$

$$b_6 \equiv [b_{61} \quad b_{62} \quad b_{63}]^T = [0.88 \quad 0.12 \quad 0]^T \qquad b_7 \equiv [b_{71} \quad b_{72}]^T = [0.89 \quad 0.11]^T$$

$$r_{8;4} = 1, r_{8;5} = 1, r_{8;6} = 1, r_{8;7} = 1 \Rightarrow r_8 = \sum_i r_{8;i} = 4$$

$$\Rightarrow (r_{8;4}/r_8) = 0.25 (r_{8;5}/r_8) = 0.25 (r_{8;6}/r_8) = 0.25 (r_{8;7}/r_8) = 0.25$$

According to formula (3)

$$P_r\{X_{81}\} = \sum_i (r_{8;i}/r_8) \sum_{j_i} a_{81;ij_i} v_{ij_i}$$
$$= 0.25 \times (0.90 \times 0.76) + 0.25 \times (0.90 \times 0.92) + 0.25 \times (0.95 \times 0.88)$$
$$+ 0.25 \times (0.90 \times 0.89)$$
$$= 0.78725$$

The threshold value of teaching strategies is 0.73; $P_r\{X_{81}\} \geq 0.73$
Rule 2 can be determined as effective. So he/she can learn the next knowledge.
For another example, the system gets his learning personalization features in the learning process are as follows:

Sadness: credibility CF 0.89; Reluctant: credibility CF 0.91;
Fatigue: credibility CF 0.85; General cognitive level: credibility CF 0.85
Low grasp status: credibility CF 0.88; Low preparation status: credibility CF 0.89

The matching rules:

Rule 1: IF sadness and reluctant and fatigue THEN negative emotion.
Rule 2: IF negative emotion AND general level of cognitive and low master status and low preparation status THEN warn the student to learn.

According to formula (3)

$$P_r\{X_{42}\} = \sum_i (r_{4;i}/r_4) \sum_{j_i} a_{42;ij_i} v_{ij_i}$$
$$= 0.25 \times (0.83 \times 0.89) + 0.5 \times (0.92 \times 0.91) + 0.25 \times (0.80 \times 0.85)$$
$$= 0.763257$$

The threshold value of emotional is 0.72; $P_r\{X_{42}\} \geq 0.72$
Rule 1 can be determined as negative emotion.
Then we analyze the teaching strategies of X_{86}
According to formula (3)

$$P_r\{X_{86}\} = \sum_i (r_{8;i}/r_8) \sum_{j_i} a_{81;ij_i} v_{ij_i}$$
$$= 0.25 \times (0.90 \times 0.76) + 0.25 \times (0.90 \times 0.85) + 0.25 \times (0.95 \times 0.88)$$
$$+ 0.25 \times (0.80 \times 0.89)$$
$$= 0.7515$$

The threshold value of teaching strategies is 0.73; $P_r\{X_{86}\} \geq 0.73$
Rule 2 can be determined as effective. So the teaching strategy is to warn the student to learn.

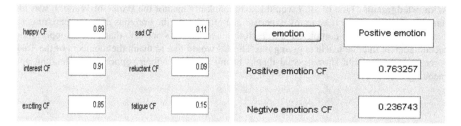

Fig. 4 Enter the parameters emotional state and the output of emotion

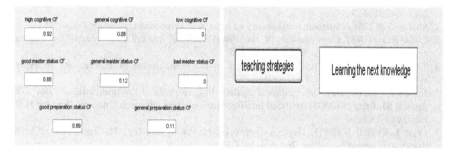

Fig. 5 Enter the parameters and the output of teaching strategies

4 Emulation

1. Enter the parameters emotional state and the output of emotion (Fig. 4)
2. Enter the parameters of cognitive level, master status and ready status and the output of teaching strategies (Fig. 5)

5 Conclusions

In this paper, the uncertainty reasoning between emotional cognitive and teaching strategies is taken by the M-DUCG model. It can not only Intuitively concisely to express a variety of complicated uncertain causal relationship within modules, but also being simplified through Deleting independent of event, letting the reasoning become more easy. Meanwhile, this thesis conducted a comprehensive analysis and reasoning between each state of the cognitive status dimension and emotional state so that the teaching strategies to be more reasonable. But it requires the event occurrence probability is 1 in the reasoning process, that is $\sum_j b_{ij} = 1$, But we calculated the probability of occurrence is only one state, while the probability of the rest state is defined as $1 - b_{ij}$, which has certain error events for the three states.

Acknowledgments First of all, I would like to thank my mentor the Professor Wang, it was his careful guidance that increasing my expertise and broadening the horizons of my researching, so that the papers could completed successfully. Second, I sincerely thank the support and encouragement that the teachers giving me. Third, I would like to thank the members of the Task Force. Finally, I would show special thanks to my family for their support understanding and encouragement.

References

1. Shortliffe EH, Buchanan BG (1975) A model of inexact reason in medicine. Math Biosci 23(3–4):351–379
2. Shafer G (1976) A mathematical theory of evidence. Princeton University Press, NJ
3. Duda RO (1978) Development of the PROSPECTOR consultation system for mineral exploration. Final Report, SRI International
4. Zadeh LA (1983) The role of fuzzy logic in the management of un-certainty in expert systems. Fuzzy Sets Syst 11(1–3):199–227
5. Jensen FV (2000) Bayesian graphical models. Encyclopedia of Environmetrics. Wiley, UK
6. Russell SJ, Peter N (2003) Artificial intelligence: a modern approach, 2nd edn. Prentice Hall, Englewood Cliffs
7. Pearl J, Russell S (2003) Bayesian networks. In: Arbib MA (ed) The handbook of brain theory and neural networks, 2nd edn, MIT Press
8. Poole D, Zhang NL (2003) Exploiting contextual independence inprobabilistic inference. J Artif Intell Res 18:263–313
9. Jensen FV (1996) An introduction to Bayesian networks. Springer, NewYork
10. Liu JH (2010) Emotional cognitive interactive in e-learning research and teaching inference system platform Beijing [D] Capital Normal University
11. Ji JZ, Liu CN, Sha Z (2003) Bayesian belief network model learning, inference and applications. Comput Eng Appl (5):24–28
12. Patricio GAA, Silvia S, Marcelo C (2007) Evaluating Bayesian networks precision for detectingstudents learning styles. Comput Educ 49:794–808
13. Zhang Q (1994) Probabilistic reasoning based on dynamic causal-ity trees/diagrams. Reliab Eng Syst Saf 46(3):209–220
14. Zhang Q (2010) DUCG-a new dynamic uncertain causal knowledge representation and reasoning. J Comput Sci 33(4):625–651
15. Li HL (2011) Emotional cognitive model of interactive online teaching strategies and reasoning of Beijing [D] Capital Normal University

Label Samples Using TC-SVDD

Gao Zhi-hua and Ben Ke-rong

Abstract In many fields, labeling samples is a time-consuming and costly work. This paper describes an automatically labeling samples method based on SVDD with transductive confidence (TC-SVDD). The new algorithm labeling samples automatically by lead transductive confidence idea into support vector data description. It gives the confidence lever about labeling result to improving the labeled samples quality. Experiment results on UCI data sets show the algorithm has advantages on label samples with high quality.

Keywords Sample labeling · TC-SVDD (Transductive Confidence-Support Vector Data Description)

1 Introduction

The current situation of many applications is unlabeled samples being easy to obtain, but labeling these samples is very hard which making been labeled samples being limited [1]. The traditional learning algorithm trains the classifier using the given labeled samples as the training set. An important step in machine learning is to generate training data. How to get high quality large amount of labeled samples become a great challenge to machine learning [2]. At the same time, how to make use of unlabeled samples to improve learning performance has become one of the popular problems in the research of machine learning [3].

The paper proposes a new algorithm, TC-SVDD, which can labeled samples automatically. This method introduces transductive confidence learning on the

G. Zhi-hua (✉) · B. Ke-rong
Department of Computer Engineering, Naval University of Engineering,
Wuhan 430033, China
e-mail: zhihuagao@126.com

Z. Wen and T. Li (eds.), *Knowledge Engineering and Management*,
Advances in Intelligent Systems and Computing 278,
DOI: 10.1007/978-3-642-54930-4_36, © Springer-Verlag Berlin Heidelberg 2014

SVDD, and labels samples according to the confidence level. This algorithm can minimize the sample labeling cost without losing the classifier performance.

2 Learning by Transduction

Transductive Confidence Machine (TCM) [4] is a machine learning algorithms to be widely used. In some applying areas, the confidence of classification prediction is more important than classification result. By TCM, the sample classification and confidence can be gained synchronously. The goal of TCM is to derive the valid confidence under the general i.i.d. (independent and identically distributed) assumption, and connected with the randomness test in the Kolmogorov's randomness theory [5]. Randomness test is not computable, but can approximatively evaluate it by the P-values. This P-value serves as a measure of how well the data supports or not null hypothesis. Users of transduction as a test of confidence have approximated a universal test for randomness by using a P-value function called strangeness measure. The strangeness measure corresponds to the ambiguity of the point being measured with respect to all the other labeling examples of a class: the higher the strangeness measure, the higher the uncertainty.

We have a labeled samples set $\{(x_1,y_1),\ldots(x_m,y_m)\}$, of m elements, where x_i is the set of feature values for example i and y_i is the classification for example i, taking values from a finite set of possible classifications, which we identify as $\{1, 2,\ldots, c\}$. The goal is to assign to every unlabeled sample one of the possible classifications.

We assign to every sample a measure called the individual strangeness measure. This measure defines the strangeness of the sample in relation to the rest of the samples. The strangeness measure is call α_i.

It is proved in [6] that the function

$$P(\alpha_{\text{new}}) = \frac{\#\{j : \alpha_i \geq \alpha_{\text{new}}\}}{m+1} \tag{1}$$

where # represents the cardinality of the set given as an argument and α_{new} is the strangeness value for the test sample. If we have a sequence $\{\alpha_1,\ldots\alpha_m\}$ and a new element α_{new} is introduced then α_{new} can take any place in the new (sorted) sequence with the same probability, as all permutations of the new sequence are equiprobable. The values taken by the above randomness test will be called P-values. The P-value for the sequence $\{\alpha_1,\ldots\alpha_m, \alpha_{\text{new}}\}$, where $\{\alpha_1,\ldots\alpha_m\}$ are the strangeness measures for the training examples and α_{new} is the strangeness measure of a new sample with a possible classification assigned to it, is the value $p(\alpha_{\text{new}})$.

The confidence learning process of TCM is as follows. Firstly, give the sample sequence $\{(x_1,y_1),\ldots,(x_m,y_m), (x_{\text{new}},y)\}$, y ($y = 1$ initially) is the classification. Calculate the strangeness values of each sample and obtain the strangeness values sequence $\{\alpha_1, \ldots \alpha_m, \alpha_{new}\}$. Then calculate the P-value of the test sample based on formula (1). The α_i is the strangeness value about sample x_i. Set $y = y + 1$, repeat calculating the P-value until $y = c$. Lastly obtain the P-values sequence about the test sample. The largest P-value P_1 is the credibility and 1-P_2 is the confidence. P_1 corresponding class is the true label for the sample.

TCM gives the unlabeled samples label with confidence. Compared with the traditional machine learning algorithms which only giving the classification results, TCM enhances the output information and has distinct characteristics. The difference of the implementing TCM algorithm mainly is how to computing the strangeness values. The strangeness measure is the foundation of calculating the P-values, there are usual two methods to strangeness measure, one method is based on the support degree, which is measured by the Lagrange operator of each sample, named TCM-SVM [7]. Another method is based on the distance, such as combing the K-Nearest Neighbors and TCM, named TCM-KNN (Transductive Confidence Machine for K-Nearest Neighbors) [8].

3 Support Vector Data Description

Support Vector Data Description (SVDD) [9] is a data description method which inspiration comes from SVM (Support Vector Machine) proposed by Tax [10]. So like SVM, The SVDD is also based on statistical learning theory, introducing the nonlinear mechanism by the kernel method, and so on. Unlike the SVM, SVDD is a one-class classification method using the closed area. The following is a brief introduction of SVDD.

The main idea of SVDD is to envelop the samples within a high-dimensional space with the volume as small as possible. Given a data set $\{x_i\}(x_i \in R^d, i = 1, 2, \ldots, n)$, the mapping $\phi : X \to H$ is implicitly done by a given Kernel which compute the inter product in the feature space. The volume of the hyper-sphere with center a and radius R contains all or most of the data objects. The SVDD solves minimizing the radius R of the hyper-sphere problem:

$$\min \quad R^2 + C \sum_{i=1}^{n} \xi_i \tag{2}$$

Subject to the constraints:

$$\|\phi(x_i) - a\|^2 \le R^2 + \xi_i \quad \xi_i \ge 0, \quad i = 1, \ldots n \tag{3}$$

where ξ_i represent the slack variables which allow the possibility that some of the training examples can be wrongly classified. C is a variable which gives the trade-off between the two error terms: volume of the hyper-sphere and the number of target objects rejected. After incorporating the constraints (2) into the function (1) by Lagrange multipliers and have:

$$L = R^2 + C \sum_{i=1}^{n} \xi_i - \sum_{i=1}^{n} \alpha_i (R^2 + \xi_i - ||\phi(x_i) - a||^2) - \sum_{i=1}^{n} \lambda_i \xi_i \qquad (4)$$

where L should be maximized with respect to Lagrange multipliers $\alpha_i \geq 0, \lambda_i \geq 0$ and minimized with respect to R, a and ξ_i.

To test if a new object x is within the hyper-sphere, the distance to the center of the hyper-sphere has to be calculated. A test object x is accepted when this distance is smaller than the radius.

$$f(x) = k(x, x) - 2 \sum_{i=1}^{n} \alpha_i k(x_i, x) + \sum_{i=1}^{n} \sum_{j=1}^{n} \alpha_i \alpha_j k(x_i, x_j) \qquad (5)$$

In training data the samples have $\alpha_i \neq 0$ and are called the support objects. These samples are outside the hyper-sphere and are considered outliers. The rest of the training data is within the description.

Hyper-spheres are built for each class when extending SVDD to multi-classes classification. Each hyper-sphere contains almost all of labeled samples in the corresponding each class.

4 TC-SVDD

This paper presents a labeling sample algorithm which using transductive confidence mechanism based on the degree of support. Because this algorithm use SVDD as classification model, we call the algorithm TC-SVDD. The strangeness measure based on the level of support is the idea of support vector. Support vector thought comes from the optimal classification hyperplane. To solve the optimization problem need to use lagrange method to convert the problem into the dual problem. And each sample has a corresponding lagrange multiplier. This is an optimization problems in inequality constrained and has the only solution. The strangeness measure based the degree of support is using every sample lagrange multiplier as its strangeness value. Then using random detection function obtains the P- value.

Suppose the number of labeled samples sets is N, the number of unlabeled samples sets is M, the number of class is c. Set the confidence threshold for η. The TCM-SVDD is described as follows:

Step 1: Select a sample x_{new} at random from unlabeled sets. Make a new sample set by combing x_{new} and labeled sampled sets. The new set sets

Table 1 Benchmark datasets used in the experiment

Dataset	No. of samples	No. of feature	No. of class
Iris	150	4	3
Wine	178	13	3
Parkinsons	195	22	2
B_Cancer	683	10	2
Glass	214	9	7

down as $S = \{(x_1, y_1), \ldots, (x_n, y_n), (x_{new}, y)\}$, and the number of the set is $n + 1$;

Step 2: Calculate the P-value of x_{new} belonging to each class. Let $y = 1$.

① Calculate the lagrange multiplier α_i of every samples in new set S using SVDD. Obtain the strangeness value sequence $\{\alpha_1, \ldots \alpha_m, \alpha_{new}\}$;

② Calculate the P-value of x_{new} belonging to class y;

③ $y = y + 1$, repeat ① and ② until $y = c_o$.

Step 3: Obtain the P-value sequence about x_{new} belonging to every class. Sort the sequence from big to small order. The final P-value sequence is $\{P_1, P_2, \ldots, P_c\}$.

Step 4: Calculate the confidence of x_{new}, which is $1-P_2$. If $1-P_2 \geq \eta$, let P_1 corresponding category as the class label of x_{new}.

Step 5: Select another new sample from unlabeled set. Repeat step 1 to step 4 until all samples in unlabeled set have been calculate the algorithm is over.

TC-SVDD algorithm use limited information from small labeled samples to achieve labeling lots of unlabeled samples automatically and remove samples which not be clear on type of class.

5 Experiment Results and Analysis

In order to verify the effectiveness of the TC-SVDD algorithm, this paper programs the code including the dd_tools [11] based on Matlab 7.0. We present our experiment performed on 5 datasets in UCI repository as shown in Table 1. For simulating been labeled samples is very few in real applications, experiment control the number of training samples strictly.

Firstly, experiment observe the label effect of TC-SVDD when test samples set having outlier samples. The outlier samples may be the noises or the new class samples. A good label samples method is not only label the samples effectually but also throw away the outlier samples in time. The paper compares TC-SVDD with normal SVDD [12]. The two algorithms set the same parameter value $\xi = 5$, $C = 0.01$. And the confidence threshold η of TC-SVDD is 0.7. We use three

Table 2 The evaluation of normal SVDD and TC-SVDD

Data sets	SVDD			TC-SVDD		
	Precision	Recall	F-measure	Precision	Recall	F-measure
Iris	1	0.8	0.8889	1	1	0.9362
Wine	0.85	0.85	0.85	1	1	1
Parkinsons	0.6333	0.6667	0.6495	1	0.7308	0.8445
B_Cancer	0.82	1	0.9011	1	0.9615	0.9804
Glass	0.8333	0.7576	0.7936	1	0.8571	0.9231

Table 3 The train sets and test sets composing

Data sets	Target class 1		Target class 2		Outlier
	Train	Test	Train	Test	Test
Iris	30	20	30	20	20
Wine	30	20	30	20	20
Glass	50	20	50	20	17

measures, Precision, Recall, F-Measure, to evaluate the performance of TC-SVDD. The measures are expressed as follows:

$$\text{Precision} = TP/(TP + FP)$$

$$\text{Recall} = TP/(TP + FN)$$

$$F - \text{measure} = (2 * \text{Precision} * \text{Recall})/(\text{Precision} + \text{Recall})$$

where TP, TN, FP and FN stand for the numbers of "true-positives", "true-negatives", "false-positives", and "false-negatives", respectively.

We select 20 % samples from the first class of each datasets as target class to train the normal SVDD and the TC-SVDD. The test sets include 50 % samples from the first class and 50 % samples from the second class of each datasets. Experiments repeat 10 times and compute the average value. Table 3 gives the experiments results.

Table 2 shows the precision of TC-SVDD all are 1. That mean the labeled samples all are target and none outlier be labeled target because TC-SVDD have transductive confidence estimation. So the F-measure of TC-SVDD is higher than normal SVDD.

The last experiment compares normal SVDD and TC-SVDD on multi-class classification. The train set select the first class and the second class as the target. The test set adds the third class (this class never be learned by classifier) as outlier. This experiment not uses datasets Parkinsons and B_Cancer because they only fit two-class classification. Table 3 gives the train sets and test sets composing from each datasets. Table 4 shows the label result of normal SVDD and TC-SVDD.

Table 4 The label result of normal SVDD and TC-SVDD

Data sets	Test number	SVDD			TC-SVDD		
		No. of labeled class 1	No. of labeled class 2	No. of outlier	No. of labeled class 1	No. of labeled class 2	No. of outlier
Iris	60	19	24	17	20	19	21
Wine	60	17	11	32	18	15	27
Glass	57	25	34	-	19	17	23

From Table 4 we can see, on iris dataset SVDD select 17 samples from outlier to throw away and other 3 outlier samples be labeled target class. Different result on wine dataset that has 12 target samples be throw away with 20 outlier samples. On Glass dataset none be throw away and all outlier be labeled target class. And be labeled samples number are more than test number that because some samples be repeat labeled when decision surface of SVDD to the two target classes being covered. So the normal SVDD is instability.

The transductive confidence idea in TC-SVDD leads to only high confidence samples can be labeled. So that most samples from the two target classes can be labeled rightly and few target samples be throw away cause low confidence. All outlier added in each datasets be recognized rightly and be throw away. The experiment illuminate TC-SVDD method insure the labeling result have high quality.

6 Conclusions

This paper proposed an automatically samples label method based on SVDD with transductive confidence. The method label samples automatically at the same time gives the confidence about the label result and it insure label quality by setting the confidence threshold. Experiment on UCI data sets shows the TC-SVDD algorithm not only can label samples from lots of datasets when train sets being limited but also insure the label result are high quality.

References

1. Chao Deng, Maozu Guo (2008) Tri-training and data editing based semi-supervised clustering algorithm. J Softw 19(3):663–673
2. Li H (2012) Machine learning new trends: learning from the human-computer interaction. Chin Assoc Artif Intell 11:1–10
3. Liao Z, Zhao L, Hu G, Wang Q (2009) Semi-supervised learning based on one-class classifier. Pattern Recognit Artif Intell 22(6):924–930

4. Nouretdinov I, Gammerman A (2003) Transductive confidence machines is universal. In: Proceedings of the 14th international conference on algorithmic learning theory, London, pp 283–297
5. Barbara D, Domeniconi C, Rogers JP (2006) Detecting outliers using transduction and statistical testing. In: Proceedings of the 12th ACM SIGKDD international conference on knowledge discovery and data mining, New York, pp 55–64
6. Vovk V (2002) Asymptotic optimality of transductive confidecne machine. In: Proceedings of the 13th international conference on algorithmic learning theory, Lubeck, pp 336–350
7. Saunders C, Gammerman A, Vovk V (1999) Transduction with confidence and credibility. In: Proceedings of the 16th international joint conference on artificial intelligence, Stockholm, pp 722–726
8. Proedru K, Nouretdinov I, Vovk V, Gammerman A (2002) Transductive confidence machines for pattern recognition. In: Proceedings of the 13th European conference on machine learning, London, pp 381–390
9. Tax DMJ, Duin RPW (2004) Support vector data description. Mach Learn 54:45–66
10. Vapnik VN (1995) The naturn of statistical learning theory. Springer, New York
11. http://www-ict.ewi.tudelft.nl/~davidt/dd_tools.html
12. Mu T, Nandi AK (2009) Multiclass classification based on extended support vector data description. IEEE Trans Syst Man Cybern Part B: Cybernetics 39(5):1206–1216

Improvement of Soluble Solids Content Prediction in Navel Oranges by Vis/NIR Semi-Transmission Spectra and UVE-GA-LSSVM

Tong Sun, Wenli Xu, Xiao Wang and Muhua Liu

Abstract The objective of this research is to improve soluble solids content (SSC) prediction in navel oranges by visible/near infrared (Vis/NIR) semi-transmission spectra and uninformative variable elimination-genetic algorithm-least squares support vector machine (UVE-GA-LSSVM). Spectra of navel oranges were acquired using a QualitySpec spectrometer in the wavelength range of $350 \sim 1,000$ nm. After applying spectral pretreatment methods, UVE-GA was used to select variables, then LSSVM with three kernel functions (RBF kernel, linear kernel, polynomial kernel) was used to develop calibration models. The results indicate that Vis/NIR semi-transmission spectra combined with UVE-GA-LSSVM has good performance on assessing SSC of navel oranges, and SSC is improved. The R^2s and RMSEPs of SSC for RBF kernel, linear kernel, and polynomial kernel in prediction set are 0.850, 0.848, 0.849 and 0.419, 0.421, 0.420 %, respectively.

Keywords Vis/NIR semi-transmission · UVE-GA · LSSVM · Soluble solids content · Navel oranges

1 Introduction

Soluble solids content (SSC) is one of the important internal quality indices in navel orange (Citrus sinensis Osbeck) that match human's taste. Visible/near infrared (Vis/NIR) spectroscopy is a nondestructive, fast, and nonpolluting technique, and it has been used for SSC assessment in a variety of fruit [1–8]. Some research work reported the application of Vis/NIR spectroscopy on internal

T. Sun · W. Xu · X. Wang · M. Liu (✉)
College of Engineering, Jiangxi Agricultural University, 1225 Zhimin Street,
Nanchang 330045, People's Republic of China
e-mail: suikelmh@sina.com

Z. Wen and T. Li (eds.), *Knowledge Engineering and Management*,
Advances in Intelligent Systems and Computing 278,
DOI: 10.1007/978-3-642-54930-4_37, © Springer-Verlag Berlin Heidelberg 2014

quality assessment of navel orange. Liu et al. [9] measured the Vis/NIR diffuse reflectance spectra of navel oranges in the wavelength range of 350 ~ 1800 nm. They used partial least squares (PLS) and principal component analysis-back propagation neural network (PCA-BPNN) combined with several spectral pretreatment methods to develop calibration models for SSC. PCA-BPNN with multiplicative scatter correction (MSC) obtained the best results, the correlation coefficient and root mean square error of prediction (RMSEP) were 0.90 and 0.68 %, respectively. Xue et al. [10] divided the diffuse reflectance spectra of navel oranges (350 ~ 1800 nm) into 15 intervals, and used genetic algorithm (GA) to select optimal combination of spectral intervals. The correlation coefficient and RMSEP for SSC prediction by GA-PLS were 0.913 and 1.26 %, respectively. Liu et al. [11] used a portable NIR device to acquire diffuse reflectance spectra of "Gannan" navel oranges. PLS and least squares support vector regression (LSSVR) combined with wavelet transformed (WT) were used to develop calibration models for SSC. The results of LSSVR were better than that of PLS. The correlation coefficient and RMSEP for WT-LSSVR were 0.918 and 0.321 %, respectively.

It can be found that most of the research work has used diffuse reflectance mode to assess SSC of navel oranges. However, the pericarp of navel orange is medium thick, which weakens the transmission of light in the diffuse reflectance mode. Therefore, in such case, the diffuse reflectance spectra may only contain the information of internal quality near the pericarp. If the SSC values of pulp in navel oranges are different from inner to outside, the prediction accuracy of SSC by diffuse reflectance mode is limited due to insufficient information of inner pulp. While this problem is nonexistent for semi-transmission mode.

The objective of this research was to improve SSC prediction in navel oranges by Vis/NIR semi-transmission spectra and UVE-GA-LSSVM. Uninformative variable elimination (UVE) combined with GA was used to select sensitive wavelength variables. By using the selected sensitive wavelength variables, least squares support vector machine (LSSVM) was used to develop calibration models for SSC of navel oranges.

2 Materials and Methods

2.1 Samples

A total of 352 navel oranges from a local terminal market were used in this study. Before spectra measurement, navel orange samples were stored in the room temperature for about 24 h to be equilibrated. The samples were randomly divided into three sets: a calibration set (n = 176), a validation set (n = 88), and a prediction set (n = 88), with the number ratio of 2:1:1. In order to ensure the adaptability of the model, the calibration set contained the samples with the minimum and maximum values of SSC. All measurements were conducted within 6 days, about 60 samples per day. SSC and spectra of the same sample were measured on the same day.

2.2 Spectra Acquisition

Figure 1 shows the schematic diagram of Vis/NIR system used in this study. This system consists of a QualitySpec spectrometer (Analytical Spectral Devices, Inc., USA), two tungsten halogen lamps (15 V/150 W), a sample holder, an optical fiber and a computer. The QualitySpec spectrometer has two detectors: an Si detector (350 ~ 1,000 nm) and an InGaAs detector (1,000 ~ 1,800 nm). Only the Si detector was used in this study. The tungsten halogen lamps were placed on both sides of the fruit. The angle of lamp-fruit-detector was set as 90°.

Vis/NIR semi-transmission spectra of navel oranges were acquired by the Vis/NIR system. Spectrometer parameters setting, spectra data obtaining and storing were carried out via the Indic software v4.0 (Analytical Spectral Devices, INC., USA). A polytetrafluoethylene (PTFE) ball, with diameter of 80 mm, was used to acquire the reference spectra prior to sample spectra acquisition. The integration time and scan number for reference spectra were 17.41 s and 1, respectively. The measurement was expressed as log (1/T). T was percent transmittance of sample spectra. The integration time and scan number for samples were 2.08 s and 1, respectively.

2.3 SSC Measurement

After spectra acquisition was done, the pulp of each sample was macerated using a manual fruit squeezer. The extracted juice was filtered, and SSC value of the filtered juice was measured by a hand-held refractometer (PR-101α, Atago, Co., Tokyo, Japan).

2.4 Data Analysis

In this research, several spectral pretreatment methods such as MSC, standard normal variate correction (SNV), and four smoothing methods were used, and the best spectral pretreatment method was chosen according to the model performance. SNV and MSC can correct both additive and multiplicative scatter effects, and eliminate the multiplicative interferences such as particle size, scatter, and the change of light distance [12, 13]. Smoothing methods can smooth and remove

Fig. 1 Schematic diagram of
Vis/NIR system

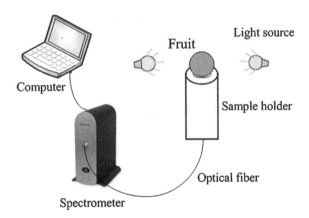

random noise from Vis/NIR spectra [14]. Spectral pretreatment methods were conducted in the Unscrambler software X 10.1 (CAMO AS, Trondheim, Norway).

UVE-GA was used to select sensitive wavelength variables from spectra after pretreated by the best spectral pretreatment method. UVE was applied first, and followed by GA accordingly. In the UVE analysis, 351 random variables (noise matrix) were added to spectral data, and the cut-off threshold was set as 99 % of maximal absolute value of noise variable. The detailed principle of UVE can be found in the literature [15]. In the GA analysis, the parameters were set as follows: population size (30 chromosomes), autoscaling, maximum number of components (15), probability of mutation (1 %), probability of cross-over (50 %), and number of runs (100). The detailed principle of GA can be found in the literature [16, 17]. The UVE and GA were conducted in the Matlab 7.6.0 (The MathWorks, Inc., USA).

For the SSC calibration model development, LSSVM with three kernel functions (RBF kernel, linear kernel, polynomial kernel) was used with selected wavelength variables. LSSVM that developed by Suykens et al. [18] is an optimized algorithm based on standard support vector machine (SVM). The detailed information of SVM can be found in [19]. In the LSSVM, simplex technique and leave one out cross-validation were applied to find the optimal parameter values. LSSVM was conducted in the Matlab 7.6.0 (The MathWorks, Inc. USA).

3 Results and Discussion

3.1 Quality Attribute Distribution

The mean and standard deviation (S.D.) of SSC of all navel orange samples were 12.42 % and 1.04, respectively. Details of SSC statistic parameters for calibration, validation, and prediction sets were presented in Table 1.

Table 1 Means, ranges, and S.D.s of navel orange samples of calibration, validation and prediction sets

Parameter	Data set	Samples	Mean	S.D.	Range
SSC (%)	Calibration	176	12.46	1.06	8.6 ~ 14.9
	Validation	88	12.45	0.95	9.5 ~ 14.7
	Prediction	88	12.27	1.08	9.4 ~ 14.5

3.2 Spectral Pre-treatment Analysis

Figure 2 shows the Vis/NIR spectra of all unpeeled navel oranges. According to the spectra, in the wavelength range of 600 ~ 950 nm, valleys appear at about 700 and 810 nm, and peaks appear at about 770 and 950 nm which correspond to the absorption peaks associated with a third overtone stretch of CH and second and third overtones of OH [20]. Therefore the wavelength range of 600 ~ 950 nm should contain information of internal qualities for SSC, and hence be useful for model calibration. Besides, there are noises on the wavelength range below 600 nm and above 950 nm, which leads to low signal/noise ratio. So only spectra in the wavelength range of 600 ~ 950 nm were used for further analysis.

Several spectral pretreatment methods such as MSC, SNV, and four smoothing methods were applied on original spectra of unpeeled navel oranges. PLS was used to develop calibration models of SSC in the wavelength range of 600 ~ 950 nm. Table 2 shows the results of PLS regression of SSC based on different pretreatment methods. For pretreatment methods of moving average (MA) and gaussian filter (GF), the results of PLS are deteriorated gradually with the increasing of segment size. For pretreatment method of median filter (MF), the results of PLS are improved gradually when the segment size increases from three to seven. In order to find the optimal segment size for MF, the segment size increased until the results are deteriorated. The optimal segment size for MF is nine. Among the four smoothing methods, the results of PLS are better when it is combined with MF than with others. And PLS with MF(9) obtains the best results among these pretreatment methods. The coefficient of determination (R^2) and root mean square error (RMSE) in the calibration, validation and prediction sets are 0.865, 0.798, 0.841 and 0.389, 0.423, 0.433 %, respectively.

3.3 UVE-GA

After spectra were pretreated by MF (9), the best pretreatment method in this study, UVE-GA, was used to select sensitive wavelength variables.

For UVE analysis, in order to avoid model overfitting, different latent variable (LV) numbers of PLS models were compared in the process of UVE [21]. The number of LVs in the optimal full-spectrum PLS model was chosen as the maximum number of LVs. Thus the numbers of LVs was ranged from 1 to 11. After

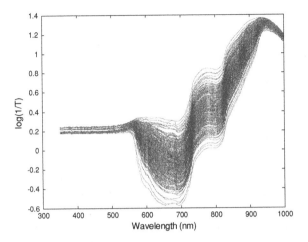

Fig. 2 Vis/NIR semi-transmission spectra of all unpeeled navel oranges

Table 2 Results of PLS regression of SSC of navel oranges at different pretreatment methods

Pretreatment method	Factors	Calibration		Validation		Prediction	
		R^2	RMSEC	R^2	RMSEV	R^2	RMSEP
Original	11	0.867	0.386	0.801	0.420	0.832	0.445
MSC	13	0.914	0.311	0.828	0.390	0.798	0.498
SNV	11	0.866	0.387	0.809	0.411	0.823	0.454
MF(3)	11	0.866	0.388	0.799	0.422	0.832	0.444
MF(5)	11	0.865	0.389	0.797	0.424	0.837	0.438
MF(7)	11	0.864	0.390	0.798	0.423	0.839	0.435
MF(9)	*11*	*0.865*	*0.389*	*0.798*	*0.423*	*0.841*	*0.433*
MF(11)	11	0.865	0.389	0.793	0.428	0.839	0.436
MF(13)	11	0.865	0.389	0.790	0.431	0.831	0.445

MSC multiplicative scatter correction; *SNV* standard normal variate correction; *MF* median filter; The number in bracket is segment size for four smoothing methods. Results of other three smoothing methods are not presented in this table

eliminating uninformative variables, leave-one-out cross-validation was performed on new spectral data, and the smallest root mean square error of cross-validation (RMSECV) value was obtained when 11 LVs were used. Figure 3 shows the stability of each variable in the PLS model with 11 LVs for SSC by UVE. The wavelength variables and random variables are separated by a solid vertical line. Two-dashed horizontal lines indicate the upper and lower cutoffs. The variables beyond the cutoffs are realized as informative variables and reserved, while the variables between the cutoffs are realized as uninformative variables and eliminated. After UVE analysis, 184 variables were selected from 351 variables for SSC.

After that, GA was conducted on the new spectral data in which uninformative variables were eliminated by UVE. Figure 4 shows the results of GA for SSC of navel oranges. Figure 4a is a frequency histogram of each selected variables. The

Fig. 3 Stability of each
variable in the PLS model
with 11 LVs for SSC by UVE

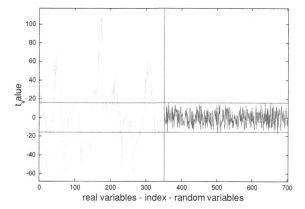

lower horizontal line indicates the cutoff for the model with the minimum
RMSECV. The upper horizontal line indicates the cutoff for the model that have
the fewest number of variables among all models which the difference between its
RMSECV and the minimum RMSECV is not significant according to an F test
($p < 0.1$). The variables above the cutoff are realized as informative variables.
From Fig. 4a, variables having higher frequency are mostly in the wavelength
range of 700 ~ 800 nm, and some of them are in the wavelength ranges of
600 ~ 650 and 850 ~ 900 nm. Figure 4b shows the changes of CV explained
variance with increasing number of variables included. At the first, the CV
explained variance ascends sharply as the number of variables increase from 1 to
7, which indicate at least seven variables should be included. After that, the CV
explained variance ascends slowly as the number of variables increase from 7 to
42. And then the CV explained variance is level off with further increasing number
of variables. Figure 4c shows the distribution of selected variables for SSC of
navel oranges. The lower line corresponds to the variables selected by the model
with minimum RMSECV, while the upper line corresponds to the variables
selected by the model with an F test RMSECV. The numbers of selected variables
for the models with minimum RMSECV and an F test RMSECV are 42 and 37,
respectively. From Fig. 4(c), it can be seen that most of the selected variables are
in the wavelength range of 700 ~ 800 nm, and some other selected variables are
in the wavelength ranges of 600 ~ 650 and 850 ~ 900 nm. This phenomenon is
accordant to the higher frequency distribution as shown in Fig. 4a. Xu et al. [22]
selected 190 variables for SSC of pears by GA. Most of selected variables were in
the wavelength range of 700 ~ 800 nm. These results agree with the finding in
this study. Figure 4d shows the changes of RMSECV value with increasing
number of variables included. The changing trend of RMSECV is inverse to the
trend of CV explained variance. The RMSECV plot drops rapidly as the number of
variables increase from 1 to 7, and then drops slowly as the number of variables
increase from 7 to 42. The RMSECV plot is level off finally with further increasing
number of variables.

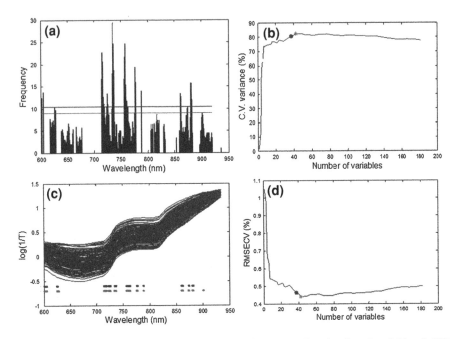

Fig. 4 Results of GA for SSC of navel oranges: **a** frequency of each selected variables; **b** CV explained variance; **c** distribution of selected variables; **d** RMSECV. The position of marker '*' and '•' indicate the minimum RMSECV and the F test RMSECV

3.4 LSSVM

Because the difference between the minimum RMSECV and F test RMSECV was not significant, the 42 variables selected by the model with minimum RMSECV were used to develop calibration models using LSSVM. In order to obtain the most suitable kernel function for SSC of navel oranges, three kernel functions (RBF kernel, linear kernel, polynomial kernel) were investigated in the LSSVM regression. Table 3 shows the results of LSSVM regression with different kernel functions for SSC of navel oranges. LSSVM models obtain better results than PLS models for SSC of navel oranges, and the model performances of three kernel functions are very close to each other, with the ranges of R^2 and RMSEPs in the prediction set 0.848 ~ 0.850 and 0.419 ~ 0.421 %, respectively. Compared to full-spectrum PLS, the number of variables used in the UVE-GA-LSSVM decreases from 351 to 42, while the R^2 in the prediction set increases from 0.841 to 0.850, and the RMSEP decreases from 0.433 to 0.419 %. Therefore, it can be stated that UVE-GA-LSSVM is an efficient method for SSC determination of navel oranges.

In order to compare the results of SSC of navel oranges with other research work, the RPD (standard deviation/RMSEP) which has been defined as a measure of model performance is introduced [23]. Compared to the results in reference [8],

Table 3 Results of LSSVM regression with different kernel functions for SSC of navel oranges

	Parameters			Calibration		Validation		Prediction	
	γ	$\sigma2/t$	Degree	R^2	RMSEC	R^2	RMSEV	R^2	RMSEP
RBF_kernel	1.5e8	4.8e4		0.879	0.370	0.810	0.414	0.850	0.419
lin_kernel	8.7e2			0.865	0.390	0.816	0.408	0.848	0.421
poly_kernel	2.3e-7	4.0e4	3	0.880	0.368	0.808	0.415	0.849	0.420

the results of SSC of navel oranges ($R^2 = 0.810$, RMSEP $= 0.680$ %, RPD $= 2.29$) are inferior to this study ($R^2 = 0.850$, RMSEP $= 0.419$ %, RPD $= 2.58$). The results of this study are superior to the results of SSC of navel oranges ($R^2 = 0.757$, RMSEP $= 0.470$ %, RPD $= 2.34$) in reference [9]. Compared to the results in reference [10], the results of SSC of navel oranges ($R^2 = 0.834$, RMSEP $= 1.258$ %, RPD $= 2.41$) are also inferior to this study. Diffuse reflectance spectra and wavelength range of 350 ∼ 1800 nm were used in the above research work, while semi-transmission spectra and wavelength range of 350 ∼ 1000 nm were used in this study. Besides, the same spectrometer was used in the above research work and our research work. Compared to our research work, the wavelength range of the above research work is much broader while the results are inferior, which may indicates that semi-transmission spectra is more suitable than diffuse reflectance spectra for assessing SSC of navel oranges. Besides, UVE-GA-LSSVM is suitable calibration method for SSC prediction.

4 Conclusions

The results demonstrate that Vis/NIR semi-transmission spectra combined with UVE-GA-LSSVM has good performance on assessing SSC of navel oranges. Compared to other similar research work, SSC prediction in our research work is improved. The results of UVE-GA-LSSVM are better than that of full-spectrum PLS. The R^2s and RMSEPs for RBF kernel, linear kernel and polynomial kernel in prediction set are 0.850, 0.848, 0.849 and 0.419, 0.421, 0.420 %, respectively.

Acknowledgment The authors gratefully acknowledge the financial support provided by the National Nature Science Foundation of China (No. 30972052), New Century Excellent Talents in Support of Ministry of Education Project (No. NCET090168) and Technology Foundation for Selected Overseas Chinese (2012).

References

1. Flores K, Sanchez MT, Perez-Marin DC et al (2008) Prediction of total soluble solid content in intact and cut melons and watermelons using near infrared spectroscopy. J Near Infrared Spec 16:91–98

2. Sun T, Lin HJ, Xu HR et al (2009) Effect of fruit moving speed on predicting soluble solids content of 'Cuiguan' pears (pomaceae pyrifolia nakai cv. cuiguan) using PLS and LS-SVM regression. Postharvest Biol Technol 51:86–90

3. Moghimi A, Aghkhani MH, Sazgarnia A et al (2011) Improvement of NIR transmission mode for internal quality assessment of fruit using different orientations. J Food Process Eng 34:1759–1774

4. Antonucci F, Pallottino F, Paglia G et al (2011) Non-destructive estimation of mandarin maturity status through portable Vis-NIR spectrophotometer. Food Bioprocess Technol 4:809–813

5. Bertone E, Venturello A, Leardi R et al (2012) Prediction of the optimum harvest time of 'scarlet' apples using Dr-UV-Vis and NIR spectroscopy. Postharvest Biol Technol 69:15–23

6. Jha SN, Jaiswal P, Narsaiah K et al (2012) Non-destructive prediction of sweetness of intact mango using near infrared spectroscopy. Sci Hortic 138:171–175

7. Tian HQ, Wang CG, Zhang HJ et al (2012) Measurement of soluble solids content in melon by transmittance spectroscopy. Sensor Lett 10:570–573

8. Liu YD, Sun XD, Ouyang AG (2010) Nondestructive measurement of soluble solid content of navel orange fruit by visible–NIR spectrometric technique with PLSR and PCA-BPNN. LWT-Food Sci Technol 43:602–607

9. Liu YD, Sun XD, Zhou JM et al (2010) Linear and nonlinear multivariate regressions for determination sugar content of intact gannan navel orange by Vis-NIR diffuse reflectance spectroscopy. Math Comput Model 51:1438–1443

10. Xue L, Li J, Liu M et al (2010) Nondestructive detection of soluble solids content on navel orange with Vis/NIR based on genetic algorithm. Laser Optoelectron Prog 47:123001

11. Liu YD, Gao RJ, Hao Y et al (2012) Improvement of near-infrared spectral calibration models for Brix prediction in 'Gannan' navel oranges by a portable near-infrared device. Food Bioprocess Technol 5:1106–1112

12. Isaksson T, Naes T (1988) The effect of multiplicative scatter correction (msc) and linearity improvement in nir spectroscopy. Appl Spectrosc 42:1273–1284

13. Barnes R, Dhanoa M, Lister J (1989) Standard normal variable transformation and detrending of near infrared diffuse reflectance spectra. Appl Spectrosc 43:772–777

14. Jamshidi B, Minaei S, Mohajerani E et al (2008) Reflectance Vis/NIR spectroscopy for nondestructive taste characterization of valencia oranges. Comput Electron Agric 85:64–69

15. Centner V, Massart DL, deNoord OE et al (1996) Elimination of uninformative variables for multivariate calibration. Anal Chem 68:3851–3858

16. Leardi R (2000) Application of genetic algorithm-pls for feature selection in spectral data sets. J Chemometr 14:643–655

17. Leardi R, Gonzalez AL (1998) Genetic algorithms applied to feature selection in pls regression: how and when to use them. Chemometr Intell Lab 41:195–207

18. Suykens JAK, Van Gestel T, De Brabanter J et al (2002) Least squares support vector machines. World Scientific, Singapore

19. Vapnik V (1995) The nature of statistical learning theory, 2nd edn. Springer, New York

20. Shao YN, Bao YD, He Y (2011) Visible/near-infrared spectra for linear and nonlinear calibrations: a case to predict soluble solids contents and pH value in peach. Postharvest Biol Technol 4:1376–1383

21. Wu D, He Y, Nie PC et al (2010) Hybrid variable selection in visible and near-infrared spectral analysis for non-invasive quality determination of grape juice. Anal Chim Acta 659:229–237

22. Xu HR, Qi B, Sun T et al (2012) Variable selection in visible and near-infrared spectra: application to on-line determination of sugar content in pears. J Food Eng 109:142–147

23. Williams PC, Sobering DC (1993) Comparison of commercial near infrared transmittance and reflectance instruments for analysis of whole grains and seeds. J Near Infrared Spec 1:25–32

An Ontology-Based Domain Modeling Framework for Knowledge Service in Digital Library

Wei Wang, Yin Zhang, Baogang Wei and Yiming Li

Abstract In Digital Library, information is often stored in unstructured and semi-structured textual form. Domain modeling techniques are used for specific domain knowledge services in order to make use of the massive amounts of textual information. Ontology is an efficient way to build specific domain model for abstraction of concepts and relations. In this , we propose an ontology-based domain modeling framework, CADAL-ODM, to structure domain model. CADAL-ODM is designed to provide multi-level and multi-granularity knowledge services in order to meet the various requirements of users in digital library instead of basic reading service. We explain our work by which knowledge is extracted automatically from the unstructured and semi-structured documents. Specific domain model is structured by specific domain ontology description and extracted knowledge. We evaluate the framework with real data sets, specifically, the medical records of TCM documents set from CADAL.

Keywords Domain modeling · Ontology · Textual analysis · Association rule · Digital library · TCM

W. Wang · Y. Zhang · B. Wei (✉) · Y. Li
College of Computer Science and Technology, Zhejiang University,
Hangzhou 310027, China
e-mail: wbg@zju.edu.cn

W. Wang
e-mail: stephenweiwang@gmail.com

Y. Zhang
e-mail: yinzh@zju.edu.cn

Y. Li
e-mail: liyiming90@163.com

Z. Wen and T. Li (eds.), *Knowledge Engineering and Management*,
Advances in Intelligent Systems and Computing 278,
DOI: 10.1007/978-3-642-54930-4_38, © Springer-Verlag Berlin Heidelberg 2014

1 Introduction

With the advent of the era of Big-Data, the explosive growth of textual and other data resources is touched in every part of social life, such as Web, Digital, Libraries, etc. A myriad of techniques including information retrieval, document classification and clustering, domain ontology and information visualization have been developing to facilitate extraction and understanding of information and knowledge which is embedded in textual documents. However, taking the ambiguity and complication of natural language textual documents into consideration, information and knowledge extraction techniques play an important role in various text analysis fields of the recent past. For example, the numerous and diverse textual documents in the digital libraries are both challenges and opportunities for information and knowledge extraction research. Domain information is founded on a massive archive of knowledge in publications, research articles, case studies, and reviews in the digital libraries. The extraction of the information and knowledge from them is highly labor intensive. Compared with the management of traditional database, semantic comprehension and knowledge extraction from collections of textual documents automatically made it possible to provide multi-level and multi-granularity knowledge services in order to meet the various requirements of users in digital library.

In this chapter, we have proposed the design of an ontology-based domain modeling framework, CADAL-ODM, for conceptualization of domain information and specialization of user query and retrieval. The ontology-based framework focuses on the key concepts and latent correlativity among the concepts.

We presented and realized the framework with real data set, specifically, the medical records of traditional Chinese medicine (TCM) documents set from China Academic Digital Associative Library (CADAL) [1]. The medical records of TCM are the records of clinical diagnosis and treatment and TCM have lasted for more than thousands of years. Hence, these accumulated TCM medical records are extremely chaotic. However, the complicated and obscure relations of medications, symptoms, and treatments which are contained in the unstructured and semi-structured documents are invaluable resources for the heritage and development of TCM. Therefore, we choose the above document sets to realize and evaluate our framework.

The main contributions of our approach can be summarized as follow: (1) We have shown that ontology-based framework could properly reflect the knowledge extracted from natural language textual documents; (2) With the ontology reasoning and inference, the framework could infer unknown information based on existing information; (3) The evaluated result shows that multi-level and multi-granularity knowledge services could be provided with our work in order to meet the various requirements of users in digital libraries instead of basic reading service.

2 Background and Related Works

As an important knowledge services platform, digital libraries could provide multi-level and multi-granularity service with related techniques. Ontology is widely used in various knowledge services for it provides methods to deal with different sources of information and knowledge. Knowledge extraction is the basis of knowledge services while information visualization is vital for conveying information effectively and positive user experience. Therefore, it is necessary for the digital libraries to apply above techniques in the knowledge services.

There are several related works that have been suggested toward the problem of domain modeling in the textual documents and digital libraries. Liu et al. [2] summarized the application of ontology in the digital libraries as four respects: knowledge representation, metadata mapping, computing supported cooperative work (CSCW), and data mining. By data modeling, ontology-based knowledge services could be provided in the digital library so as to meet the people's needs of knowledge. Hu et al. [3] used ontology to express semantic information in digital libraries. Two interconnected ontological classes have been structured to describe users' interests and knowledge resources. Therefore, users' demands could be reflected to some extent which have reference value for knowledge services. Bhujade and Janwe [4] provided a completed system, EART, to extract effective association information and knowledge in textual documents. The EART system performs an efficient algorithm for association rule mining for complicated natural language textual documents. Lin et al. [5] proposed a domain ontology development procedure to extract knowledge from domain handbooks. Moreover, the procedure could be used for domain-specific modeling from domain information data sets.

Above researches are of great academic values. However, our work is different from the researches above. Our work is focused on the completed process from unstructured digital books to efficient and convenient knowledge services. As faced with huge amounts of resources, how to provide an advanced knowledge service is the future development orientation of digital library. This chapter will apply ontology modeling method with information and knowledge extraction and retrieval techniques to dispose information and knowledge of different sources in order to provide multi-level and multi-granularity knowledge services in digital library.

3 CADAL-ODM Architecture

In this section, we present the complete architectural detail of CADAL-ODM which consists of four phases: Text Preprocessing Phase, Association Rule Mining Phase, Ontology Modeling Phase, and Application Phase (as shown in Fig. 1). The design and working principles of above phases are presented in the following subsections.

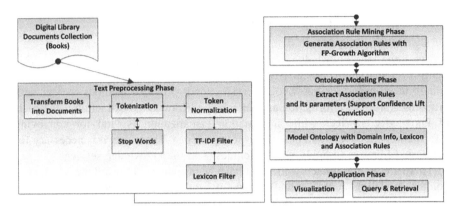

Fig. 1 CADAL-ODM architecture

3.1 Text Preprocessing Phase

The text preprocessing phase is aimed at optimizing the performance of the latter phases.

3.1.1 Transform Books into Documents

Our data set, the medical records of TCM documents set, comes from CADAL. The medical records of TCM are derived from books, clinical cases and Web, etc. And physical resources have been digitalized by related techniques such as: scanning and OCR [6] including books, clinical cases, etc. As the multi-granularity of these resources, we first need to transform these resources into medical records, specifically, the books.

Taking "The Quintessence of TCM Cases" as an example, it's a series of six volumes TCM clinical cases. In order to unify data granularity, regular expression matches have been used to transform the books into documents. Although the clinical cases in "The Quintessence of TCM Cases" contain lots of information, each clinical case begins with Chinese characters number and ends with editors' comments. Therefore, the regular expression could be structured as follow (the Chinese characters are translated into UTF-8 in the real code):

([一二三四五六七八九十]、(.*)\[按语\])

Its corresponding English expression is:

([1234567890] (.*)\[editors' comments\])

With the regular expression, the books have been segmented into TCM medical records; each TCM medical record is an entire medical diagnosis and treatment record, contains the patient's personal information with the first name withheld to protect patient privacy, signs, and symptoms, medical diagnosis and treatment, etc.

3.1.2 Tokenization and Related Operations

Tokenization and related operation is used to segment Chinese characters in the TCM medical records and optimize the segmentation results. As a matter of fact, Chinese word segmentation and syntax analysis is difficult than several other languages for its close correlation with the context and there is no space between words. Usually, there are various errors when segmenting the Chinese words which significantly degrade the quality of tokenization. As our goals are extracting related subjects around medicines, prescriptions and symptoms and diseases, we are not concerned with some of the details about diagnosis and treatment. Therefore, as domain lexicons play an important role for extraction in multiple phases, we first have to construct domain lexicons. The domain lexicons are based on Chinese Pharmacopoeia, related National Standard and various kind of Local or Hospital Standard, etc. The specific domain lexicons are used in both text pre-processing and ontology modeling phases which will be described in latter sections.

Tokenization is achieved through the specific domain lexicons to segment the words in TCM medical records. With the domain lexicons, the quality of tokenization is significant improved, especially for key words. Meanwhile, stop words have been eliminate from the segmentation results including meaningless words, void value words, etc. A series of related operations are applied to optimize the tokenization results including token normalization, stemming, or lemmatization to establish equivalent relation between different tokens.

3.1.3 Filters

The filters contain frequency filter and lexicon filter using TF-IDF and specific domain lexicon, respectively. The tokenization results are filtered by removing the unimportant or unconcerned words from the tokenized content. Specifically, TF-IDF filter focuses on the distribution of words while lexicon filter focuses on whether the words are domain vocabulary or not. After the filter process, the TCM medical records have been transformed into significative sequences of words which largely maintained the topics of original documents.

Part of the result of tokenization and filters for an example of TCM medical records are shown in Table 1. We can see only the words which are related to diagnosis and treatment are shown on the left side in Table 1. To make the result in Table 1 easier to follow, we translate the words into English on the right side.

3.2 Association Rule Mining Phase

The association rule mining phase describes a general method for extracting association rules from document sets automatically.

Table 1 Result of
tokenization and related
operations for an example of
TCM medical records

Token	Corresponding english version
盗汗	Night sweat
持续高热	Persistent fever
抽搐	Hyperspasmia
竹叶	Bamboo leaf
偏瘫	Hemiplegia
腹胀	Abdominal distension
黄岑	Radix scutellariae
…	…

3.2.1 Association Rule Mining and Related Parameters

In data mining and related fields, association rule mining is a popular and well-researched method for discovering interesting patterns in large data sets. It is intended to identify strong association rules discovered in data sets using different measures of interestingness [7]. Association rule is used to express these positive or negative relations which indicate things appearing simultaneously or not.

According to the original definition by Agrawal et al. [8, 9], the problem of association rule mining is defined as follow: Let $I = \{i_1, i_2, …, i_n\}$ be a set of n binary attributes called items. Let $D = \{t_1, t_2, …, t_n\}$ be a set of transactions called data set. Each transaction in D has a unique transaction ID and contains a subset of the items in I. Association rule is defined as an implication in the form of $X \Rightarrow Y$ where $X, Y \subseteq I$ and $X \cap Y = \emptyset$.

The related parameters of association rule mining including Support, Confidence, Lift and Conviction. Each parameter is brief introduced in the following:

- **Support**: The support $\text{supp}(X)$ of an item set X is defined as the proportion of transactions in the data set which contains the itemset.
- **Confidence**: The confidence $\text{conf}(X \Rightarrow Y)$ of a rule is defined as $\text{conf}(X \Rightarrow Y) = \text{supp}(X \cup Y)/\text{supp}(X)$.
- **Lift**: The lift $\text{lift}(X \Rightarrow Y)$ of a rule is defined as the ratio of the observed support to that expected if X and Y are independent or $\text{lift}(X \Rightarrow Y) = \text{supp}(X \cup Y)/\text{supp}(X) \times \text{supp}(Y))$.
- **Conviction**: The convention $\text{conv}(X \Rightarrow Y)$ of a rule is defined as the ratio of the expected frequency that X occurs without Y, if X, and Y were independent divided by the observed frequency of incorrect predictions or $\text{conv}(X \Rightarrow Y) = (1-)\text{supp}(Y))/(1-\text{conf}(X \Rightarrow Y))$.

3.2.2 FP-Growth Algorithm

FP-Growth algorithm [10] is based on a frequent pattern tree (FP-tree) [10] structure, which is an extended prefix-tree structure for storing compressed, crucial information about frequent pattern. With the structure of FP-tree, the transactions could be compressed and do not need to generate myriads of candidate sets.

To illustrate the result of association rule mining, we interpret some of the association rules. For example, the first association rule is shown as follow:

[柴胡 = t, 小柴胡汤 = t]:49 ⇒ [黄芩 = t]:47

conf:(0.96) *lift*:(5.1) *lev*:(0.01) *conv*:(13.26)

As the above example shown, the association rule of the example is {radix bupleuri, Xiaochaihu decoction} ⇒ {radix scutellariae}. Moreover, radix bupleuri and Xiaochaihu decoction appear simultaneously 49 times while radix scutellariae appears 47 times in the data set of TCM medical records where *conf* = 0.96, *lift* = 5.1, *conv* = 13.26 and *supp* could be calculated by the times of items appeared and the total amount of transactions.

3.3 Ontology Modeling Phase

The ontology modeling phase is intended to modeling the knowledge model so as to make use of the valuable knowledge and principle of TCM extracted in former phases.

3.3.1 Fetch Effective Information

As we illustrated in Sect. 3.1, fetch effective information from the former results of association rules and its parameters also use regular expression matches which transform books into documents. The only difference is that the regular expression should be structured as required. Therefore, association rules and its parameters could be transformed into the format which is facilitated for knowledge modeling.

3.3.2 Ontology Modeling

Domain ontology which consists of both concepts and entities, along with their relations and properties, is a new medium for the storage and propagation of domain-specific knowledge [11]. TCM ontology which we built defines the layered concepts of TCM and the relations among the concepts.

When TCM medical records are used as experimental data set in the CADAL-ODM, three basic classes consist of medicine, prescription and symptom-disease, and five basic properties consist of conjugation, similarity, sharing, and function basesd on the internal and external relationships among the classes have been established in order to reflect the obscure and complex relations among the TCM medical records. The properties of conjugation, similarity, and sharing are internal relations in the symptom- disease, prescription and medicine respectively. And the property of function describes the relation from class prescription to class symptom and disease while the property of bases describes the relation from class prescription to class medicine. The five basic properties could properly reflect the basic relations in TCM. Moreover, the parameters of association rules and the URI of the original documents set have been also add the TCM ontology in order to measure the strength of the association.

The Protégé [12] API has been used in the CADAL-ODM for ontology modeling.

3.4 Application Phase

The application phase describes the application of the CADAL-ODM including knowledge visualization and knowledge services in multi-level and multi-granularity instead of basic reading service in digital library.

In specific applications, the ontology-based domain model which is extracted from domain data sets provides domain specific knowledge service. With ontology application tools, the domain modeling result could be used for knowledge query and retrieval which the entities, relations, parameters and URIs could be effectively used. The knowledge network which is generated by domain model and users' query is also provided in the knowledge service. The knowledge network is an important tool in knowledge service and management. The entities of domain model are linked to the detailed description in CADAL. And the URIs in the domain modeling result is linked to the index documents, original TCM documents and original books so as to provide multi-level and multi-granularity for the digital library.

4 Experimental Evaluation

We evaluate the framework CADAL-ODM with real data sets, specifically, the medical records of TCM documents set from CADAL. According to our data set of TCM medical records, an extra sublayer is added between the Model and Meta Data layer so as to transform books into documents. In the digital library CADAL, there are over ten thousand medicines, one hundred thousand prescriptions and five thousand symptoms and diseases. However, only a small part useful data appear in the TCM medical records document sets. In order to optimize the performance of CADAL-ODM, only useful entities and properties have been added into the TCM domain model.

Figure 2 shows the query and visualization result of the disease Alzheimer. In Fig. 2, the disease Alzheimer is shown as "痴呆" in Chinese. The relations in blue connect the entities of the class prescription and symptom-disease. The relations in wine are of the property function while the entities connected by the relations in wine are of the class prescription. The relations in yellow are of the property

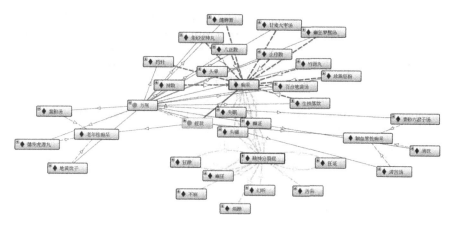

Fig. 2 Example of the disease Alzheimer

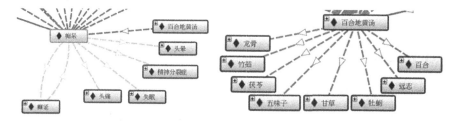

Fig. 3 Corresponding relationship in the linked original TCM documents of Alzheimer

conjugation while the entities while the entities connected by the yellow relations are of the class symptom-disease.

In the Fig. 3, the disease Alzheimer and other symptoms are shown in green and entities connected by relations in yellow including dementia, schizophrenia and insomnia, etc. The composition of lily bulb and rehmannia decoction which is shown in red consists of lily, concha ostreae and liquorice, etc. which are connected by relations in wine. And the URIs in the domain modeling result could link to original TCM documents.

As Figs. 2 and 3 shown, the CADAL-ODM could properly reflect the knowledge extracted from natural language textual document. Moreover, multi-level and multi-granularity knowledge services could be provided with our work in order to meet various user requirements for digital library.

5 Conclusion and Future Work

This chapter has proposed the design of an ontology-based domain modeling system, CADAL-ODM, to model domain knowledge from specific domain data sets. The system is a domain-independent framework so it is flexible to apply to arbitrary domains in digital library. The system uses linguistic and semantic analysis to extract knowledge from the unstructured and semi-structured documents sets. As the result of experimental evaluation shown, our ontology-based system could properly reflect the domain knowledge and its complicated and obscure relations which are extracted from natural language textual documents. Moreover, CADAL-ODM is designed to automatically model-specific domain ontology so as to provide multi-level and multi-granularity knowledge services in the digital library instead of basic reading service.

Since the research is still in progress, the order information of token sequences, complicated concepts, and relations of TCM ontology and complex ontology reasoning and inference have been omitted here [13, 14]. However, the preliminary results indicated a promising future of our work. The omitted parts will be implemented in our future work.

Acknowledgments This research was supported by Chinese Knowledge Center of Engineering Science and Technology (CKCEST) and China Academic Digital Associative Library (CADAL).

References

1. China Academic Digital Associative Library (CADAL), Zhejiang University, http://www.cadal.cn
2. Liu W, Li D, Xia C (2004) Ontology-based metadata application for digital libraries. Libr J 157(6):120–125
3. Hu C, Zhao Y (2007) An ontology-based framework for knowledge service in digital library. In: International conference on wireless communications, networking and mobile Computing. WiCom 2007, pp 5345–5348
4. Bhujade V, Janwe NJ (2011) Knowledge discovery in text mining technique using association rules extraction. In: International conference on computational intelligence and communication systems. CICN 2011, pp 498–502
5. Lin H-T, Hsirh S-H, Chou K-W, Lin K-Y (2009) Construction of engineering domain ontology through extraction of knowledge from domain handbooks. Comput Civ Eng, pp 207–216
6. Schantz, Herbert F(1982) The history of OCR: optical character recognition, [Manchester Center, Vt.]: Recognition Technologies Users Association, ISBN-9780943072012
7. Piatetsky-Shapiro G (1991) Discovery, analysis, and presentation of strong rules. In: Piatetsky-Shapiro G and Frawley WJ (eds) Knowledge discovery in databases. AAAI Press, Menlo Park, pp 229–248
8. Agrawal R, Imielinski T, Swami A (1993) Database mining: a performance perspective. IEEE Trans Knowl Data Eng 5:914–924
9. Agrawal R, Imielinski T, Swami A (1993) Mining association rules between sets of items in large databases. In: Proceedings of ACM SIGMOD international conference management of data. Washington, pp 207–216
10. Han J, Pei J, Yin Y (2000) Mining frequent patterns without candidate generation. In: Proceedings of ACM SIGMOD international conference on management of data (SIGMOD'00). Dallas, pp 1–12
11. Hsieh S-H, Lin H-T, Chi N-W, Chou K-W, Lin K-Y (2011) Enabling the development of base domain ontology through extraction of knowledge from engineering domain handbooks. Adv Eng Inf 25(2):288–296

12. Protégé, Stanford University, http://protege.stanford.edu
13. Ying G, Jun W, Xiao-yan Z, Meihong Y (2011) Domain service acquisition and domain modeling based on feature model. In: 2011 IEEE 14th international conference on computational science and engineering (CSE), pp 26–33
14. Rong P, Xiaozhen Z (2008) Domain model evolutionary approach based on semantic association. In: 2008 international conference on computer science and software engineering, pp 239–243

A New Two-Dimension Model of Evaluating the Quality of Distance Education

Jiang-Chun Song and Yan Wang

Abstract This paper established and designed evaluation index system based on "Information Technology application fields—service quality promotion" two dimensions model. The Information Technology application fields included academic and non-academic student support services. The service quality promotion included tangibles, reliability, responsiveness, assurance and empathy. This two dimensions framework evaluation method is helpful to understand the performances on specific aspects of student support services, and then do targeted improvement for the distance educational institutions.

Keywords Distance education · Two-dimension model · Quality of service · Evaluating

1 Introduction

With the rapid development of distance education, the problems caused by lacking of student support services and declining of student satisfaction are more and more prominent. With the continuous development and wide application of Information Technology (IT), our real world is influenced and changed by digital network profoundly, and gradually formed a new social structure—the network society. Distance education students almost no more than 30 years old, can use IT skillfully, so it is a great significance for improving student satisfaction and teaching quality to improve distance education student support services system by the combination of IT and student support services.

Researches in student support services are gradually refined, and many modern service ideas and management methods are applied to student support services.

J.-C. Song (✉) · Y. Wang
Information and Engineering School, Shenzhen Open University, Shenzhen, China
e-mail: 563794919@qq.com

Z. Wen and T. Li (eds.), *Knowledge Engineering and Management,*
Advances in Intelligent Systems and Computing 278,
DOI: 10.1007/978-3-642-54930-4_39, © Springer-Verlag Berlin Heidelberg 2014

Parasuraman et al. [1] who studied perceived service quality deeply, established SERVQUAL perceived quality evaluation method from which five dimensions of service quality was established. Those five dimensions are reliability, responsiveness, empathy, assurance and tangibles. SERVQUAL perceived quality evaluation method has been widely used in educational fields. For example, Fan and Jin [2] gave a case study for the evaluation of distance study support services quality from three aspects, that is study resource services, study process services and technology support services. To a certain extent, these researches prove that it is practical to evaluate distance education quality through common services and customer satisfaction theory. And some researchers investigate the satisfaction of service quality through case studies. Lee [3] used factor analysis and structural equation model techniques to research on the online education of Korean and American students. He found that perception of online support service quality was a significant predictor of online learning acceptance and satisfaction. Helgesen and Nesset [4] emphasized that service quality affected student loyalty through a case study of a Norwegian university college. Yukselturk and Yildirim [5] realized that interaction, online support, course structure and flexibility were the contributing factors to student satisfaction in an online certificate program.

Although more scholars using SERVQUAL [1] have studied student support services from different points of view, their studies, without clearing the core areas of student support services, still focus on superficial understanding of student support services, and there are some defects in the evaluation of student support services quality. This article will explore the enhancing role of IT on the quality of student support services in two aspects including academic student support services and non-academic student support services. Academic student support services and non-academic student support services proposed by Ormond Simpson [6] is well recognized by academia. Academic student support services which will run through the whole learning process, including teaching, learning, counseling and evaluating, is to provide all kinds of helps to distance learners to help them solving academic problems; non-academic student support services which mostly focus on the management and emotional aspects, is to provide helps to distance learners to help them solving all problems except academic problems. In short, it is clearly that we can understand the enhancing role of IT on the quality of student support services if we evaluate the quality of student support services both from academic and non-academic aspects.

2 IT Application in the Distance Education Student Support Services

With the rapid development of distance education promoted by the development of IT, It has brought profound impacts on academic student support services and no-academic student support services in distance education, such as shortening teaching time, expanding teaching scope, and etc. These changes have impacted on the distance teaching model, changing from teacher-centered learning and course-centered learning to student-centered learning.

2.1 Academic Student Support Services

With the rapid development of IT, new information technologies have integrated into distance education. It is mainly used in tracking students' learning process, diversified learning resources, interactive teaching and learning for teachers and students with IT.

2.1.1 Tracking Students' Learning Process

Whether in traditional education or in distance education, students analysis is the beginning of teaching. Since a student registers into distance education, we should analyze student's learning motivation, initial educational level, study style and objective. Then we track the student's learning process, get to know his situation, and give appropriate learning strategy to him. This is the first step of distance education, also is the foundation of student support services. Furthermore, it is an essential factor for ensuring study quality of distance education and raising the rate of register. By the help of IT, it is easier to form students' learning archives. With the support of Management Information System (MIS) and technology of intelligent analysis, we can track students' learning process, alarm the absence of student and provide list of students who rarely participate in teaching. In addition, teacher also can help students solving their learning problems, encourage them continuing to study.

2.1.2 Diversified Learning Resources

Learning resources such as textbooks and courseware, which is master factor of course, have played an important role in distance education. To develop high quality of diversified learning resources is a basic work of distance education, also is a necessary condition for ensuring the quality of distance education. Diversified learning resources developed with IT, such as network courses, network videos, network audios, CAI, online live classrooms, electronic course documents, digital books, digital journals, little multimedia courseware, and etc., have been widely used in distance education to fulfill all kinds of study requirements of different students.

2.1.3 Interactive Teaching and Learning for Teachers and Students with IT

Teaching counseling, a study service related with student studying courses, is the most fundamental and important service among all services of student support. It provides teaching counseling services and teaching guides to students helping them study courses, its basic features include: ① explanation, counseling and answering of course content, especially inspiration and guidance for solving practical problems with discipline theory; ② common study method guidance related with course discipline and course content; ③ correction of assignment, evaluation of test and guidance of exam; ④ guidance for experiment, other practical teaching and training project; ⑤ organization for class discussion and collaborative learning; ⑥ other teaching helps related with course study and etc.

Interactive teaching and learning for teachers and students was powerfully supported by IT. With the support of computer and multimedia technology, teachers can provide much more information to students in class and stimulate students' interest on the course; students now can finish their homework and tests online with the help of development of software technology; students can do their experiments online by using virtual simulation technology; instant message technologies, such as tencent QQ, MSN, OICQ, fetion and etc., make it possible for teachers and students to communicate online without restrictions. At the same time, with the support of computer network technology, teachers and students can discuss on BBS provided by school's teaching platform, all discussions will be saved on the server for everyone authorized reviewing, which formed a interactive platform both for teachers and students.

2.2 Non- Academic Student Support Services

Non-academic student support services put forward by Ormond Simpson [6] are as following: consulting, providing information, explaining questions, giving advices; evaluating, giving feedback on students' abilities and skills of non-academic; assisting, giving practical helps and assistances to improve study and work efficiency of students; proposing, making sample case of economical support, writing a recommendation; improving, benefiting students from innovation of educational institutions; managing, organizing and managing study support services. At present, consultation, management and ancillary services of distance education have been improved by IT obviously.

2.2.1 Study Consulting and Psychological Counseling Services

Study consulting services are consulting services for the whole study process, including recruitment, enrollment, assignment, examination, graduation and etc.

we can provide these services by phone, e-mail, BBS, short message, instant message tools (QQ, MSN, OICQ and fetion) and social network tools (BBS, micro-blogging, micro-letter), to answer questions asked by distance learner. Furthermore, we can also update and publish kinds of information to learners helping them know well about newest information and adjust their study plan in time.

Psychological counseling services are also important in distance education. Most distance education learners often feel nervous and anxious caused by the pressure of study, overburden of family and job, which block the progress of study. It is good for distance educational institutions to form a psychological counseling area online through BBS, QQ group, micro-blogging group, micro-letter group and other IT technologies, to provide psychological counseling and help to learners encountering psychological problems.

2.2.2 Learning Process and Personalized Enrollment Management Services

To provide students with personalized, user-friendly, high quality, ease of management services is significant content of non-academic student support services. On one hand, distance educational institutions need to develop enrollment management system for newly enrolled learners to establish enrollment archive, to update information of enrollment, to organize exam and perfect student score management. On the other hand, it is necessary to establish a national enrollment transaction subsystem connected by network from which we can help individual student to complete mutual recognition of credits, exemption, transfer to another school, change major and etc. Because different students came from different culture level and family, they will have different needs of learning. To meet their needs by using enrollment transaction system can best embody personalized and individualized services.

To monitor and manage learning process is an important factor for ensuring study quality. Since most distance learners are working adults, who always ignore some important contents in learning process due to busy work, distance educational institutions should be able to provide tracking services to students. It can record learning process of students and the implementations of study plans, estimate learning progress of students, visit those slower learners and remind them of study, help students solving problems, stimulate the study motivation of learners and reduce the dropout rate. There should be fixed teachers to keep in touch with those potential dropouts, who hasn't registered, taken exam, paid in a few semesters. They can enquire the requirements of students and provide timely feedback and help to them. Active tracking services reflect services of humanity fully, and it is clear that excellent study tracking system and personalized management is an important foundation for quality management.

2.2.3 Educational Administration Information and Ancillary Services

Many MIS such as educational administration management system, online course-chosen system and educational administration information querying system, are widely used in distance educational institutions either in intranet or in internet. With helps of those MIS, Students can register for course, register for course examination, query score, select course online, query a variety of educational administration information and review graduation on the network. It makes students' learning style more flexibility and teachers' support services more effectively, and the level of teaching management and services also increase significantly.

Application of IT on ancillary services except learning study content, including online bill payment, frequently asked questions, online technical support and ancillary services, etc., enhances the overall service capacity of distance educational institutions greatly.

3 The Evaluation Index System on Enhancing Role of it on the Quality of Student Support Services

As mentioned above, we hope to use SERVQUAL [1] scale proposed by PZB to research the enhancing role of IT on the quality of student support services affected by tangibles, reliability, responsiveness, assurance and empathy, in two aspects including academic student support services and non-academic student support services.

3.1 Influencing Factors Analysis of Student Support Services' Quality

3.1.1 Reliability of Student Support Services' Quality

Reliability is an ability to execute service agreements reliably and accurately. The reliable services desired are to execute all agreements timely in the promised way without any error. In fact, the reliability asks distance educational institutions to avoid errors in the process of service, because the errors is not only means direct economic losses to distance educational institutions, but also means losing a lot of potential students.

The evaluation factors of the reliability of academic student support services' quality are as following:

F1   Teachers can periodically tracking survey on students' learning status, and give appropriate guidance on learning strategy;
F2   Diversified learning resources of school can be accessed conveniently and successfully;

F3   Students can easily take counseling, tutorials, and panel discussions and other learning activities through a variety of information tools, such as curriculum BBS, QQ group, Fetion, micro-blogging and etc..

The evaluation factors of the reliability of academic student support services' quality are as following:

F4   Students could efficiently consult and learn through a variety of information channels, such as school home page, phone, QQ and Fetion and other instant message tools, mail, forums, SMS and etc.;
F5   Students can get effective psychological counseling services by a variety of information channels;
F6   Students can easily deal with the matters related to enrollment transaction in the country, such as mutual recognition of credits, exemption, transfer to another school, change major and etc.;
F7   School monitors students' learning process by perfect monitoring system, such as classes, assignments, examinations and other learning-related reminders, early warning prompts and returning of unfinished study planning and etc.;
F8   School provides convenient and reliable educational administration management system, such as course registration, registration of course examination, score query, online course, a variety of teaching information query, review of graduation and etc.;
F9   School offers convenient and reliable acting auxiliary services, such as online bill payment, frequently asked questions, online technical support and etc..

3.1.2 Assurance of Student Support Services' Quality

Assurance needs the teacher has knowledge, courtesy, and expressed a confident and credible capacity. It can enhance the students' confidence of distance educational service quality and the sense of security. When students are talking with teachers, teachers' friendly, kind, warm attitude and vast knowledge can give students confidence and sense of security. These can promote students to take part in the distance educational institute. Friendly attitude and business skills are both indispensable.

The evaluation factors of the assurance of academic student support services' quality are as following:

F10   Teachers can use information tools familiar to students, to give them advices with the attitude of respect;
F11   Teachers can teach students advanced and modern knowledge content though a variety of information channels;
F12   Teachers can take many different measures to communicate with student professionally, warmly and tolerantly.

The evaluation factors of the assurance of non-academic student support services' quality are as following:

F13   Students could obtain efficient and polite consulting services and learning contents through a variety of information channels;
F14   Students can get psychological counseling services by a variety of information channels with privacy protecting;
F15   School monitors students' learning process and remind appropriately, which has to be polite and do not interfere with the normal life of students;
F16   School is starting from the perspective of students to provide various ancillary services.

3.1.3 Responsiveness of Student Support Services' Quality

The responsiveness is to help students and to express the desire providing services quickly and effectively. Allowing students to wait without any instructions or explanation for the students will have unnecessary negative impact on the quality perception. Students might feel that they are not respected and have no basic right to know. When service failure occurs, sincerely apology and solving problems quickly can calm angry sentiments of students and have a positive impact on the quality perception. Whether distance educational institutions could satisfy the various requirements of students will indicate the service orientation of the distance educational institutions, that is, whether the interests of students are put in the first place. At the same time, the efficiency of service delivery is also a reflection of the service quality of the distance educational institutions. Studies have shown that the time of students waiting for service is the important influence factor for the feeling, impression and satisfaction of students, and the formation of the image of the distance educational institutions in the service delivery process. So shortening the waiting time and improving the efficiency of service delivery can effectively improve the service quality of the distance educational institution.

The evaluation factors of the responsiveness of academic student support services' quality are as following:

F17   Learning resource update timely;
F18   Diversified learning resources of school can be accessed quickly;
F19   Students can quickly take counseling, tutorials, and panel discussions and other learning activities through a variety of information tools.

The evaluation factors of the responsiveness of non-academic student support services' quality are as following:

F20   Students could quickly consult and learn through a variety of information channels, and teachers can answer as soon as possible;

F21 Students can get quick psychological counseling services by a variety of information channels;

F22 Students can quickly deal with the matters related to enrollment transaction in the country;

F23 School monitors students' learning process and supply many reminder functions timely, such as homework reminder, exam reminder, pre-warning for uncompleted study plan and etc.;

F24 School provides a fast educational administration management system;

F25 School offers fast acting auxiliary services, and problems can be promptly corrected.

3.1.4 Tangibles of Student Support Services' Quality

Tangibles refer to the quality of physical facilities, equipment, teacher resources and communication materials of the distance educational institutions. The hardware of the Open and Distance Education Institutions is essential. Modern equipments and supporting laboratory equipments can enhance students' confidence and also help to access better teaching effects.

The evaluation factors of the tangibles of academic student support services' quality are as following:

F26 Teaching resources content (such as textbooks and courseware) is reasonable, and conducive to learning;

F27 Learning resources support platform is conducive to learning;

F28 Online counseling and answering system can be used conveniently.

The evaluation factors of the tangibles of non-academic student support services' quality are as following:

F29 School website provides clear information of learning and processing guide;

F30 Educational administration management system is conducive to students' self- learning management;

F31 Website and related systems platform operating normally, and can be accessed.

3.1.5 Empathy of Student Support Services' Quality

Empathy is putting the institutions themselves in the students' position and giving personalized attention to students. Empathy exhibits that distance educational institutions could communicate actively with the students and understand the needs of students. The institutions could adjust teaching content and methods according to the needs of students in the teaching process, and provide different services for different students.

The evaluation factors of the empathy of academic student support services' quality are as following:

F32   Targeted study suggestions are provided to appropriate students;
F33   Diversified learning resources of school focus on creating and sustaining students' study motivation;
F34   Online counseling and answering system supplies individual counseling and Q & A;
F35   Online live classrooms provide targeted time, study contents, and teaching material to students considering the situation of students.

The evaluation factors of the empathy of non-academic student support services' quality are as following:

F36   Through IT channels, schools can provide targeted learning relevant advice for needy students;
F37   Through IT channels, schools can provide targeted psychological counseling services for needy students;
F38   Through enrollment transaction system, schools can change enrollment for needy students;
F39   Schools build individual learning monitoring file for every student and provide personalized learning evaluation feedback and learning abnormalities remind;
F40   Online educational management administration system, payment system, and technical issues can always give personalized support.

3.2 Evaluation Index System: "IT Application Fields-Service Quality Promotion" Two Dimensions Framework

Those five quality influencing factors mentioned above complement and influence each other, although each representing a certain aspects of service quality, they work together to influence students' evaluation to service quality. This paper establish evaluation index system framework from IT application and service quality promotion dimensions, select and correct evaluation index according to the advices from professional and practical persons. According with physical truth and easy measuring is the chosen principle of evaluation index. After cutting non appropriate indexes, we get 40 evaluation indexes listed as Table 1.

Table 1 Evaluation index system: "IT application fields-service quality promotion" two dimensions framework

IT application fields		Service quality promotion				
		Reliability	Assurance	Responsiveness	Tangibles	Empathy
Academic student support services	Tracking students' learning process	F1	F10	–	–	F32
	Diversified learning resources	F2	F11	F17, F18	F26, F27	F33
	Interactive teaching and learning for teachers and students with IT	F3	F12	F19	F28	F34, F35
Non-academic student support services	Study consulting and psychological counseling services	F4, F5	F13, F14	F20, F21	F29	F36, F37
	Learning process and personalized enrollment management services	F6, F7	F15	F22, F23	F30	F38, F39
	Educational administration information and ancillary services	F8, F9	F16	F24, F25	F31	F40

4 Conclusion

This paper established and designed evaluation index system based on "IT application fields - service quality promotion" two dimensions model. The IT application fields included academic and non-academic student support services. The service quality promotion included tangibles, reliability, responsiveness, assurance and empathy. This two dimensions framework evaluation method is helpful to understand the performances on specific aspects of student support services, and then do targeted improvement for the distance educational institutions.

Acknowledgment The work presented in the paper is supported by e-commerce core courses teaching and learning under the cloud platform environment (Project No. Q3003A).

References

1. Parasuraman A, Zeithmal VA, Berry LL (1988) SERVQUAL: a multiple-item scale for measuring consumer perceptions of service quality. J Retail 64(1):12–40
2. Fan WQ, Jin HF (2010) Case study on evaluation of student perceived quality of learning service in distance education. 20:86–90
3. Lee JW (2010) Online support service quality, online learning acceptance, and student satisfaction. Internet High Educ 13:277–283
4. Helgesen O, Nesset E (2007) Images, satisfaction and antecedents: drivers of student loyalty? a case study of a Norwegian university college. Corp Reputation Rev 10(1):38–59
5. Yukselturk E, Yildirim Z (2008) Investigation of interaction, online support, course structure and flexibility as the contributing factors to student satisfaction in an online certificate program. Educ Technol Soc 11(4):51–65
6. Ormond S (2002) Supporting students in online, open and distance learning, 2nd edn. Rout ledge Falmer, London, p 226

Research on the Representation Methods for Rough Knowledge

Zhicai Shi, Jinzu Zhou and Chaogang Yu

Abstract Now there exists a lot of information from Internet and the information implies abundant knowledge. But the information is usually uncertain, imprecise, incomplete, coarse, and vague. So it is very difficult to acquire knowledge from the miscellaneous information. One of the effective methods to process this kind of information is Rough Set theory. As an important method of the acquisition and process for knowledge, Rough Set theory has given the algebraic representation for knowledge by means of the equivalence relations of algebra and the inclusion relations of set theory. But this representation makes knowledge understanding difficult. In order to overcome this problem the concept of granulating is proposed and the granular representation for knowledge is suggested. The granular representation for knowledge makes complicated problems simplified. It is nearer to the thinking habits of mankind and it can describe the roughness of knowledge quantitatively. This chapter discusses the algebraic and granular representations for knowledge, respectively. Some definitions, properties, and theorems under two different representations are analyzed. The research results justified that two different representations for knowledge are equivalent, but the granular representation is more direct and easier to be understood.

Keywords Rough set · Granular computing · Knowledge representation

1 Introduction

With the advent of information society and the wide application of computer network, there exists a lot of useful information from Internet. So, it seems to be very important how to acquire the information and process them so as to abstract

Z. Shi (✉) · J. Zhou · C. Yu
School of Electronic and Electrical Engineering, Shanghai University of Engineering Science, Shanghai, China
e-mail: szc1964@163.com

Z. Wen and T. Li (eds.), *Knowledge Engineering and Management*,
Advances in Intelligent Systems and Computing 278,
DOI: 10.1007/978-3-642-54930-4_40, © Springer-Verlag Berlin Heidelberg 2014

knowledge and to make decisions effectively. Knowledge is the most advanced representation form of information and the knowledge representation methods decide the efficiency of information processing, so the research for the knowledge representation methods is very meaning and it has become a hot topic in some fields related to data mining, knowledge discovery, and decision analysis. But the structure of the information from Internet is miscellaneous. The information is often uncertain and it is usually vague, incomplete, coarse, and imprecise. The information has many different characteristics and it is not processed effectively by means of the traditional methods. Some special methods have been proposed to process the information. The typical methods are Fuzzy Set theory, Rough Set theory and Granular computing. Because Rough Set theory is not dependent on any prior knowledge and Granular computing is close to the thinking habit of mankind they are researched and applied widely. In this chapter we use them to research the representation methods for knowledge, and discuss the difference and relationship between the different knowledge representations.

The remaining part of this chapter is arranged as follows: In Sect. 2, we simply review some related research works. In Sect. 3 we introduce Rough Set theory and use it to represent knowledge, which is called the algebraic representation for knowledge. In Sect. 4, we discuss how to use granular computing to represent knowledge and give the granular representation for knowledge. Section 5 is the focus of this chapter, we discuss the relationship between two different representations for knowledge and their equivalence. This chapter concludes in Sect. 6 with remarks for the works.

2 The Related Works

In the objective world, there exists a lot of useful information and this information is ascertain and precise. The knowledge abstracted from the information is deterministic generally. Researchers proposed first-order predicate logic, production rule, state space, semantic network, and other traditional methods to process the information so as to acquire and represent knowledge. But in natural world there also exists a lot of uncertain, coarse, or rough information. This kind of information is usually vague, imprecise, and incomplete, they cannot be processed effectively by means of the traditional methods. So many researchers have focused their efforts on researching and developing some new methods to process this kind of uncertain information effectively. They proposed fuzzy Set theory, D-S evidence theory, and other new methods. Among these new methods Rough Set theory and Granular computing are the most important methods.

Rough Set theory is one of the most methods to be used for identification and recognition of common patterns from a large amount of information, especially some uncertain information. Rough Set concept was introduced by Pawlak [1, 2]. It provides a technique to analyze data dependencies, to identify fundamental factors and to discover rules from data, both deterministic and nondeterministic rules [3]. It used the equivalence relations of algebra and the inclusion relations of set theory to explain and define the concepts and calculations for Rough Sets, which is called the algebraic representation for knowledge. But some concepts and calculations about this representation are so complicated that it is difficult to understand the essence of knowledge. In some cases, e.g., attribute reduction, this representation has not given any effective algorithm to process knowledge [4]. On the basis of information theory, Duoqian and Jue [5] thought that the knowledge might be considered as some random variables composed by some subsets of the universe on σ-algebraic and used information entropy to represent knowledge, which is called the information representation for knowledge. Under this representation the entropy of the knowledge is bigger then it will provide more information and its classification ability is stronger. This representation method explains the essence of the knowledge more directly. But its calculation concerns some random variables and their probabilistic distributions, so this representation method for knowledge is complicated.

Granular computing is a very important method in the design and implementation of intelligent information systems. It includes computing with words, Rough Set, and the quotient space theory. A basic concept of granular computing is the granulating of the universe [6, 7]. A granule normally consists of elements which are drawn together by indistinguishable degree, similarity, or proximity. There are many reasons for the study of granular computing. The simplicity for problem solving is perhaps one of the most important reasons that granular computing is widely used. When a problem involves some incomplete, uncertain, vague, or imprecise information, it may be difficult to differentiate distinct elements. Although sometimes detailed information may be available, very precise solutions may not be required for many practical problems. The use of granulating generally leads to simplification of practical problem solving. The granulating is very near to the thinking habit of mankind brain. If granules could be used to represent knowledge then it may describe knowledge more directly and precisely so as to be understood easily. Some research works [8] used the base of a set to define the information granularity and gave some concepts and properties of the granularity. But many descriptions and definitions are incomplete, some properties, theorem and corollary are not proved. The formulation representation methods for knowledge need to be improved and perfected.

3 The Algebraic Representation for Rough Knowledge

As one of the most effective methods to process uncertain, imprecise, coarse, and incomplete information Rough Set theory is researched and applied widely since it was proposed. Rough Set theory is different from traditional Fuzzy Set theory because it is not dependent on any prior knowledge. This theory expresses vagueness not by means of membership but by employing the boundary region of a set and it has been applied to many different fields such as knowledge acquisition, decision analysis, data mining, and pattern recognition. Then on the basis of Rough Set theory we apply equivalence relations and some concepts of set theory to discuss the algebraic representation for rough knowledge.

For the algebraic representation for rough knowledge, let R be a two-tuple equivalence relation on the set U, U/R represents the set which is composed of all equivalence classes of R and $[x]_R = \{y | (x, y) \in R, x \in U, y \in U\}$ represents the set of all elements which are equivalent to x in U. A Knowledge Base is often represented as a relation system $K = (U, P)$, where U is a finite and nonempty set called the universe and P is a cluster of equivalence relations on U, $P \subseteq U \times U$. Let Q be another cluster of equivalence relations, $Q \subseteq P$ and $Q \neq \Phi$, then for any Q, $\cap Q$ is also an equivalence relation, we call it an indistinguishable relation on U, denoted by $\text{Ind}(Q)$. $U/\text{Ind}(Q)$ is the quotient set after U is partitioned by the equivalence relation $\text{Ind}(Q)$ and it represents some knowledge related with the equivalence relation Q, we call it the basic knowledge of Q on U, which is simply called the knowledge Q [4, 8].

Definition 3.1 Let $K = (U, P)$ and $K1 = (U, Q)$ be two Knowledge Bases, if $\text{Ind}(P) = \text{Ind}(Q)$, then we call that K is equivalent to $K1$ or P is equivalent to Q, denoted by $K \cong K1$ or $P \cong Q$.

Definition 3.2 Let U be the universe and P be the cluster of equivalence relations which are defined by U. For $\forall R \in P$, if $\text{Ind}(P - \{R\})$ is equal to $\text{Ind}(P)$ we call that R is dispensable or redundant, otherwise we call that R is indispensable.

Definition 3.3 Let U be the universe and P be the cluster of equivalence relations which are defined by U. For $\forall R \in P$, R is indispensable in P, we call that P is independent, otherwise we call that P is dependent each other.

Definition 3.4 Let U be the universe and P be the cluster of equivalence relations which are defined by U. The set which is composed of all indispensable equivalence relations among P is called the kernel of P, denoted by $\text{Core}(P)$.

Definition 3.5 Let U be the universe, P and Q be two clusters of equivalence relations which are defined by U and $Q \subseteq P$. If Q is independent and $\text{Ind}(Q)$ is equal to $\text{Ind}(P)$, then we call that Q is a reduction of P, denoted by $\text{Red}(P)$.

Theorem 3.1 *The kernel of P, denoted by $\text{Core}(P)$, is the intersection of all reductions of P, this can also be described as $\text{Core}(P) = \cap \text{Red}(P)$.*

Definition 3.6 Suppose that $K = (U, R)$ is a Knowledge Base, $Q \subseteq R$, $P \subseteq R$, P and Q represent two kinds of knowledge, then:

  (i) if $\text{Ind}(P) \subseteq \text{Ind}(Q)$, we call that Q is dependent on P, denoted by $P \Rightarrow Q$.
  (ii) if $P \Rightarrow Q$, $Q \Rightarrow P$, we call that P is equivalent to Q, denoted by $P \equiv Q$.
  (iii) if there is neither $P \Rightarrow Q$ nor $Q \Rightarrow P$, we call that P and Q are independent each other, denoted by $P \neq Q$.

where $\text{Ind}(P) \subseteq \text{Ind}(Q)$ does not mean that $\text{Ind}(P)$ is a subset of $\text{Ind}(Q)$. It states that for $\forall A \in U/\text{Ind}(P)$, then $\exists B \in U/\text{Ind}(Q)$ makes $A \subseteq B$.

Definition 3.7 Let P be the cluster of all equivalence relations which are defined by the universe U, $R \in P$, $S \in P$. for $\forall u, v \in U$, if $uSv \Leftrightarrow uRv$ then we call that S is equal to R, denoted by $S = R$. For $\forall u, v \in U$, if $uSv \Rightarrow uRv$ then we call that S is finer than R or R is coarser than S, denoted by $S \leq R$. If $S \leq R$ and $S \neq R$, then we call that S is strictly finer than R, denoted by $S < R$.

Theorem 3.2 *Let U be the universe, R and S be two equivalence relations which are defined by U, let $U/R = \{X1, X2, \ldots, Xn\}$ and $U/S = \{Y1, Y2, \ldots, Ym\}$, then there are the following relations:*

(i) $S = R \Leftrightarrow m = n, Xi = Yi, i \in \{1, 2, \ldots, n\}$
(ii) $S \leq R \Leftrightarrow \forall Yi \in U/S, \exists Xj \in U/R, Yi \subseteq Xj, i \in \{1, 2, \ldots, m\},$
    $j \in \{1, 2, \ldots, n\}$
(iii) $S < R \Leftrightarrow \forall Yi \in U/S, \exists Xj \in U/R, Yi \subseteq Xj, i \in \{1, 2, \ldots, m\}, j \in \{1, 2,$
    $\ldots, n\}$ and $\exists i_0 \in \{1, 2, \ldots, m\}, \exists j_0 \in \{1, 2, \ldots, n\}, Yi_0 \subset Xj_0$

4 The Granular Representation for Rough Knowledge

It is known to the public that the knowledge represents the understanding of mankind to the objective world. The understanding degree can be described quantitatively by the granularity. The knowledge granule is finer states that the knowledge is more precise and its distinguishability to the universe is stronger. Otherwise, the knowledge granule is bigger states that the knowledge is coarser and its distinguishability to the universe is weaker. In the following part of this section the cardinality of a set is used to describe the granularity of knowledge quantitatively [8, 9], which is called the granular representation for rough knowledge.

Definition 4.1 Suppose that $K = (U, P)$ is a Knowledge Base, $R \in P$, R is an equivalence relation. The granularity of the knowledge R is defined as $GD(R) = |R|/|U|^2$, Where $|R|$ is the cardinality of the set R, $R \subseteq U \times U$.

Theorem 4.1 *Let R is a knowledge on the Knowledge Base $K = (U, P)$, $U/R = \{X1, X2, \ldots, Xn\}$, then The granularity of the knowledge R is described by $GD(R) = |R|/|U|^2$, Where $|R| = \sum_{i=1}^{n} |X_i|^2$ represents the cardinality of R, $R \subseteq U \times U$.*

Proof For $\forall (u, v) \in R$, then $\exists Xi \in U/R$ and $(u, v) \in Xi$. For $\forall Xi \in U/R$, $i \in \{1, 2, \ldots, n\}$, Xi is composed of any two elements among Xi, hence the cardinality of Xi is $|Xi|^2$, So $|R| = \sum_{i=1}^{n} |Xi|^2$ and $GD(R) = |R|/|U|^2 = \sum_{i=1}^{n} |Xi|^2/|U|^2$.

Corollary 4.1 *Let $K = (U, P)$ be a Knowledge Base, P be a cluster of equivalence relations which are defined by U, $U/\text{Ind}(P) = \{X1, X2, \ldots, Xn\}$, the granularity of the knowledge P is defined as $GD(P) = GD(\text{Ind}(P)) = \sum_{i=1}^{n} |Xi|^2/|U|^2$.*

It is easy to conclude that Theorem 4.1 is identical to Corollary 4.1. Because for Theorem 4.1 the equivalence relation R can also be regarded as a cluster of equivalence relations which only consists of one equivalence relation R.

Property 4.1 *Let R be a knowledge in the Knowledge Base $K = (U, P)$, if R is the equal relation then $GD(R) = 1/|U|$. If R is the universe relation then $GD(R) = 1$.*

Proof Let R is an equal relation, then $U/R = \{\{u\}|u \in U\}$, so $GD(R) = \sum_{i=1}^{n} |X_i|^2/|U|^2 = \sum_{i=1}^{n} 1^2/|U|^2 = |U|/|U|^2 = 1/|U|$.

Let R is an universe relation, then $U/R = \{U\}$, so $GD(R) = \sum_{i=1}^{n} |X_i|^2/|U|^2 = |U|^2/|U|^2 = 1$.

In general, for knowledge R defined the universe U its granularity is from $1/|U|$ to 1.

It is obvious that for knowledge R, its minimum granularity is $1/|U|$ and its maximum granularity is 1. When the granularity of R is the minimum value the corresponding partition to U is the finest and the distinguishing ability is the strongest. When the granularity is the maximum value the corresponding partition to U is the coarsest and the distinguishing ability is the weakest. So the granularity of knowledge describes its distinguishing ability to the knowledge quantitatively.

Property 4.2 *Suppose that R and S are two equivalence relations which are defined by the universe U, let $U/R = \{X1, X2, \ldots, Xn\}$ and $U/S = \{Y1, Y2, \ldots, Ym\}$, if $S = R$ then $GD(S) = GD(R)$.*

Proof If $S = R$ then $m = n$ and $Xi = Yi$, $i \in \{1, 2, \ldots, n\}$, hence $GD(R) = \sum_{i=1}^{n} |X_i|^2/|U|^2 = \sum_{j=1}^{m} |Yj|^2/|U|^2 = GD(S)$.

Property 4.3 *Suppose that R and S are two equivalence relations which are defined the universe U, let $U/R = \{X1, X2, \ldots, Xn\}$ and $U/S = \{Y1, Y2, \ldots, Ym\}$, if $|U/S| = |U/R|$, $\exists f : U/S \to U/R$, $|f(Yi)| = |Yi|$, then $GD(S) = GD(R)$, where f is a one-to-one mapping.*

Proof Let $U/R = \{X1, X2, \ldots, Xn\}$ and $U/S = \{Y1, Y2, \ldots, Ym\}$, owing to $|U/S| = |U/R|$, then $m = n$. By the known condition $\exists f : U/S \to U/R$ and f being a one-to-one mapping, for $\forall Yi \in U/S$ there exists $Xj \in U/R$ and $f(Yi) = Xj$, $i \in \{1, 2, \ldots, m\}$, $j \in \{1, 2, \ldots, n\}$. By another known condition $|f(Yi)| = |Yi|$, we have $|Xj| = |Yi|$ for $i \in \{1, 2, \ldots, m\}$, $j \in \{1, 2, \ldots, n\}$, which implies $\sum_{j=1}^{n} |Xj|^2 = \sum_{i=1}^{m} |Yi|^2$, hence by Theorem 4.1 $GD(S) = GD(R)$.

Property 4.4 *Let R and S are two equivalence relations which are defined by the universe U, let $U/R = \{X1, X2, \ldots, Xn\}$ and $U/S = \{Y1, Y2, \ldots, Ym\}$, if $S < R$, then $GD(S) < GD(R)$.*

Proof By Theorem 3.2, if $S < R$, then $\forall Yi \in U/S$, $\exists Xj \in U/R$, $Yi \subseteq Xj$, $i \in \{1, 2, \ldots, m\}$, $j \in \{1, 2, \ldots, n\}$ and $\exists i_0 \in \{1, 2, \ldots, m\}$, $\exists j_0 \in \{1, 2, \ldots, n\}$, $Yi_0 \subset Xj_0$, which implies $|Yi| \leq |Xj|$ and $|Yi_0| < |Xj_0|$ for $i \in \{1, 2, \ldots, i_0 - 1\} \cup$

$\{i_0 + 1, \ldots, m\}$ and $j \in \{1, 2, \ldots, j_0 - 1\} \cup \{j_0 + 1, \ldots, n\}$. Hence we have $\sum_{i=1}^{m} |Yi|^2 < \sum_{j=1}^{n} |Xj|^2$, by Theorem 4.1 we have $GD(S) < GD(R)$.

Corollary 4.2 *Let R and S are two equivalence relations which are defined by the universe U (i) if $S > R$, then $GD(S) > GD(R)$. (ii) if $S = R$, then $GD(S) = GD(R)$.*

By Property 4.4 and Corollary 4.2, we can see:

(i) $S < R$ states that S is finer than R so its granularity is smaller than the granularity of R.
(ii) $S > R$ states that S is coarser than R so its granularity is bigger than the granularity of R.
(iii) $S = R$ states that S has the same distinguishing ability as R so they have the same granularity.

Property 4.5 *Let R be an equivalence relation which are defined by the universe U, $U/R = \{X1, X2, \ldots, Xi, \ldots, Xn\}$, choose $Xi \in \{X1, X2, \ldots, Xn\}$ randomly, let $Xi = Xi1 \cup Xi2$ and $Xi1 \cap Xi2 = \phi$, we will get a new equivalence relation R' and $U/R' = \{X1, X2, \ldots Xi1, Xi2, \ldots, Xn\}$, then $GD(R') \leq GD(R)$.*

Proof Let $U/R = \{X1, X2, \ldots, Xi, \ldots, Xn\}$, choose $Xi \in \{X1, X2, \ldots, Xn\}$ randomly, let $Xi = Xi1 \cup Xi2$ and $Xi1 \cap Xi2 = \phi$, $U/R' = \{X1, X2, \ldots, Xi1, Xi2, \ldots, Xn\}$. By Theorem 4.1, we have:

$$GD(R) = \left(\sum_{j=1}^{i-1} |Xj|^2 + (|Xi1| + |Xi2|)^2 + \sum_{j=i+1}^{n} |Xj|^2 \right) \bigg/ |U|^2$$

$$GD(R') = \left(\sum_{j=1}^{i-1} |Xj|^2 + |Xi1|^2 + |Xi2|^2 + \sum_{j=i+1}^{n} |Xj|^2 \right) \bigg/ |U|^2$$

$$(|Xi1| + |Xi2|)^2 \geq |Xi1|^2 + |Xi2|^2$$

So $GD(R') \leq GD(R)$.

Corollary 4.3 *Let R be an equivalence relation which is defined by the universe U, $U/R = \{X1, X2, \ldots, Xn\}$, choose Xi and Xj randomly, $Xi, Xj \in \{X1, X2, \ldots, Xn\}$, $i \neq j$, let $Xi' = Xi \cup Xj$, we will get a new equivalence relation R' and $U/R' = \{X1, X2, \ldots, Xi \cup Xj, \ldots, Xn\}$, then $GD(R') \geq GD(R)$.*

Property 4.5 and Corollary 4.3 states that the change of the knowledge granules will affect its granularity. With the decomposition of the equivalent classes the granularity will become smaller and the related knowledge will become finer. With the union of the equivalence classes the granularity will become bigger and the related knowledge will become coarser.

By the above analysis, it can be concluded that the roughness of knowledge decreases monotonously with its information granularity becoming small.

Property 4.6 *Let R and S are two equivalent relations which are defined by the universe U, if $S \leq R$ and $GD(S) = GD(R)$, then $S = R$.*

Proof Let $S \neq R$, by the known condition $S \leq R$, then $S < R$. By Property 4.4 we can conclude $GD(S) < GD(R)$, but it is in contradiction with the known condition $GD(S) = GD(R)$, so $S = R$.

5 The Relationship Between Two Different Representations for Rough Knowledge

As the algebraic representation for knowledge, the granular representation for knowledge can also describe the roughness of the knowledge quantitatively. This kind of representation method is more direct and easy to be understood. Then someone may feel puzzled about what relationship there exists between two different representations for rough knowledge? Are they to equivalent each other? Which is the best? Then we will discuss these questions.

Theorem 5.1 *Let $K = (U, R)$ be a Knowledge Base, P and Q be two clusters of equivalence relations which are defined by U. If $\mathrm{Ind}(P) = \mathrm{Ind}(Q)$, then $GD(P) = GD(Q)$.*

Proof The known condition: $\mathrm{Ind}(P) = \mathrm{Ind}(Q)$ means $U/\mathrm{Ind}(P) = U/\mathrm{Ind}(Q)$, let $U/\mathrm{Ind}(P) = \{X1, X2, \ldots, Xn\}$ and $U/\mathrm{Ind}(Q) = \{Y1, Y2, \ldots, Yn\}$, then $Xi = Yi$, $i \in \{1, 2, \ldots, n\}$, by Theorem 4.1 and Corollary 4.1 we can prove $GD(P) = GD(Q)$.

Theorem 5.1 states that under the algebraic representation two same knowledge have the same granularity.

Theorem 5.2 *Let $K = (U, R)$ be a Knowledge Base, P and Q be two clusters of equivalence relations which are defined by U. If $Q \subseteq P$ and $GD(P) = GD(Q)$, then $\mathrm{Ind}(P) = \mathrm{Ind}(Q)$.*

Proof (i) By the known condition $Q \subseteq P$, we can conclude $\mathrm{Ind}(P) \subseteq \mathrm{Ind}(Q)$. (ii) Next we prove $\mathrm{Ind}(P) \supseteq \mathrm{Ind}(Q)$. We suppose that $\mathrm{Ind}(P) \supseteq \mathrm{Ind}(Q)$ is false, then we have $\mathrm{Ind}(P) \subset \mathrm{Ind}(Q)$. Let $U/\mathrm{Ind}(P) = \{X_1, X_2, \ldots, X_m\}$ and $U/\mathrm{Ind}(Q) = \{Y_1, Y_2, \ldots, Y_n\}$, then by definition 3.6 for $\forall k \in \{1, 2, \ldots, m\}$, $\exists l \in \{1, 2, \ldots, n\}$ makes $X_k \subset Y_l$. Let $Y_l = X_k \cup (Y_l \backslash X_k)$. For $\forall i \in \{1, 2, \ldots, m\}$ and $i \neq l$, $\exists j \in \{1, 2, \ldots, n\}$ makes $X_i \subseteq Y_j$ and $|Y_j|^2 \geq |X_k|^2$, so $GD(Q) = \sum_{j=1}^{n} \left(|Y_j|^2 / |U|^2 \right) = \sum_{j=1}^{l-1} \left(|Y_j|^2 + |Y_l|^2 + \sum_{j=l+1}^{n} |Y_j|^2 \right) / |U|^2 = \sum_{j=1}^{l-1} \left(|Yj|^2 + |X_k \cup (Y_l \backslash X_k)|^2 + \sum_{j=l+1}^{n} |Y_j|^2 \right) / |U|^2$, by $|X_k \cup (Y_l \backslash X_k)|^2 > |X_k|^2$ we get $GD(Q) > GD(P)$. But this is in contradiction with the known condition $GD(P) = GD(Q)$, then we can conclude $\mathrm{Ind}(P) \supseteq \mathrm{Ind}(Q)$. So we can conclude $\mathrm{Ind}(P) = \mathrm{Ind}(Q)$.

Theorem 5.2 states that if knowledge includes another knowledge and they have the same granularity, then we can conclude that they should have the same distinguishing ability and their algebraic representations are equivalent also.

Theorem 5.3 *Let U be the universe and P be a cluster of equivalence relations which are defined by U. For $R \in P$, R is dispensable in P if and only if $GD(P - \{R\}) = GD(P)$.*

Proof (i) Let R is dispensable for P, then $\text{Ind}(P) = \text{Ind}(P - \{R\})$. By Property 4.2, $GD(P - \{R\}) = GD(P)$. (ii) Suppose that R is indispensable for P, then $P - \{R\} \subset P$ and $GD(P) < GD(P - \{R\})$, this is in contradiction with the known condition $GD(P - \{R\}) = GD(P)$, so R is dispensable for P.

Theorem 5.3 states that a dispensable relation is added into a cluster of equivalence relations and the granularity of the related knowledge is not changed. According to Theorem 5.3 we can conclude Theorem 5.4.

Theorem 5.4 *Let U be the universe and P be a cluster of equivalence relations which are defined by U. For $\forall R \in P$, P is independent if and only if $GD(P - \{R\}) > GD(P)$.*

Proof (i) Assume that P is independent, then $\forall R \in P$, R is indispensable for P, $P - \{R\} \subset P$ and $\text{Ind}(P) \subset \text{Ind}(P - \{R\})$. By Theorem 4.1 and Corollary 4.1 we have $GD(P - \{R\}) > GD(P)$. (ii) For $\forall R \in P, GD(P - \{R\}) > GD(P)$ means that $\forall R \in P$ is deleted from P so as to reduce the granularity of the knowledge about P and the related distinguishing ability, so R is indispensable for P, P is independent.

Theorem 5.5 *Let U be the universe, P and Q be two clusters of equivalence relations which are defined by U, $Q \subseteq P$, Q is a reduction of P if and only if: (i) $\forall S \in Q, GD(Q - \{S\}) > GD(Q)$. (ii) $GD(Q) = GD(P)$.*

Proof Let $Q \subseteq P$, by Definition 3.5 Q is a reduction of P if and only if $\text{Ind}(Q) = \text{Ind}(P)$ and Q is independent. By the known condition $Q \subseteq P$ and Theorem 5.2, if and only if $GD(Q) = GD(P)$, then $\text{Ind}(Q) = \text{Ind}(P)$. By Theorem 5.4, if and only if $\forall S \in Q, GD(Q - \{S\}) > GD(Q)$, then Q is independent.

Theorem 5.5 states that for knowledge reduction the granular representation for knowledge is equivalent to its algebraic representation. Knowledge reduction can be also completed by granular computing.

Theorem 5.6 *Let $K = (U, R)$ be a Knowledge Base, P and Q be two clusters of equivalence relations defined by U, $Q \subseteq R, P \subseteq R$, there exit the following conclusions: (i) if $P \Rightarrow Q$, then $GD(P) \le GD(Q)$. (ii) if $P \equiv Q$, then $GD(P) = GD(Q)$.*

Proof (i) If $P \Rightarrow Q$, then $\text{Ind}(P) \subseteq \text{Ind}(Q)$ and $|\text{Ind}(P)| \le |\text{Ind}(Q)|$, $GD(P) = GD(\text{Ind}(P)) = |\text{Ind}(P)|/|U|^2$, $GD(Q) = GD(\text{Ind}(Q)) = |\text{Ind}(Q)|/|U|^2$, so $GD(P) \le GD(Q)$. (ii) $P \equiv Q$ means that $P \Rightarrow Q$ and $Q \Rightarrow P$, then $GD(P) \le GD(Q)$ and $GD(Q) \le GD(P)$, so $GD(P) = GD(Q)$.

6 Conclusion

The processing of the massive information or big data is becoming more important in the modern society. The information is usually coarse, imprecise, uncertain, and vague. Rough Set theory is a kind of method often used to deal with this sort of information or data. It uses the equivalence relations of algebra and inclusion of relations of set theory to represent knowledge. This chapter has discussed the algebraic representation for knowledge in detail and given some definitions, properties, theorems, and their proofs. But the algebraic representation for knowledge cannot make the essence of knowledge understood easily. In order to understand knowledge directly and easily, we have proposed the granular representation for knowledge and use granulating to describe the procedure that mankind recognize the natural world. We have discussed the relationship between two different representations for knowledge and their equivalence. The research results justify that some functions, e.g., attribute reduction, can also be completed effectively by granular computing, these functions are accustomed to be completed by Rough Set theory previously. Granular computing is nearer to the traditional thinking habits of mankind and it is more effective method to acquire and recognize knowledge.

Acknowledgment The work about this chapter is supported by National Natural Science Foundation of China (No. 61272097), The Science Research Project (No. 2011XY16) and Discipline Developing Foundation (No. XKCZ1212) of Shanghai University of Engineering Science.

References

1. Pawlak Z (1991) Rough set: theoretical aspect of reasoning about data. Kluwer Academic Publishers, Boston
2. Pawlak Z (1998) Rough set theory and its application to data analysis. Cybern Syst 29(7):661–688
3. Mollestad T, Skowron A (1996) A rough set framework for data mining of propositional default rules. In: The 9th international symposium on intelligent systems, pp 448–457
4. Qinrong F, Duoqian M (2009) Knowledge representation using partition granularity. Chin J Pattern Recognit Artif Intell 22(1):64–69
5. Duoqian M, Jue W (1999) An information representation of the comcepts and operation in rough set theory. Chin J Softw 10(2):113–116
6. Guan J, Bell D (1998) Rough computational methods for information systems. Artif Intell 105:77–103
7. Yao YY (1998) A comparative study of fuzzy sets and Rough Sets. Inf Sci 109:227–242
8. Li X (2011) Knowledge granular representation of concepts and operations in Rough Set theory. Chin J Comput Eng Appl 47(11):34–36
9. Zadeh LA (1997) Towards a theory of fuzzy information granulation and its centrality in human reasoning and fuzzy logic. Fuzzy Sets Syst 19:111–127